T0312828

FUTURE SUSTAINABLE ECOSYSTEMS
Complexity, Risk, and Uncertainty

CHAPMAN & HALL/CRC
APPLIED ENVIRONMENTAL STATISTICS

Series Editors

Doug Nychka
Institute for Mathematics
Applied to Geosciences
National Center for
Atmospheric Research
Boulder, CO, USA

Richard L. Smith
Department of Statistics &
Operations Research
University of North Carolina
Chapel Hill, USA

Lance Waller
Department of Biostatistics
Rollins School of
Public Health
Emory University
Atlanta, GA, USA

Published Titles

Michael E. Ginevan and Douglas E. Splitstone, **Statistical Tools for Environmental Quality**

Timothy G. Gregoire and Harry T. Valentine, **Sampling Strategies for Natural Resources and the Environment**

Daniel Mandallaz, **Sampling Techniques for Forest Inventory**

Bryan F. J. Manly, **Statistics for Environmental Science and Management, Second Edition**

Bryan F. J. Manly and Jorge A. Navarro Alberto, **Introduction to Ecological Sampling**

Steven P. Millard and Nagaraj K. Neerchal, **Environmental Statistics with S Plus**

Wayne L. Myers and Ganapati P. Patil, **Statistical Geoinformatics for Human Environment Interface**

Nathaniel K. Newlands, **Future Sustainable Ecosystems: Complexity, Risk and Uncertainty**

Éric Parent and Étienne Rivot, **Introduction to Hierarchical Bayesian Modeling for Ecological Data**

Song S. Qian, **Environmental and Ecological Statistics with R**

Thorsten Wiegand and Kirk A. Moloney, **Handbook of Spatial Point-Pattern Analysis in Ecology**

FUTURE SUSTAINABLE ECOSYSTEMS
Complexity, Risk, and Uncertainty

Nathaniel K. Newlands

CRC Press
Taylor & Francis Group
Boca Raton London New York

CRC Press is an imprint of the
Taylor & Francis Group, an **informa** business

A CHAPMAN & HALL BOOK

CRC Press
Taylor & Francis Group
6000 Broken Sound Parkway NW, Suite 300
Boca Raton, FL 33487-2742

© 2017 by Taylor & Francis Group, LLC
CRC Press is an imprint of Taylor & Francis Group, an Informa business

No claim to original U.S. Government works

Printed on acid-free paper
Version Date: 20160531

International Standard Book Number-13: 978-1-4665-8256-9 (Hardback)

Library of Congress Cataloging-in-Publication Data

Names: Newlands, Nathaniel K., 1973-
Title: Future sustainable ecosystems : complexity, risk, and uncertainty / Nathaniel K. Newlands.
Description: Boca Raton : CRC Press, 2017. | Series: Chapman & Hall/CRC applied environmental statistics ; 11 | Includes bibliographical references and index.
Identifiers: LCCN 2016018594 | ISBN 9781466582569 (alk. paper)
Subjects: LCSH: Biotic communities. | Habitat conservation. | Environmental sciences. | Environmental protection.
Classification: LCC QH541 .N49 2017 | DDC 577.8/2--dc23
LC record available at https://lccn.loc.gov/2016018594

**Visit the Taylor & Francis Web site at
http://www.taylorandfrancis.com**

**and the CRC Press Web site at
http://www.crcpress.com**

Printed and bound in the United States of America by Publishers Graphics, LLC on sustainably sourced paper.

Contents

List of Figures

List of Tables

Preface

There has been rapid development of statistical theory, methods, and applications in recent years. Many leading universities and research institutions are initiating new and/or expanding existing interdisciplinary and integrative science programs. This book aims to showcase interdisciplinary and integrative approaches to environmental problem solving using statistical methods. Environmental science and management students, sustainability practitioners and professionals need guidance, and a synthesis of fundamental theory and applied modeling and application principles to better frame, understand, and assimilate highly specialized and technical congruent findings and insights that are emerging across scientific domains. This book addresses the need to connect theoretical understanding and real-world investigation. It bridges natural and social science context from a public science (i.e., broad, integrated, and interdisciplinary) perspective.

The term "real-world" in the title of this book refers to scientific assessment, modeling, and forecasting of "real" complex, adaptive ecosystems (i.e., the ocean/marine, coastal/ocean-land interface, or land/terrestrial-based forestry, agroecosystems, for example) integrated with human impacts and interactions. Our interactions with these systems vary from weak to strong, their boundaries change from open to closed, among other defining characteristics. By "real-world" I am contrasting such problems with problem solving and obtaining solutions and insights regarding more theoretical or more "idealized" or abstracted problems. "Real-world" problem solving recognizes the inherent interconnectivity, heterogeneity of ecosystems, and the ability of these systems to adapt. It aims to bridge the divide between our qualitative and quantitative understanding of more tractable solutions and defined problems to those that are more integrative, extended, and complex. This demands more of our current scientific theory, methodologies, and knowledge, to advance the frontier of science toward a future that is sustainable for both ourselves and the systems we depend on.

In *Complexity Theory for a Sustainable Future* (2008), J. Norberg and G. Cumming detail how advances in complexity theory, and understanding of complex adaptive ecosystems and socio-economic systems contribute new insights for adaptive management of these systems involving the concepts of resilience, scaling, asymmetries, self-organization, and regime shifts. Their book focuses on bridging the gap between theoretical and applied perspectives in the management of complex adaptive systems. While my book shares the same central aim to further the theoretical and applied understanding of complex dynamic systems, it expands on how statistics bridges this theory-data gap and showcases the real-world application of statistics to complexity, risk, and uncertainty.

The integrated approach, while at first may appear more heterogeneous and hybrid, has far more realistic assumptions, typically incorporates a broad set of uncertainties, provides greater methodological flexibility, and has a higher potential for inclusion in practical, operational, and industrial applications. Through my own interdisciplinary training, teaching, and public-good research, it became apparent to me that not enough emphasis and attention is being given to teaching or applying integrative problem solving in addressing environmental problems. Such integration requires more than just putting researchers or practitioners in a room together with a problem statement. It requires a sufficient appreciation, qualitative and quantitative understanding of ecosystems and knowledge of the wide variety of approaches, methods and datasets available for advancing a more holistic, interdisciplinary perspective and 'systems-level' understanding of environmental problems in terms of how we rely on and interact with ecosystems.

The broader motivation behind writing this book was to detail how statistical concepts, methodologies and applied insights are an essential requirement for translating sustainability concepts, principles, and frameworks into meaningful action. Scientific-based sustainability frameworks that go beyond broad concepts and generalities need to be devised, tested, and put into action. International and national institutions need to establish new sustainable metrics capable of integrating theoretical and applied knowledge and providing sufficient flexibility to refine them. These frameworks must be well-structured, salient, credible, and legitimate to adequately support and instill urgent action by better engaging organizations and citizens and helping them individually or collectively to respond - increasing societal adaptation capacity.

I see an urgent need to accelerate the application of statistics to ecosystems in providing enhanced methods and integrative technology-transfer solutions to address problems related to food, water, energy scarcity and climate change risks, vulnerability, impacts, and adaptation. Such problems include: pesticide transport and soil water contamination, greenhouse gas emissions, climate change impacts on crop production, crop growth and phenology dynamics, biofuel supply-chain modeling and energy supply risk assessment, remote sensing applications, and forecasting of extreme climatic events. The book highlights substantial, upcoming challenges linked with major aspects of modeling ecosystems that require further research.

This book is directed toward technical practitioners, scientific professionals, advisors and researchers, as well as upper undergraduate- and graduate-level students. It requires a fundamental grounding in theory and practice related to environmental science, and a basic level of understanding in building and applying quantitative methods for exploring data and quantitative modeling of spatial and temporal patterns and processes. Nonetheless, it provides an up-to-date reference to many significant scholarly and government literature references that provide a rich synthesis for those interested in further clarification, self-learning, and more in-depth investigation.

This book has many unique elements and features:

- It showcases the integrative, interdisciplinary scientific perspective

- It contains self-contained, state-of-the-art expositions of the core issues with extensive cross-referencing to help guide the reader through details and aspects in problem solving involving complex systems

- It connects the latest ecosystem concepts, hypotheses, and theories to applied understanding and insights

- It provides a synthesis and discussion of the latest context and knowledge of Earth system dynamics, resource depletion, and sustainability

- It includes relevant material and latest reported estimates, condensed from a broad set of regional, national, and international scientific and governmental reports

- It summarizes the current understanding and knowledge gaps related to major environmental problems and challenges, providing a foundation for putting statistics at the core of emerging sustainability science and industrial ecology

- It highlights applications involving a diverse range of data types, and the integration of new observational technology and platforms, such as satellite remote-sensing and wireless sensor networks

Chapters 1 and 2 provide a broad, integrated, perpective on sustainability and integrated risk, followed by a background and synthesis of the multi-scale changes and impacts taking place within ecosystems worldwide. Chapter 3 focuses on complex adaptive systems theory and alternative ways it can be useful in solving sustainability problems. Chapter 4 details many of the challenges in real-world problem solving and how geospatial approaches and statistical attributes of integrated risk can be statistically framed as an integrated stochastic sustainability control problem. Building on fundamentals detailed in the first three chapters, this chapter provides a synthesis of the latest measurement and analytical techniques, including research conducted by the author. Here, uncertainty arising from the strength of cumulative impacts and sensitivity between environmental, social, and economic trade-offs are key to obtaining robust solutions. Chapter 5 provides the future outlook in terms of statistical-based forecasting and the increasing use of artifical intelligence techniques in tracking and guiding our sustainable development.

Our sustainable future will require adaptive policy, science, and institutional arrangements, operational mechanisms, and sustainability tools. Such flexibility will enable us to track, integrate, and respond to environmental, economic, and societal risks more rapidly and comprehensively. I hope this book informs, motivates, and offers useful and meaningful theoretical and practical guidance, as part of a scientific evidence-based prescription for furthering global sustainable development.

N.K. Newlands (Kelowna, British Columbia, Canada)

Acknowledgments

I dedicate this book to my wife, Dr. Tracy Ann Porcelli and to my son, Tepairu Kenneth James Porcelli-Newlands for their love and compassion. I thank my teachers, mentors, colleagues, post-doctoral fellows and students around the world, who have contributed to shaping my worldview, perspective, and learning.

About the Author

Dr. Nathaniel Kenneth Newlands was born in Toronto, Canada. He has lived in North America, Europe and Africa, receiving an education within both private and public schools. He completed a B.Sc. in mathematics and physics (University of Guelph, Ontario, Canada), an M.Sc. in astrophysics (University of Calgary, Alberta, Canada), a Ph.D. in resource management and environmental studies (University of British Columbia (UBC), British Columbia, Vancouver, Canada). He completed post-doctoral training through the Institute of Applied Mathematics (IAM) and the Pacific Institute of the Mathematical Sciences (PIMS) at UBC. He is a strong lateral learner and systems thinker. He has worked as an scientific consultant, university lecturer, and since 2004, as a research scientist in environmental health within the Government of Canada (Agriculture and Agri-Food Canada). He is an associate faculty member in geography at the University of Victoria (UVic), Canada. He is a member of the Statistical Society of Canada (SSC) and The International Environmetrics Society (TIES), an editor for *Frontiers in Environmental Science*, and a regular reviewer for grant programs and is regularly solicited as a journal reviewer, with reviews conducted for over 33 different journals spanning a broad range of scientific disciplines. He has delivered talks at major international conferences, workshops, and broader public events, and has published his research across a variety of media: books, journals, newsletters and popular articles. He has participated in major, multidisciplinary, collaborative research and development projects of varying size and complexity within the water-energy-food nexus that have achieved high-quality, high impact public-good outcomes. In 2016, he was a recipient of a prestigious Government of Canada national award, the Public Service Award of Excellence in Innovation.

1

Sustaining Our Ecosystems

CONTENTS

"Let us be good stewards of the Earth we inherited. All of us have to share the Earth's fragile ecosystems and precious resources, and each of us has a role to play in preserving them. If we are to go on living together on this Earth, we must all be responsible for it."

"Let Us Be Good Stewards Of The Earth We Inherited" by Kofi Annan, 7th Secretary-General of the United Nation, Nobel Peace Prize Winner, UN Meetings Coverage and Press Releases (May 24, 2001) ©2001 United Nations. Reprinted with the permission of the United Nations.

"We must connect the dots between climate change, water scarcity, energy shortages, global health, food security and women's empowerment. Solutions to one problem must be solutions for all."

"We the Peoples" by Ban Ki-moon, 8th Secretary-General of the United Nations, Address to the 66th General Assembly (September 21, 2011) ©2011 United Nations. Reprinted with the permission of the United Nations.

1.1 Risk and Our Sustainable Future

We depend on renewable and non-renewable natural resources provided by managed ecosystems (i.e., marine and terrestrial) to provide water, food, energy, land, mate-

1

rial and other interrelated ecosystem services—now and in the future. Agriculture—
the cultivation of animals, plants, fungi, and other life forms for food, fiber, fuel,
medicine and other products is essential to sustain and enhance human life and societal well-being. However, human societal demands are rapidly increasing—being
driven, predominately, by the world's accelerating human population that is expected to reach 9.3 billion in 2050 (UN, 2011). Combating hunger and poverty
worldwide remains a tremendous challenge. Perhaps the greatest challenge of the
21st century is to meet and sustain society's growing needs for resources without
harming the environment over the long-term. More affluent societies are becoming keenly aware of the consequences of their reliance on increasingly scarce, non-
renewable natural resources. Major societal advancements, all play a role in increasing worldwide resource demands, including: modern rates of educational advancement, improved human health and longevity, higher expectations for prosperity and
affluence, utilization, expanded trade and wider availability of resources and products, urbanization, and the broader use of latest technologies. The rapid transportation of goods, mobility of people, and expansion of social communication networks
continue to increase our global societal awareness, interaction, and interdependence.

1.1.1 Risk era and societies—urgency versus uncertainty

We are living within the Holocene ("entirely recent") epoch (i.e., a geological time-
scale) that first began 11,700 years ago, after the last major ice age (International
Union of Geological Sciences (IUGS)). Human activities are having cumulative,
lasting impacts on the Earth's ecosystems and are leading to mass extinctions of
plant and animal species, polluted and perturbed oceans, land, and air. Many argue
that we now live in a new epoch called the "Anthropocene," from anthropo, for
"man" , and cene, for "new." From longer time-periods of "epochs" to the shorter
"eras," human society has undergone several major transformations as a result of
major technological advancements in agriculture, industry, computers and digital
information. While agriculture is believed to have first started 12,000 years ago,
involving irrigation and crop rotation, the Modern Agricultural Revolution that
occurred in the mid-1700s to mid-1800s led to unprecedented increases in agricultural
production and land productivity by the use of synthetic fertilizers and pesticides,
selective breeding, and mechanization. The Industrial Revolution that started circa
the 1780s and expanded through the mid-1800s, led to another technological-driven
transformation involving large-scale industrial production and new manufacturing
processes (e.g., textiles), requiring greater natural resource exploitation, supply, and
distribution. Human capabilities then continued to expand with the Information
Revolution, driven by the discovery and design of the microchip, computers, and the
production of huge amounts of digital information. Computing power, information
storage and access, technological miniaturization with rapid global communications
and networking continue to transform human societies. The Information Revolution
added a "knowledge" economy to a resource-driven one. The cumulative changes of
these recent eras have largely designed the modern capitalistic society within which
we now live.

Human behavior has been shaped so much by technology that our dependence on it has overridden our awareness of the damages and risks we face as a result of severely exploiting our ecosystems. We still know very little about how resilient ecosystems are to the broad, cumulative impacts of human activity. Nonetheless, evidence to date indicates that human activities are very likely out-pacing the ability of ecosystems to respond and adapt in a resilient and sustainable way. While still a matter of great debate, some argue that human activities are now having such a large impact on the Earth's ecosystems and its atmosphere that we may be at a high probability of extinction due to climate change. If one also considers that these changes will persist far into the future and may also be irreversible, then we may also be underestimating global catastrophic and human extinction risks (Bostrom and Cirkovic, 2011). Irrespective of how high our current extinction risk is, accumulated evidence to date on broad, cumulative impacts of climate change, are a wake-up call, and alert us that we must not only be aware and acknowledge that such change is occurring, but also that we must mitigate and adapt (IPCC, 2014b). A Fourth Industrial Revolution comprising rapid communication, innovation, and integration of smart algorithms and technologies is emerging in the face of multiple, exponential pressures and catastrophic tipping points; involving not only keeping sea-level rise below 6 meters globally (associated with global temperature rise between 1–2°C), but also addressing water scarcity, and extreme weather shocks[1,2].

In this book, I argue that modern society is in the process of being transformed, possibly irreversibly, through another new Era that I term the "Risk Era," for which the Fourth Industrial Revolution represents the first of many future development progressions or collective/societal behavioral shifts. The Risk Era is led not by resources or knowledge alone, but by our ability to identify, understand, anticipate, communicate, and take real, meaningful action to minimize known and unknown risks due to climate and global change. This challenge may be beyond the current capabilities of human intelligence, requiring us to extend our intelligence and capabilities by a broad integration of artificial, machine-based intelligence to inform societal learning, and decisions guided by human-intervention with computer-based decision support tools with machine intelligence. As argued by many leading authorities and visionaries, such as Nick Bostrom (Director of the Future of Humanity Institute at Oxford University) in his new book entitled, *Superintelligence: Paths, Dangers, Strategies*, re-conceptualization is an essential societal (i.e., human collective) function that, today, requires human intelligence to be more fully integrated and coexist with artificial, machine intelligence and learning. Human societies and technology are becoming so complex, integrated, and interactive, that answers and solutions that involve a closer human-machine symbiosis and co-learning may offer new opportunities, capabilities, and possibilities that would otherwise not be realized nor be possible to achieve.

---

[1]World Economic Forum 2016,
http://www.weforum.org/agenda/2016/01/revolution-warming-planet
[2]The Biosphere Code, "A manifesto for algorithms in the environment", The Guardian, 5 October, 2016, http://www.theguardian.com/science/political-science/2015/oct/05/a-manifesto-for-algorithms-in-the-environment

Recently, Nobel Laureates, the world's leading scientists and experts, key opinion leaders, policy makers, and the general public came together as part of a Nobel Week Dialogue focused on "The Future of Intelligence", discussing that the way we interact, learn, share our expertise, and contribute to a better world is anticipated to dramatically and fundamentally change.[3] There are clearly both beneficial and detrimental potential impacts and risks associated with the way that human societies decide to use new technologies as they transform and adapt. But as our long-term survival is being threatened, if humans alone cannot sufficiently understand and solve the complexity of highly integrated and interactive ecosystem response processes to ensure the resiliency and sustainability of ecosystems, then our capacity and ability to develop and guide solutions that utilize machine super-intelligence may be our only hope. Despite significant advancements and investments in digital technologies (i.e., the Internet, mobile phones, and all the other tools used to collect, store, analyze, and share information/knowledge digitally), the broader development benefits of such technologies (termed information and communication technologies or ICTs) have lagged behind, producing a so-called digital dividend or digital divide. The Internet still remains unavailable, inaccessible and unaffordable to a majority of the world's population.[4] Closing this digital divide is anticipated to help remove policies that promote inequality, increase social responsibilities, spur innovation, and strengthen sustainable development practices globally.

Collectively, we are becoming far more conscious of extreme events and emerging trends of inequality and different levels of perceived environmental and socio-economic vulnerabilities and risks. Adaptation and mitigation are two complementary options for responding to climate change. *Adaptation* is a process by which individuals, communities, and countries seek to cope with the consequences of climate change, and adjust to actual or expected climate and its effects, in order to either lessen or avoid harm or exploit beneficial opportunities. While the process of adaptation is not new, the idea of incorporating future climate risk into policy-making is (Burton et al., 2004). *Mitigation* is the process of reducing known risks (e.g., sources or enhancing sinks of Greenhouse Gas Emissions (GHG)) so as to limit future climate change impacts (IPCC, 2014b). While defined here in the context of anticipated climate change and its impacts and risks, these definitions can be applied more broadly to include other Earth system changes (i.e., global change). The urgency for adaptation action in response to climate risks is rapidly growing and climate change mitigation efforts alone are now insufficient to avoid further, and often negative, impacts. As the famous marine biologist, Rachel Carson, who wrote *Silent Spring* observed more than 40 years ago, modern science and technology so strongly rely on "...specialists, each of whom see his own problem and is unaware of or tolerant of the larger frame which it fits" (Walker, 1999). Interdisciplinary learning and training (i.e., both theoretical and applied), while anticipated to contribute

---

[3]December 9, 2015, Gothenburg, Sweden, http://www.nobelweekdialogue.org/

[4]The World Development Report (2016) of The World Bank estimates 4 billion people worldwide still have no access to the Internet (WBG, 2016).

substantially to major scientific advancement this century, still faces, however, many substantial scientific, policy and societal challenges (Sideris and Moore, 2008).

1.1.2 Risk communication and awareness

Humanity is entering into a renaissance—a dramatic change from one that perceived people and nature as separate actors to one that sees them as interdependent social-ecological systems that are dynamic and inter-connected, and subject to both gradual and abrupt changes (Folke and Gunderson, 2012). Whether this renaissance and our heightened awareness and perception of impacts, vulnerabilities and risks will, over the longer term, translate into meaningful, lasting human welfare and sustainable developmental benefits, is not clear. Regardless, we continue to be forced to react to a myriad of possible foreseen and unforeseen situations and events. It is up to us whether we react in an informed, unbiased, and precautionary way. We also continue to interact, utilize, and exploit ecosystems to an extent we never have before, with increasing sophistication, intensification, and expansion. If human development is to continue and ecosystems are to remain resilient under such external demands and pressures, and we are to sustain our development over the long-term, we must find new and better ways to ensure natural and industrial production systems can function and adapt congruently.

Naomi Klein, a social activist known for her political analyses and criticism of corporate globalization, in her new book, *This Changes Everything: Capitalism vs. The Climate*, argues, we *can* design, deploy and adopt a better economic system that can ensure environmental, economic, and social sustainability, by "reining in corporate power, rebuilding local economies, and reclaiming democracies" (Klein, 2014). She argues that facing climate change head on requires changing capitalism, as it is irreconcilable with a livable climate, and because the laws of nature can't change. While natural laws don't change, how we interact with nature can. Market-based insurance instruments (e.g., weather-index based property, production, or other types of insurance), government intervention, subsidies and disaster funding relief may be, on their own, insufficient over the longer-term. Science and technology play a major role in driving socio-economic transformations and adaptations. For this reason, it is very likely that new sustainability and resilience decision-making tools at the community level, which integrate economic, environmental, and social considerations, will drive the next transformation and adaptation of our human activities around the globe towards a sustainable future[5]. Nonetheless, consistency, influence, and equity in decision-making will be necessary, involving far more than the needs and desires of the world's wealthy corporate and political elite. A 2014 report on the economic risks of climate change in the United States from the "Risky Business Project" provides a climate risk assessment, leveraging recent advances in climate modeling, econometric research, private sector risk assessment, and scalable cloud computing (processing over 20 terabytes of climate and economic data) now available[6], with

---

[5]US Resilience Toolkit, `toolkit.climate.gov`

[6]`climateprospectus.rhg.com`

interactive maps and content[7] (Gordon, 2014). Three general collective actions are prescribed: 1) business adaptation – changing everyday business practices to become more resilient, 2) investor adaptation – incorporating risk assessment into capital expenditures and balance sheets, and 3) public sector response – instituting policies to mitigate and adapt to climate change.

Explaining risk and how it changes within and across different societies is extremely challenging. Moreover, while public apathy over climate change and one's inability to comprehensively assess risk is often attributed to knowing too little science, or a lack of technical reasoning, there is a lack of support for this viewpoint. Instead, there is a dichotomy in risk between the personal interest of *individuals* linked with social or cultural groups that they belong to, and broader, societal *collective* interests (Kahan, 2012)[8]. Such "cultural polarization" around assessing risk has been measured to be greatest in those who have the highest level or degree of science literacy and technical reasoning capacity.

Collective action can be modeled as a public good game of cooperation, where contributions depend on the risk of future losses, and the window of opportunity for cooperation to thrive, changes in the presence of risk. In such games, the net risk of collective failure provides a possible escape from the Tragedy of the Commons (Santos and Pacheco, 2011). Appraisal theory provides an integrative framework explaining how we perceive risk according to a wide array of cognitive and affective approaches, whereby individuals elicit different emotions and action tendencies (Keller et al., 2012; Watson and Spence, 2007). Figure 1.1 outlines five dimensions of outcome (i.e., desirability, agency, fairness, certainty, and coping potential) of environmental risk perception, whereby one's appraisal differs between various risks (e.g., climate change). The cumulative or combined effect of individual appraisals within and between emergent cultural groups in society, therefore, plays a major role in our capacity and ability to address environmental risk (Kahan et al., 2012). A society that is so strongly driven to minimizing risks, however, also possesses a collective anxiety to complexity and interrelated hazards (e.g., disasters, epidemics, catastrophes), which can disengage groups from understanding the consequences of individual and societal, collective decision-making. Paradoxically, much of our collective anxiety stems from an ability to determine the causality and precise triggers and root causes of various threats (Walker (1999) and references therein).

1.1.3   Risk assessment and modeling

Risk assessment and modeling is directed toward understanding the processes and interrelationships that generate different types of risk. A simple equation for environmental risk is,

$$R = f(H, V, C) \tag{1.1}$$

where R denotes "risk," generally defined as "the probability of a future loss" or more precisely as, "the chance of harmful effects to human health or to ecological systems

---

[7]Risky Business, `riskybusiness.org`
[8]`www.culturalcognition.net/kahan/`

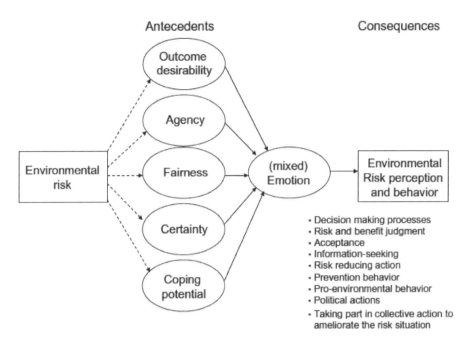

FIGURE 1.1: An integrative model based on appraisal theory in the context of environmental risk (Keller et al., 2012; Watson and Spence, 2007).

resulting from exposure to an environmental stressor"[9] (United States Environmental Protection Agency (EPA)), or similarly, "the probability (or likelihood) of that event combined with the severity of its consequences" (Byrd and Cothern, 2000; Gordon, 2014; Beer, 2003). A stressor is any physical, chemical, or biological entity that can induce an adverse response. Stressors may adversely affect specific natural resources or entire ecosystems, including plants and animals, as well as the environment with which they interact. Risks associated with climate change, for example, may include sea-level rise causing large-scale losses of coastal property and infrastructure, increased frequency of high rainfall events causing floods and/or Atlantic hurricanes/Pacific cyclones causing extreme heat waves affecting livestock and crop survival, severe damage to infrastructure and energy systems, or long-term impacts to human and environmental health (i.e., soil health, water and air quality) (Gordon, 2014).

A formal definition of risk is that it is a function, f, of a given threat or hazard, H. A threat or hazard is an event, condition, or circumstance that could lead to an unplanned or undesirable outcome/s. They can arise from to a sequence of interrelated causes and cumulative impact/s, or a single event with broad impact/s. Hazards have several main defining characteristics, namely: waiting time, duration/persistence, frequency and intensity. In Equation 1.1, V denotes "vulnerability" as a measure of the statistical likelihood of success of a particular hazard, and C is a function of the expected consequence or total cost (i.e, environmental,

---

[9]www.epa.gov/risk_assessment/basicinformation.htm

economic and social) of the impact of a particular hazard experienced by a vulnerable target. Vulnerability (V) itself can be further expressed as a response function,
g, involving a measure of an exposure, E of a system to hazards, its intrinsic sensitivity (or system robustness), $S(E)$, to an exposure E, and its adaptive capacity
(or system resistance), A, expressed as,

$$V = g(E, S(E), A). \tag{1.2}$$

Random variables are those that depend on chance and can be discrete (finite number
of values) or continuous (unbroken chain of values or outcomes with infinite number
of possibilities). For the discrete case, risk is a function involving the probabilities,
$p(i)$ associated with all potential events, i, linked or attributed to a given hazard,
H, and their expected outcomes or consequences, expressed as,

$$R = \sum_i p(i) \cdot c(i). \tag{1.3}$$

For the continuous case, we define $p(t)$ to be the probability density function (pdf)
of events, where t denotes time. This probability density is often derived from
available data that may have been historically observed, or model scenario output,
or a combination of both. The above discrete equation can then be re-expressed
as *integrated* risk associated with a given hazard, in relation to a set of continuous
probability distribution functions for E, S, A, and C, where,

$$R = \int_E \int_S \int_A p(A|E,S)p(S|E)p(E) \cdot c(E,S,A)dAdSdE \tag{1.4}$$

with c denoting the total cost or consequence of a *sequence* of potential events
$i = (1, ..., n)$, not just a single event, i.

Let a function, $f(t)$, be defined as the pdf of waiting times, T, until the occurrence of an event, assumed to be a non-negative and continuous random variable. Its
corresponding *cumulative* distribution function (cdf), is defined as $F(t) = P(T < t)$,
and is the probability that an event has occurred by time t. The probability of survival just before time t (i.e., duration t), or likewise, the probability that an event
has not occurred by duration t, is then (Lawless, 2002),

$$S(t) = P(T \geq t) = 1 - P(T < t) = 1 - F(t) = \int_t^\infty f(x)dx \tag{1.5}$$

Threat or hazard waiting times can also be expressed in terms of the *rate* of occurrence of an event after a duration t; as the ratio of the density of events at t, divided
by the probability of surviving to that time without the event occurring, given by,

$$\lambda(t) = f(t)/S(t) = -\frac{d}{dt}logS(t) \tag{1.6}$$

One can obtain a single equation that links the survival and hazard rate functions.
By applying boundary conditions $S(0) = 1$ (event will not have occurred by duration

0), and integrating over time from 0 to t, one obtains the probability of surviving to duration t, expressed as a function of hazards for durations up to time t, or,

$$S(t) = exp\left(-\int_0^t \lambda(x)dx\right) = exp(-\Delta(t)), \quad (1.7)$$

where $\Delta(t)$ is defined as the *cumulative* risk over time duration $[0, t]$. Further assumptions regarding the intensity of a threat or hazard, and ranking the severity of risk outcomes are also important, including their consistency in reporting. In some cases, "likely" may describe outcomes with at least a 67% (or 2-in-3) chance of occurring. Likewise, for extreme outcomes, different levels of chance related to a benchmark or baseline threshold may be assumed. Such benchmarks may be based on the historical conditions, whereby a 1-in-20 chance (or 5%) of being worse than (or better than) a particular threshold, or 1-in-100 outcomes as a 1% chance of occurring.

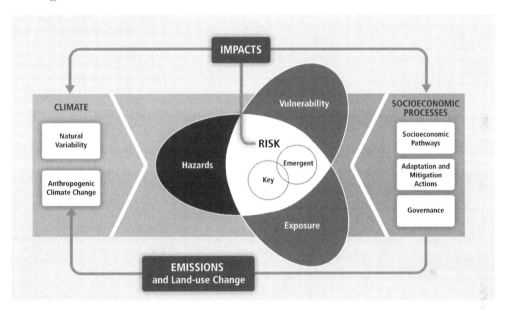

FIGURE 1.2: Schematic of the complex interaction among the physical climate system, exposure and vulnerability producing risk. Vulnerability and exposure, are, as the figure shows, largely the result of socio-economic development pathways and societal conditions. Changes in both the climate system (left) and developmental processes (right) are key drivers of the different core components (vulnerability, exposure, and physical hazards) that constitute risk (IPCC (2014a), Chapter 19, Section 19.2, Figure 19-1, pg. 1046).

Figure 1.2 provides a schematic of the complex interaction between the physical climate system, exposure, and vulnerability that produce *integrated risk*. The survival and hazard rate functions above assume homogeneous spatial and temporal conditions in terms of explanatory variables related to climate, environment, threats and hazards, vulnerability, exposure, sensitivity, and adaptive capacity. In reality, all these variables are heterogeneous and dynamic, and are not governed by the same survival or hazard functions.

The Cox proportional hazards model, while often applied in the analysis of event-time data, assumes a nonlinear relationship between a hazard function and covariates, but avoids having to specify an exact form of a baseline hazard function. This model, however, breaks down in the presence of multiple, competing risks, because of a lack of direct interpretation in terms of survival probability, and due to its censoring assumption/s. Censoring occurs when the information on survival time is incomplete, where random and non-informative censoring (i.e., censoring is not related to probability of events occurring) is often assumed to reduce model bias. Here, stratified approaches are required that involve the concept of *sub-distribution hazard*, defined as the hazard of failing from a given cause in the presence of competing events, given survival or failure due to different causes (Scrucca et al., 2010).

Thus, in real-world systems, leading predictors of risk (i.e., the set of key explanatory variables) may keep switching and changing, according to variation in their strength and causal interdependence in time and space. More general approaches for survival modeling, such as accelerated life, proportional, and general rate hazard, exponential and Weibull, time-varying covariates, and time-dependent effects models, all explore greater complexity between explanatory variables to assess, model and forecast *direct*, *integrated* or *cumulative* risk.

1.1.4   Risk management—tolerance, reward, volatility

There are four major types, categories or so-called quadrants of risk, namely: hazard and/or operational (classed as "pure" risks), and "speculative" risks that are financial and/or strategic. Different institutions consider, and different threats involve, one or more of these risk types. Figure 1.3 provides an overview of the four categories (Byrd and Cothern, 2000). Hazard risks arise from property, liability, or personnel loss exposures, and are mitigated through insurance. The Basel Committee is a forum that provides cooperation on banking supervisory matters and enhances the understanding of operational risk faced by institutions alongside addressing key supervisory issues and improving the quality of banking supervision worldwide (Perera, 2012)[10]. They define *operational* risk as, "any loss resulting from inadequate or failed internal processes, people and systems or from external events." Operational risks are outside the hazard risk category and are typically considered to be mitigated by simple changes in behavior. Financial risks arise from the effect of market forces on financial assets or liabilities, and include market, credit, liquidity, and price risk. Market risk is the uncertainty about an investment's future value because of potential changes in a market. Liquidity risk is the risk that an asset cannot be sold on short notice without incurring a loss. With the motivation to increase profits for stakeholders, many companies focus on these economic risks, devoting time to identifying and then monitoring to mitigate them. Broader, integrative risks, termed *strategic risks*, arise from trends in the economy and society, including changes in the economic,

---

[10]The Basel Committee, `www.bis.org`

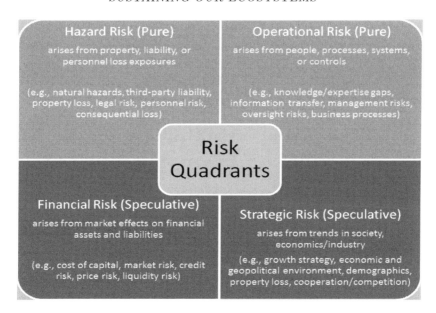

FIGURE 1.3: General categories of risk.

political, competitive environments, and demographic shifts. These are often the risks that many organizations neglect to comprehensively consider.

Broad, integrated threats like climate change impact all four risk quadrants to varying extents. Many organizations may inadequately assess strategic and hazard risks, spending most of their efforts on tracking operational and financial risks with high likelihoods of occurrence and a varying degree of direct consequences for the organization itself. In the past, many organizations have simply relied on government or other collective forms of intervention if large-scale disasters and threats take place. Today, many organizations are adopting broader sustainability goals whereby they are driven not only to generate profits and increased economic sustainability, but also to serve customer demands for increased environmental and social sustainability. However, as the occurrence of extreme events and threats continue to rise, a "new normal" is establishing itself, whereby organizations must themselves assess, model, and forecast broader, integrated threats and hazards. There is a clear need to develop and provide sustainability decision-support tools to a wide array of different organizations, to enable them to reliably assess, model, and forecast *integrated* environmental risk.

Environmental risk-based decisions depend on risk tolerance, which can vary greatly, because it depends on both qualitative and quantitative factors related to how an individual or collective perceives, understands, and assesses risk. Risk tolerance must account for the time and differences in consumer demand (i.e., tolerance heterogeneity). This is because environmental and climate change risks involve high near or short-term mitigation and adaptation costs, as well as uncertain long-term returns and benefits. Individuals, groups, and organizations decide if they should receive or profit less on an immediate return on an investment in exchange for less risk, so as to be risk adverse. Generally, the more risk adverse the individual and/or the lower their risk discount rate, the higher their willingness to pay. In terms of the assessment side, there exists a critical reliance on risk tolerance on assessing vulnerability (V) and the system (e.g. ecosystem) response function comprising exposure, sensitivity, and adaptive capacity. The stability of this vulnerability response function is termed "sustainability" and its sensitivity is termed "resilience."

The Capital Asset Pricing Model (CAPM) (Sharpe, 1964) is a simple model that illustrates the fundamental trade-off between risk and return, defining the "risk reward to volatility ratio." Despite the long history of actual risk-bearing and risk-sharing in organized financial markets, the Capital Asset Pricing Model was developed at a time when the theoretical foundations of decision-making under uncertainty were relatively new and when basic empirical facts about risk and return in the capital markets were not yet known (Perold, 2004). This model assumes investors maximize economic utility, have homogeneous expectations, are rational and risk adverse, possess a diversified portfolio of investments that are liquidable and divisible, have no influence on market prices, incur no transaction or taxation costs, and rely on the same available market information at the same time.

An optimal monetary reward (i.e., not necessarily sustainable) portfolio involving a set of feasible choices, involves a risk-free return rate (R_f), and an expected return on individual capital asset, $E(R_i)$, in relation to an expected market return, $E(R_m)$. The model assumes that there exists an equilibrium (called the securities markets line or SML) between an individual asset's risk premium (i.e., $E(R_i) - R_f$) with a market's risk premium $(E(R_m) - R_f)$, factoring in a risk scaling or conversion factor for a specific asset, β_i, such that,

$$E(R_i) - R_f = \beta_i(E(R_m) - R_f), \tag{1.8}$$

Here β_i is termed the "risk sensitivity" or "risk reward to volatility ratio." The optimal risk reward to volatility ratio (associated with an optimal risk strategy) can be obtained by setting the rate of change in an investment goal or utility function (i.e., derivative $E(R_m)/dR_m$) to zero, and solving for β_i. Considering the following goal function,

$$U = R_f + \beta_i(E(R_m) - R_f) - \frac{1}{2}\kappa\beta_i^2\sigma^2(R_m), \tag{1.9}$$

where κ and $\sigma^2(R_m)$ denote risk tolerance level and risk volatility, respectively, the following optimal solution is obtained,

$$\beta_i^* = \frac{E(R_m) - R_f}{\kappa\sigma^2(R_m)} \tag{1.10}$$

This equation shows that the risk reward to volatility ratio (β) is directly proportional to market risk premium and inversely proportional to the level of risk tolerance or aversion, and level of risk. More generally, this ratio is termed the "Sharpe ratio" (Sharpe, 1964),

$$\beta_i = \frac{Cov(R_i, R_m)}{\sigma^2(R_m)}, \tag{1.11}$$

and is the ratio of the *covariance* between individual and market return rate, and the *variance* or volatility in market return rate itself. Such variance can be extended to include environmental and social variance contributions to economic variance.

The Capital Asset Pricing Model (CAPM) predicts the optimal return or discount rate; the rate at which future cash flows generated by an asset should be discounted depending on risk. $\beta > 1$ indicates greater than average riskiness (similarly, $\beta < 1$ is lower than average risk). Investments with higher risk are "discounted" at a higher rate. However, even with higher discount rates, large future environmental or climate change impacts have negligible effect on current decisions, unless a constant discount rate is replaced by a declining rate (i.e., hyperbolic discounting) (Karp, 2005). The CAPM model, while useful in highlighting how assessing risk can benefit return on investments under a prescribed goal or utility, has very idealistic assumptions that are not realistic in the real-world, whereby economic, environmental, or social variables are heterogeneous in time and space, and dynamic, exhibiting both fast shocks, and slower changes due to internal, *non-equilibrium* behavior. In real-world markets, for example, investment returns are typically not normally-distributed, investors have heterogeneous information and biased expectations, and transaction and taxation costs are incurred in relation to incentives. There is also no special requirement for the broad consideration of short/near versus longer-term investment decisions, such that larger, observed variation or volatility in expected return would result—much higher than the CAPM predicts. The basic assumptions of the first versions of the CAPM are, therefore, inadequate for identifying optimal, sustainable investment portfolios that must consider a wider array of economic, environmental, and social risks linked with technology and scientific discovery, environmental impacts, and policy changes.

The CAPM has been recently extended for heterogeneous beliefs, the elimination of risk-free lending and borrowing, non-marketable assets, and dynamic investment opportunities, including other effects (Perold, 2004). For example, Breeden (1979) has extended the CAPM with endogenous risk by devising a Consumption Capital Asset Pricing Model (CCAPM). Here, *endogenous risk* refers to the risk from shocks that are generated and amplified *within* a system, whereas *exogenous risk*, refers to shocks from *outside* or external to a system (e.g., extreme climatic and/or weather events such as droughts, heat-waves, floods). Financial markets are subject to both types of risk (Danielsson, 2013; Danielsson and Shin, 2002).

There are many arguments that favor a low rate of return on long-term investments, either when high uncertainty is considered in future market interest rates or in an assumed risk-free rate. Long-term investment is also favored to reduce endogenous risk and enhance social welfare that can yield immediate socio-economic consequences (protection) and alter the possibility landscape of future consequences (e.g., insurance) in mitigating or adapting to future environmental or climate-change hazards and threats (Sandsmark and Vennemo, 2007; Weitzman, 2001). As highlighted by Levin, "we all discount our own futures, enjoying todays certain benefits against the uncertainty of what will happen tomorrow" (Levin, 2012). Accordingly, "how utility is measured and what discount rate is chosen requires an awareness and understanding of temporal allocation, of pro-sociality as concern for the welfare of others, of cooperation, and of the social norms that sustain cooperative behavior." Furthermore, he emphasizes that we live in a global commons where the collective or cumulative consequences of individual actions have externalities and social costs. Conventional market mechanisms are not adequate to address such aspects, which are magnified as the scale of organization increases, especially beyond individual nations that serve as the primary functional unit of global decision-making.

Using the extended CCAPM model of optimal portfolio selection involving correlation between rates of return and levels of endogenous risk, Sandsmark and Vennemo (2007) have further explored alternative investment hypotheses that are aimed at reducing environmental and climate-change related risks. Such investments are justified in cost-benefit accounting based on lower discount rates, despite low, short-term expected returns. To illustrate this, let $U_s(c_t)$ represent the social utility of consumption (i.e., consumption goal), and $\phi(t)$ be the discount factor for consumption (e.g., $\phi(t) = exp(-\delta(\tau - t))$. The associated benefit or "payoff" at time t is,

$$\int_t^\infty \phi(\tau)U(c_{t+\tau})d\tau = \int_t^\infty U(c_{t+\tau})e^{-\delta(\tau-t)}d\tau \qquad (1.12)$$

such that the present value of \$1 additional consumption τ time units in the future is $\phi(\tau)U'(C_{t+\tau})$. The *social discount rate* is the amount one is willing to spend; less than a dollar now to prevent a dollar's worth of damage in a year, or a decade in the future, to avert climate change. Let $r(t)$ denote the social discount rate. This rate is then the negative of the rate of change of the present value in a future marginal utility of consumption, expressed as (Karp, 2005),

$$r(t) = \frac{-d[ln(\phi(t))U'(c_t)]}{dt} = \eta(t) + \nu(c_t)\frac{dc(t)/dt}{c(t)}, \qquad (1.13)$$

where $\eta(t)$ is a rate of time preference for discounting, and $\nu(c)$ is the elasticity of a future marginal utility of consumption. In hyperbolic discounting, one can consider the rate of time preference to decline over time, or the social discount rate as a whole to decline.

Let K_i represent savings invested in a climate stabilizing asset i over an initial period, whereby $K_i = x_i(W - C_1)$, and W denotes wealth and C_1 consumption over this period of time. Further, let C_2 be a random variable with distribution function, $F(C_2, K_i, \eta)$, of consumption in a second period following this initial one, where η as an *exogenous* climate change signal such as frequency of extreme events, or an increase in temperature. Under these assumptions, a modified form of Equation 1.8 for expected return under the assumptions of the CCAPM (i.e.,CAPM with consumption), is given by (Sandsmark and Vennemo, 2007),

$$E(R_i) = R_f + \beta_i(E(R_m) - R_f) - \frac{\int u'(C_2)F(C_2, K_i, \eta)dC_2}{E(u'(C_2))} \qquad (1.14)$$

where, similar to Equation 1.11, $\beta_i = Cov(C_2, R_i)/\sigma^2(C_2)$ for a given asset i. Here, the variance of C_2 or $Var(C_2) = \sigma^2(C_2)$ represents the correlation between a climate investment and consumption (so-called "self-insurance"). The third term in the above equation is like a penalty function, and measures the "self-protection" level in relation to a low optimal expected rate of return on the environmental asset, assuming that investing in the environmental asset would increase the probability of higher consumption. In summary, self-protection is the increasing probability of favorable outcomes from investment that reduces risk, while reducing risk of the consequences or severity of risk is termed self-insurance. This model represents risky investing, explaining a lower discount rate as a balance between investments that reduce the future, longer-term, consequences of risks in relation to shorter-term risk linked with current investments.

It is widely acknowledged that climate change will have a broad impacts on economies and financial markets in the future. However, there are major knowledge gaps on the investment implications of climate change at a portfolio level. This is linked to uncertainty about how climate change might impact the underlying drivers of the major asset classes around the world, and how institutional investors will respond (Mercer, 2011). An economic model used in a recent study conducted by the International Finance Corporation (World Bank Group) (IFC)'s Advisory Services in Sustainable Business explores the implications of climate change scenarios for the allocation of assets to address strategic risks. It highlights three major issues for "early warning" risk management processes and systems: 1) traditional approaches to modeling strategic asset allocation must account for climate change risk, 2) new approaches to Strategic Asset Allocation are required to tackle fundamental shifts in the global economy, and 3) climate policy could contribute as much as 10% to overall portfolio risk according to the IFC's *TIP* framework, that assesses three variables for climate change risk: the rate of development and opportunities for investment into low carbon technologies (*Technology*), the extent to which changes to the physical environment will affect investments (*Impacts*), and the implied cost of carbon and emissions levels resulting from global policy developments (*Policy*).

In the IFC report, four future scenario projections were considered that span the range of likelihood from highest to lowest, respectively. This report compares alternative impacts of climate change policy (i.e., carbon cost, and greenhouse-gas emission drivers as leading explanatory variables). The *first scenario* involves regional divergent and unpredictable action (carbon cost of US $110/t $CO_{2,e}$[11] and carbon emission reductions of 50 Gt $CO_{2,e}$/yr in 2030 as -20% of the Business-as-Usual or BAU benchmark scenario). The *second scenario* characterizes *delayed* action and late (after 2020) stringent policy measures (carbon cost of US $15/t $CO_{2,e}$ by 2020 then dramatic rise to US $220/t $CO_{2,e}$ and 40 Gt $CO_{2,e}$/yr in 2030 (i.e., -40% BAU). The *third scenario* envisions strong, transparent, internationally co-ordinated action. Such action is advocated for and outlined in the "Stern Report," a comprehensive study on the "Economics of Climate Change" conducted by Sir Nicholas Stern, Head of the UK Government Economic Service, and a former Chief Economist of the World Bank (Stern, 2007). In this third scenario (the so-called

---

[11]$CO_{2,e}$ denotes carbon-dioxide net equivalent of all greenhouse gas emissions.

Stern Action scenario), carbon cost is US \$110/t $CO_{2,e}$ globally and 30 Gt $CO_{2,e}$/yr in 2030 (i.e., -50% BAU). In a *fourth* scenario, involving climate breakdown and no mitigation efforts beyond current BAU, carbon cost is US \$15/t $CO_{2,e}$, relying on or limited to the European Union (EU) Emissions Trading Scheme[12] (and other regional schemes), with emissions of 63 Gt $CO_{2,e}$/yr in 2030 (equivalent BAU).

The IFC analysis also explores diversification of investment (i.e., the portfolio mix) for a 7% level of return, considering: cash, sovereign fixed income, and credit. It also explores both developed/emerging markets and private equity markets with renewables and renewable-themed equities. Forest, agriculture, and real-estate investment are also considered. Overall, the main findings indicate that the economic *costs* of new, more stringent climate policy aimed at reducing our reliance on carbon, will require global economic markets to absorb an estimated US \$8 trillion cumulatively, by 2030. It further indicates that additional investment in technology will increase portfolio risk (for a representative set of portfolio mixes that were considered) by about 1%, with global investment accumulating to US \$4 trillion by 2030. Such investment is expected to be beneficial for many institutional portfolios. Having some economic perspective on risk and investment that address climate change, we now turn our focus to the multi-functional context of ecosystems, their resilience and sustainability, and the importance of ecosystem goods and services to economic security and our survival.

1.1.5 Sustainability and resiliency

A general qualitative definition for "sustainability" is, "an act, a process or a situation, which is capable of being upheld, continued, maintained or defended" (De Vries, 2013). Sustainability science aims to better understand (and quantify) the dynamics of Coupled Socio-Economic and Environmental Systems (SES) and changes in their processes, emergent patterns, and evolution. This necessitates an integrated, interdisciplinary approach to problem solving. It focuses on interactions between resources, its users, and the governance required to sustain ecosystems while also delivering what people need and value. The multifunctional perspective of agriculture and the inescapable interconnections of agriculture's different roles and functions. "Integrated" can be defined in broad terms as, "The interdisciplinary combination or coupling of theory, empirical data, techniques, methods, and tools in devising, testing, and deploying problem solving research and development, scientific and technological solutions for real-world operational application."

The concept of multifunctionality recognizes agriculture as a multi-output activity producing not only commodities (food, fodder, fibers, and biofuels), but also non-commodity outputs such as ecosystem services, landscape amenities, and cultural heritages. This definition of multifunctionality is the one adopted by the Organisation for Economic Co-operation and Development (OECD), and used by the International Association of Agricultural Science and Technology for Development (IAASTD). It acknowledges that: (1) multiple commodity and non-commodity

---

[12]`ec.europa.eu/clima/policies/ets/index_en.htm`

outputs are jointly produced by agriculture; and (2) some of the non-commodity outputs may exhibit the characteristics of externalities or public goods, such that markets for these goods function poorly or are nonexistent (IAASTD, 2009a). Sustainable development of agricultural production systems (i.e., agroecosystems), for example, involves an integrated development pathway of environmental, economic, social and institutional goals, and achievable targets (IAASTD, 2009a). Whether for the agricultural or any other resource-based or resource-dependent sector, sustainable development encompasses social, economic, environmental, and institutional dimensions; the first of these address key principles of sustainability, while the final dimension addresses key institutional policy and capacity issues (Figure 1.4)[13] (see also Huang et al. (2015) on a review on ecosystem multifunctionality).

All ecosystems, as open systems, are exposed to gradual changes in climate, nutrient loading, habitat fragmentation, and biotic exploitation. Nature is usually assumed to respond to gradual change in a smooth way. However, studies on lakes, coral reefs, oceans, forests, and arid lands have shown that smooth change can be interrupted by sudden drastic switches to a contrasting state. Although diverse events can trigger such shifts, recent studies show that a loss of resilience usually paves the way for a switch to an alternative state. This suggests that strategies for sustainable management of such ecosystems should focus on maintaining resilience (Scheffer et al., 2001). One definition of *resilience* is, "the capacity of a system to absorb disturbance, undergo change and still retain essentially the same function, structure, identity, and feedbacks". Accordingly, one metric of the resilience of a system is the, "full range of perturbations over which the system can maintain itself" (De Vries, 2013). Theoretical findings on system resilience offer insights on the effect of different kinds of disruptions, but lack specific context and relevance to specific types of disturbances and interactions associated with various industries, resource sectors, and ecosystem services (Biggs et al., 2012). Understanding how specific types of disturbances and kinds of interactions detrimentally affect the long-term sustainability of managed ecosystems is challenging, because ecosystems are continually evolving and adapting in response to cumulative impacts and disturbance effects. However, the ability of SES systems to self-organize indicates that they may all rely on a core set of critical probabilistic or stochastic processes (i.e., processes that involve chance) to create and maintain their functioning (Holling, 2001, 1973).

Our current understanding of resilience is that it involves non-linear dynamics, thresholds, uncertainty and a lack of predictability (i.e., surprise) such that "resilience is not always a good thing" (Folke, 2006). Currently, there are four resilience concepts spanning different system size and extent of cross-scale interplay or level of integration across spatial and temporal scales. Each of these concepts of resilience has their own defining characteristics, empirical support, and application context, namely: engineering, ecological/ecosystem, and coupled social-ecological resilience. In more isolated systems with one identifiable steady state, or for dynamics in the vicinity of a global stable equilibrium, the concept of engineering resilience can quan-

---

[13] Multifunctional perspetive of agriculture,
`www.grida.no/graphicslib/detail/a-multifunctional-perspective-of-agriculture_1097`

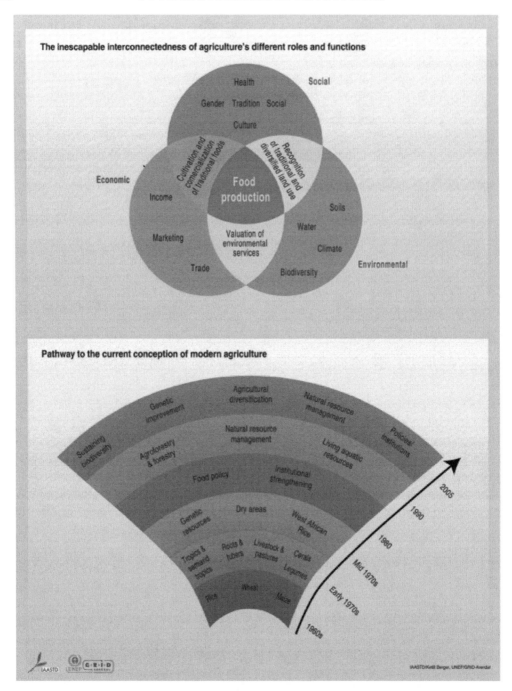

FIGURE 1.4: The multifunctional perspective of agriculture and the inescapable interconnections of agriculture's different roles and functions. The concept of multifunctionality recognizes agriculture as a multi-output activity producing not only commodities (food, fodder, fibers and biofuels), but also non-commodity outputs such as ecosystem services, landscape amenities and cultural heritages. Sustainable development of agriculture therefore involves an integrated development pathway of environmental, economic, social and institutional goals, and achievable targets (IAASTD, 2009a,b) (Cartographer/Designer: Ketill Berger, GRID-Arendal).

tify and explain the persistence/constancy, and recovery/renewal, regeneration, and reorganization after a disturbance in their observed dynamics.

Persistence is the ability of a stable steady state to stay within a stability domain, "domain of attraction" or "basin of attraction" under external perturbations. For large ecological systems involving a higher degree of randomness and pure random, independent interactions, the probability of persistence is close to zero (Allesina and Tang, 2012; May, 2001, 1972). The frequency and strength of trophic consumer-resource relationships (i.e., predator-prey) interactions may play a leading role in ensuring stability in ecosystems under destabilizing effects arising from mutualistic and competitive interactions (Allesina and Tang, 2012). In this way, individual- and group-level interactions and their strength may more strongly drive the intrinsic dynamic properties that we observe when they emerge at the population or broader system level (e.g., possible dynamic states and their attraction domains/resilience limits, switching between sets of leading variables, asymmetry of interactions etc.). In the original concept of resiliency, a system restores its equilibrium after a disturbance occurs with a given speed of return or "return time." Formally, return time is not the exact time it would take to return to an equilibrium, as this would theoretically be infinite, but instead is defined as the inverse of the time in which deviations from the equilibrium state shrink by a factor of $1/e$ (Grasman et al., 2005). This is similar to the relaxation time in physics, where a reference reduction factor of $1/2$ (half-time in radioactive decay) is adopted. In addition to return time, the size of a stability domain can be used as a measure of engineering resilience.

Given that many different variables are typically involved, a system that undergoes continual disturbance may move between different stability domains rather than between steady states or reaching a single one. In this way, understanding the variability in a system by applying statistics, rather than necessarily identifying all its possible steady states, may offer greater knowledge and insights on how systems continually adapt to increase their resiliency. Transitions between multiple equilibria and stable steady states, that evolved ecosystems exhibit, may provide enhanced system robustness and persistence when responding to uncertain, external shocks and changes in their ability to maintain internal functions. The existence of multistable states in a system's behavioral plasticity leads to departures from standard equilibrium-based assumptions. Ecosystems may rely on nonstationarity as it creates an array of extended possibilities, in addition to a core suite of dynamics. Stochastic assessment, modeling, and forecasting methodologies may offer significant insights on their boundaries or the thresholds of stability that are far-from-equilibrium.

While SESs exhibit a wide plethora of functional responses to disturbances, involving the interplay of integrated feedbacks and cross-scale interactions in time and space, they also, at a fundamental level, are continually reorganizing, transforming and adapting their structures. Accordingly, just as there are resilience *functional aspects*, there are also *structural aspects*: latitude, resistance, precariousness, and panarchy (Norberg and Cumming, 2008). These aspects are defined as follows: *Latitude* is the maximum amount (threshold) that a system can be reversibly or irreversibly changed, *resistance* is the ease or difficulty of a structural or process change in a system, *precariousness* is a measure of how close a system is to a limit or

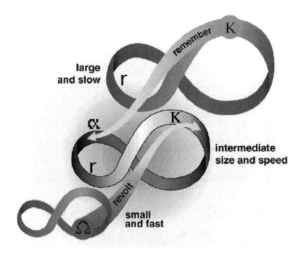

FIGURE 1.5: Panarchy as a heuristic model of nested adaptive renewal cycles with cross-scale interplay (Folke (2006), as modified from Gunderson and Holling (2002)).

threshold, and *panarchy* is the degree that a certain hierarchical level of an ecosystem is influenced by other levels.

A heuristic (or conceptual) model of panarchy, formed from two nested adaptive renewal cycles associated with the fast and slow dynamics of Complex Adaptive Systems (CAS) and their cross-scale interplay is depicted in Figure 1.5 (Folke, 2006; Gunderson and Holling, 2002; Gunderson and Pritchard, 2002). This model describes four main development phases: 1) r-phase (exploitation and exponential change), 2) K-phase (conservation, stasis and rigidity), 3) Ω-phase (re-adjustment and collapse), and 4) α-phase (re-organization and renewal). Gradual or slower changes accompany rapid transitions, whereby disturbance and development are closely intertwined. This cycling sustains patterns and processes at the same time that it develops and adapts them. As highlighted by Folke (2006), while much attention is often placed on systems undergoing the r and K phases, far less attention is typically given to the "backloop" (i.e., Ω and α phases).

Understanding resilience involves understanding the fundamental characteristics or essential elements of CAS system dynamics. Levin (1998) identifies several essential elements: 1) sustained diversity and individuality of components, 2) localized interactions between components, 3) an ability to select structural components prescribed by different sets of local interactions and variables, and 4) replication and enhancement. Such essential elements are congruent to properties of CAS that consist of heterogeneous groups of individual agents that interact locally, and evolve in their genetics, behaviors and spatial distributions based on the outcome of their interactions (so-called path dependency). Others identify six elements, namely: 1) aggregation, 2) nonlinearity (generates path dependency), 3) hierarchical organization, 4) continual adaptation, 5) perpetual novelty (linked with the elements of diversity and individuality), and 6) far-from-equilibrium dynamics (Folke, 2006). The essential elements of CAS can be conceptual or theoretically related to leading *statistical* characteristics of ecosystem patterns and processes (Newlands, 2006). Figure 1.6 provides an overview of these statistical elements. This views systems from an inherently statistical perspective involving continual variability within and across domains of attraction. Such stochastic elements distinguish fundamental features that are associated more closely with spatio-temporal emergent patterns, from those closely associated with underlying ecological processes, namely: *pattern* (i.e., heterogeneity (aggregation), interaction (nonlinearity), scaling (hierarchical organization) and *process* (i.e., self-organization (adaptation), asymmetry (novelty/diversity), stochasticity (far-from-equilibrium)), as outlined in Figure 1.6. Vasconcelos et al. (2011) have recently explored a way to find the optimal set of predictors or "principal axes of stochastic dynamics" for CAS in terms of the above essential elements. The statistical approach they employ finds eigenvalues and eigenvectors for the stochastic dynamics applicable to systems of arbitrary dimension. The eigenvalues measure the amplitude of a stochastic force and their corresponding eigenvectors show the direction toward which such force acts. They show how CAS systems can be represented as a system of coupled Langevin equations (i.e., the Fokker–Planck equations), whereby drift and diffusion functions can be related to measurements of a system without prior knowledge on its dynamics. However, they assume that the underlying stochastic process is stationary, such that the drift and diffusion coefficients do not depend explicitly on time. Research is now exploring non-stationary assumptions and ways to derive link functions for quantifying the sensitivity in the response of such eigenvectors and eigenvalues to each of the essential elements of CAS depicted in Figure 1.6. Further testing of real-world systems of different size and complexity, where multi-variate monitoring time-series data is available, is required to validate such statistical approaches.

While desirable states of a system may exist and influence its dynamics (i.e., change in space and time), the number and characteristics of these states also change (panarchy heuristic model). Humans are continually disturbing and influencing ecosystems, driving them to be in perpetual non-equilibrium, whereby their resilience does not coincide with a distinguishable or measurable attraction domain. A more meaningful and robust measure of resilience, other than return time or size of an attraction domain, is then the distance of the state of the system from an attraction domain boundary, whereby resilience is smaller nearer such a boundary (Martin et al., 2011). The statistical distribution of this distance metric to various boundaries may offer clues to forecasting ecosystem dynamics, where there is sufficient memory (or temporal autocorrelation) between an ecosystem's upcoming or new dynamic state in relation to states it has exhibited in the past.

A dynamic model is next presented and discussed to illustrate how one can obtain mathematical equations for the resilience measures of return time, attraction domain size, and distance to an attraction domain boundary (Martin et al., 2011; Anderies et al., 2002). This model focuses on the growth of grass shoots (i.e., the portion of a plant representing new upward growth following seed germination in advance of leaf development) in relation to their crowns (i.e., all the above-ground plant biomass consisting of stems, leaves, and reproductive structures) and competition from shrubs for soil nutrients and water, and animal/cattle/livestock grazing within

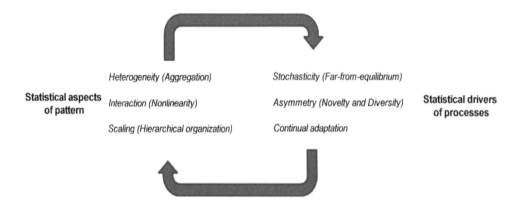

Heterogeneity (Aggregation)　　　　　　Stochasticity (Far-from-equilibrium)

Statistical aspects　Interaction (Nonlinearity)　　　　　Asymmetry (Novelty and Diversity)　　**Statistical drivers**
of pattern　　　　　　　　　　　　　　　　　　　　　　　　　　　　　**of processes**

Scaling (Hierarchical organization)　　　Continual adaptation

FIGURE 1.6: Statistical elements of pattern and process, and their interaction within CAS.

a managed perennial grassland/rangeland ecosystem or community cattle pasture. A plant canopy represents multiple crowns growing within a specific area. Some broader context is first needed before presenting the dynamic model in greater detail.

A major challenge in managed perennial grassland or rangeland systems is to trade-off increasing short-term profit while preserving the long-term sustainability of the rangeland. Increasing the stocking rate (number of animals per unit area) of the rangeland increases short-term gain, but can lead to effectively irreversible land (soil, plant, and water) degradation of the rangeland and detrimental effects on livestock nutrition, weight, and reproductive status, if the resultant grazing pressure is too high. A low stocking rate leads to income loss. In some cases, government tax deferral provisions or schemes are available to help farmers/ranchers by offsetting or buffering against high income losses in periods of excess soil moisture/flooding and prolonged drought by enabling them to sell part of their breeding herd and defer a taxable portion of sale proceeds to the following year. Rangeland or pasture management then seeks to adjust the stocking rate in relation to grazing records that balances these competing environmental and socio-economic sustainability goals (Martin et al., 2011; Diaz-Solis et al., 2006).

Let $s(t)$ and $c(t)$, $w(t)$ represent shoot, crown and shrub biomass at time t, respectively. Shrub biomass is assumed to be constant over time (i.e., $w(t) = w$). Let r_c , and the crown growth rate. Let a_c be the shoot biomass growth rate per unit crown biomass as the portion of shoot growth that is dependent on crown growth (so-called tiller potential), and r_s be the shoot biomass growth rate per unit crown per shoot biomass (i.e., a rate that depends on the interaction of shoot and crown). When shoots are present, $r_c s$ is the growth rate of crowns, and when no shoots are present, crowns are assumed to grow at a constant rate of $\delta_c c$. Let s_{max} be the maximum shoot biomass per unit area limited by soil nutrients and available water, and w_{max} be the maximum shrub biomass per unit area. Let γ_g be the fraction of shoots that are removed by cattle grazing (grazing pressure). Grazing pressure is defined as the ratio of demand to supply of dry matter requirements for livestock to the quantity of forage available for them within a given pasture area at a specific time. As it increases, the likelihood of over-grazing other preferred plant species also increases.

In this model, grazing pressure is assumed to be constant over time, even though it is expected to vary substantially between pasture areas and over time (years). Finally, let $\alpha_{ws}(w/w_{max})\beta$ be a nonlinear, competitive plant-community effect of the shrubs on grass, where β is a constant. A system of ordinary differential equations that represents this ecosystem under the above simplifying assumptions, is given by,

$$\frac{dc(t)}{dt} = r_c s - \delta_c c \tag{1.15}$$

$$\frac{ds(t)}{dt} = c(a_c + r_s s)\left(1 - \frac{s}{s_{max}} - \alpha_{ws}\frac{w}{w_{max}}\beta\right) - \gamma_g s \tag{1.16}$$

The above system is termed "autonomous," as time, t, does not appear on the right-hand side of either of these equations. To solve this system when shrubs are at their maximum growth potential and crown biomass is at an equilibrium, we set $dc(t)/dt = 0$, $w = w_{max}$ and $\beta = 1$. This yields $c = r_c s/\delta_c$ by solving the first of these equations. Substituting this into the second equation and collecting terms in s, s^2 and s^3, yields

$$0 = f(c, s) = s\left(\frac{r_c a_c}{\delta_c}(1 - \alpha_{ws}) - \gamma_g\right) + s^2\left(\frac{r_c}{\delta_c}(r_s(1 - \alpha_{ws}) - \frac{a_c}{s_{max}}\right) - s^3\frac{r_c r_s}{\delta_c s_{max}} \tag{1.17}$$

Here, we define f as a function of the time-dependent variables s and c. Dividing all terms by the common factor of s, and substituting parameter values of $r_c = 1$, $\delta_c = 1$, $a_c = 0.3$, $r_s = 3$, $s_{max} = 1$, $\alpha_{ws} = 0.50$ into the above equation, yields a *phase equation* for generating a *phase plot or diagram* of s versus γ_g. This reveals three equilibria: $s \geq 0$, $s = 0$ and $ds(t)/dt = 0$, given by,

$$(0.15 - \gamma_g) + 1.2s - 3s^2 = 0 \tag{1.18}$$

For $s = 0$, grazing pressure is unstable below 0.15, and stable above this value. For $s > 0.20$ there is a stable equilibrium, and for $0 < s \leq 0.20$, there is an unstable equilibrium. Applying linear stability analysis by evaluating the partial derivative of f with respect to s (i.e., the Jacobian, $J(s) = \partial f/\partial s$) within the local neighborhood of the solution $s = 0$, yields,

$$J(s) = 0 = \left(\frac{r_c a_c}{\delta_c}(1 - \alpha_{ws}) - \gamma_g\right) + 2s\left(\frac{r_c r_s}{\delta_c}(1 - \alpha_{ws}) - \frac{a_c}{s_{max}}\right) - 3s^2\frac{r_c r_s}{\delta_c s_{max}} \tag{1.19}$$

Evaluating the Jacobian under the same parameter values yields the *characteristic polynomial* of solutions,

$$J(s) = (0.15 - \gamma_g) + 2.4s - 9s^2. \tag{1.20}$$

Solving the above equation equilibrium shoot growth, s_e yields,

$$s_e = 0.2 + \sqrt{0.09 - \frac{\gamma_g}{3}}, \quad \gamma_g \in (0, 0.27). \tag{1.21}$$

When $\gamma_g \leq 0.27$, an equation for resiliency with respect to the equilibrium solution, r_e, is $R(\gamma_g) = -s_e(1.2 - 6s_e)$, and when $\gamma_g > 0.27$, $R(\gamma_g) = 0$ which indicates that no asymptotically stable solution exists. The resultant expression for the domain of attraction measure of resiliency (Martin et al., 2011) is,

$$R(s_e) = \frac{min(s_e, 2\sqrt{0.09 - \gamma_g/3})}{s_e - min(s_e, 2\sqrt{0.09 - \gamma_g/3})}. \tag{1.22}$$

Based on this resiliency measure, when there is only one global stable equilibrium, $R(s_e)$ is infinite and when $\gamma_g > 0.27$, resiliency is zero (see Figure 1.7).

Additional statistical modeling could, in the future, expand this model assuming maximum growth (i.e., s_{max} and w_{max}) and/or alternatively shoot, crown and shrub growth rates are all dependent on regional climate variability in temperature and precipitation. This would enable exploring the effect of prolonged inter-annual drought, and a prediction of the resiliency of this ecosystem integrating longer-term variability and its impact. Ecosystem management is also important to factor into such model dynamics, such that the deterministic model could be made stochastic by adding statistical factors/terms involving additional random variables and uncertainty.

More complex dynamic simulation models exist for guiding stocking rate decisions that are represented in this illustrative model example. More complex models incorporate aspects of livestock/animal production (voluntary intake, digestibility, net energy intake), plant/grass growth, losses from trampling and feces deposition, and variation in plant/crop biomass due to changing nutrient levels, temperature, rainfall and irrigation. Optimal production is then determined in relation to different stocking rates, pasture, and climate characteristics (Diaz-Solis et al., 2006). However, as such models are simulation-based and involve a large number of variables, it is a challenge to determine multi-equilibria, as well as to obtain analytical expressions for resilience measures. Network resilience that takes into account network topology and interactions (Smith et al., 2011), resilience measures for individual-based, cellular automata and multi-agent-based simulation models (Batt et al., 2013; Ortiz and Wolff, 2002; Janssen and Carpenter, 1999) and simulated industrial supply chains (WEF, 2013) are all currently being further explored. More on resiliency, stability analysis, and computational approaches for higher-dimensional dynamic systems will be discussed in Chapter 3.

The need for realistic assumptions and the real-world operational application of such resilience models of land-use and land productivity over all land-types and their management is critical. For instance, a recent study on the protection and conversion of California's rangeland ecosystems (1984–2008) based on aerial observational imagery, indicates close to half has been converted for residential/commercial development, with additional land conversion to agriculture (Cameron et al., 2014). In protecting biodiversity and critical wildlife habitat and other ecological benefits of

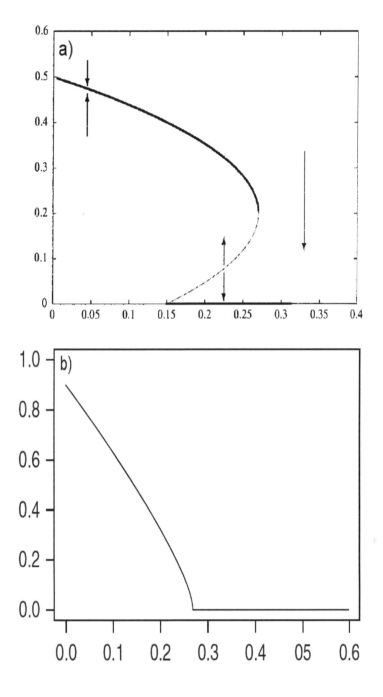

FIGURE 1.7: a) Phase diagram showing equilibria of the rangeland ecosystem (i.e., Shoot biomass versus grazing pressure), $s = 0$ (unstable for $\gamma_g < 0.15$, stable for $\gamma_g > 0.15$), and $s > 0.2$ (stable), $s \in (0, 0.2)$ (unstable) (the arrows of motion show the transition in the system's dynamic behavior from old and new steady states, following a shock), b) Resiliency $R(\gamma_g)$ versus grazing pressure.

rangelands, an agricultural tax incentive program with voluntary enrollment report-edly has provided widespread protection. While this program has protected 37% of the remaining rangeland from commercial conversion, it has not deterred intensive agricultural conversion. Remaining rangeland (24%) is protected by private conservation organizations or public agencies, through land or easement ownership, while 38% had no protection status at all.

There is an urgent need for a better understanding of the dynamic, adaptive behavior of complex systems and their resilience in the face of disruptions, recognizing that steady-state sustainability models are simplistic or reductionist. Investigation of more real-world complex systems with less restrictive or idealistic assumptions would enable a more integrated approach to systems analysis, beneficial intervention, and improvement of resilience (Fiksel, 2006). Ecosystems that face high risks with high resilience are expected to perform "better" than those facing similar risks but with low resilience. Similarly, systems with low risk but also low resilience may perform the same as, or even worse than, systems with high risk and high resilience (Linkov et al., 2014). While this seems conceptual, and straight forward, it is idealistic, because in real-world CAS, risks are dynamic and interchangeable. Moreover, systems adapt to varying levels of risk, with different levels of resilience, involving a complex interplay of dynamic interactions. As highlighted by Fiksel (2006), for system models to be reliable and useful for the integrated assessment, modeling and forecasting of sustainability and resiliency, they must capture emergent behaviors and dynamic relationships of real CAS. System representations need to go beyond just assuming that system components can be simply linked together. Instead, they need to focus on representing and reproducing observed CAS behavior, given that it is not necessarily governed by a simple sum of their individual parts. Fundamental aspects related to system dynamics are presented next, to provide sufficient background for understanding how ecosystems evolve and transform over time.

In general, a dynamic system can be represented as a linear/nonlinear system of differential equations, which expressed in vector form is,

$$\frac{dX^n(t)}{dt^n} = \mathbf{G}\left(t, X(t), \frac{dX}{dt}, \dots, \frac{dX^{n-1}}{dt^{n-1}}\right). \tag{1.23}$$

For a first-order system, $n = 1$, and the above system reduces to,

$$\dot{\mathbf{X}} = \frac{dX(t)}{dt} = \mathbf{G}(X(t), t), \tag{1.24}$$

where $X(t)$ is the state of the system at a time (i.e., unknown continuous functions of the real-valued variable t, representing time), and $\mathbf{G}$ is a vector-valued function of both the dependent variable (i.e., state) and an independent variable (i.e., time) (i.e., $n+1$ variables. This system is termed *non-autonomous*. As an initial value problem, $X(t_0) = X_0$, so that the system will have a solution defined for time displacements $|t - t_o| < \delta$, for some $\delta > 0$. The time interval of existence of a solution depends on the functional form of $\mathbf{G}$ and the initial conditions. System solutions may be bounded $\alpha < t < \beta$ or be a global solution that is infinite $-\infty < t < \infty$. Readers are referred to two reference sources that provide additional theoretical background and context in applying mathematics (i.e., calculus and differential equations) in the study of biological system dynamics (Edelstein-Keshet, 2005) and climate system dynamics (Shen, 2015).

A non-autonomous system representation can be transformed into an *autonomous* one by removing the time dependency of the evolution or changes in state, $X(t)$, so that its dynamics depend only a function $\mathbf{F}$ of location (i.e., *path* dependent not *time* dependent). Let the state and rate vectors, $\mathbf{X}$ and $\mathbf{F}$, comprise n components, denoted as, $\mathbf{X} = (x_1, x_2, \ldots, x_n)^\mathsf{T}$, and $\mathbf{F} = (f_1, f_2, \ldots, f_n)^\mathsf{T}$, respectively (Note: $\mathbf{X}^\mathsf{T}$ indicates the transpose of a vector $\mathbf{X}$). For a first-order linear/nonlinear autonomous system of ordinary differential equations (ODEs) we obtain,

$$\begin{aligned}
\dot{x}_1 &= f_1(x_1, x_2, \ldots, x_n) \\
\dot{x}_2 &= f_2(x_1, x_2, \ldots, x_n) \\
&\vdots \qquad \vdots \qquad \vdots \\
\dot{x}_n &= f_m(x_1, x_2, \ldots, x_n)
\end{aligned} \tag{1.25}$$

The set of all points that satisfy $f_j(x_1, x_2, \ldots, x_n) = 0$ is called the $x_j - nullcline$, and the intersection of all nullclines is termed an equilibrium or fixed-point of the system (i.e., not to be construed with a stationary, critical, or turning point). A fixed or invariant point of a function is an element of the function's domain that is mapped to itself by the function, where $\mathbf{X} = f(\mathbf{X})$. Whereas, a stationary or critical point is where all the partial derivatives are zero (equivalently, the gradient is zero), such that $\partial \mathbf{F}/\partial \mathbf{X} = 0$, where $\mathbf{F}$ is a differentiable function of several variables. A *turning point* is a point at which the derivative changes sign. If the function is differentiable, then a turning point is a stationary point; however, not all stationary points are turning points. If the function is twice differentiable, stationary points that are not turning points, are horizontal inflection points.

A *multivariate* Taylor series expansion of the right-hand side of the above differential equation, yields,

$$\dot{\mathbf{X}} = \mathbf{F}(\mathbf{X}^*) + \left.\frac{\partial \mathbf{F}}{\partial \mathbf{X}}\right|_{\mathbf{X}^*} (\mathbf{X} - \mathbf{X}^*) + \frac{1}{2}\left.\frac{\partial^2 \mathbf{F}}{\partial \mathbf{X}^2}\right|_{\mathbf{X}^*} (\mathbf{X} - \mathbf{X}^*)^2 + \ldots \tag{1.26}$$

At an equilibrium point, $\mathbf{F}(\mathbf{X}^*) = 0$, and for $\mathbf{X}$ sufficiently close to $\mathbf{X}^*$, the higher order terms will be very close to zero, and one obtains the approximation,

$$\begin{aligned}
\dot{\mathbf{X}} &= \mathbf{F}(\mathbf{X}^*) + \left.\frac{\partial \mathbf{F}}{\partial \mathbf{X}}\right|_{\mathbf{X}^*} (\mathbf{X} - \mathbf{X}^*) + \ldots \\
&= \left.\frac{\partial \mathbf{F}}{\partial \mathbf{X}}\right|_{\mathbf{X}^*} (\mathbf{X} - \mathbf{X}^*) + \ldots
\end{aligned} \tag{1.27}$$

The $m \times n$ matrix of partial derivatives is called the *Jacobian* matrix, denoted $\mathbf{J}$, given by,

$$\mathbf{J} = \frac{\partial \mathbf{F}}{\partial \mathbf{X}} = \frac{\partial(f_1, f_2, \ldots, f_m)}{\partial(x_1, x_2, \ldots, x_n)} = \begin{pmatrix} \frac{\partial f_1}{\partial x_1} & \frac{\partial f_1}{\partial x_2} & \cdots & \frac{\partial f_1}{\partial x_n} \\ \frac{\partial f_2}{\partial x_1} & \frac{\partial f_2}{\partial x_2} & \cdots & \frac{\partial f_2}{\partial x_n} \\ \vdots & \vdots & \ddots & \vdots \\ \frac{\partial f_m}{\partial x_1} & \frac{\partial f_m}{\partial x_2} & \cdots & \frac{\partial f_m}{\partial x_n} \end{pmatrix}. \tag{1.28}$$

The Jacobian summarizes the effects of slight changes in a variable on itself and all the other variables. In the case $m = n$, $\mathbf{J}$ is a square matrix, and its determinant is the *Jacobian determinant* of $\mathbf{F}$. If this determinant is non-zero, then the function $\mathbf{F}$ has an inverse that is differentiable within a local neighborhood of the equilibrium $\mathbf{X}^*$. Given $\delta\mathbf{X} = (\mathbf{X} - \mathbf{X}^*)$ for a small displacement near an equilibrium point $\mathbf{X}^*$ (i.e., as per linear perturbation or stability analysis), then, $\dot{\delta\mathbf{X}} = \mathbf{J}^* \delta\mathbf{X}$, where $\mathbf{J}^*$ is the Jacobian evaluated *at* the equilibrium point.

The behavior of the system near an equilibrium point is related to the eigenvalues of $\mathbf{J}(\mathbf{X}^*)$. The eigenvalue equation is given by,

$$\mathbf{J}\mathbf{V} = \lambda\mathbf{V} \tag{1.29}$$

where $\mathbf{V}$ are eigenvectors, and λ are eigenvalues. Involving the identity matrix, $\mathbf{I}$, one obtains,

$$(\mathbf{J} - \lambda\mathbf{I})\mathbf{V} = \mathbf{0} \tag{1.30}$$

and the eigenvalues are the roots of $det(\mathbf{J} - \lambda\mathbf{I})$ (i.e, the "characteristic polynomial" where $\Delta(\mathbf{J}) = det(\mathbf{J} - \lambda\mathbf{I}) = 0$). In the case of a 2×2 square matrix,

$$\lambda^2 - \tau(\mathbf{J}) + \Delta(\mathbf{J}) = 0 \rightarrow \lambda_{1,2} = \frac{\tau(\mathbf{J}) \pm \sqrt{\tau^2(\mathbf{J}) - 4\Delta(\mathbf{J})}}{2} \tag{1.31}$$

where $\tau()$ is the trace and $\Delta()$ the determinant of a matrix, respectively. The determinant is a property of any square matrix that describes the degree of coupling between the equations. The determinant equals zero when the system is not linearly independent, meaning one of the equations can be decomposed into a linear combination of the other equations. The eigenvalues of the Jacobian matrix are, in general, complex numbers. Let $\lambda_j = \mu_j + i\nu_j$, where μ_j and ν_j are the real and imaginary parts of the eigenvalue.

$$e^{\lambda_j t} = e^{\mu_j t} e^{\nu_j t} \tag{1.32}$$

$$e^{\nu_j t} = \cos(\nu_j t) + i\sin(\nu_j t) \tag{1.33}$$

The system solution is a superposition of terms $e^{\lambda_j t}$, where λ_j are eigenvalues of the Jacobian. When the eigenvalues of a fixed point are complex, the point is called a focus. The dominant eigenvalue determines the long-term behavior of a system governed by the Jacobian. If eigenvalues all have negative real parts, then the system is stable near the stationary point, and if any eigenvalue has a real part that is positive, then the point is unstable. In cases where the largest real part of the eigenvalues is zero, $\mathbf{J}$ does not allow for an evaluation of the stability.

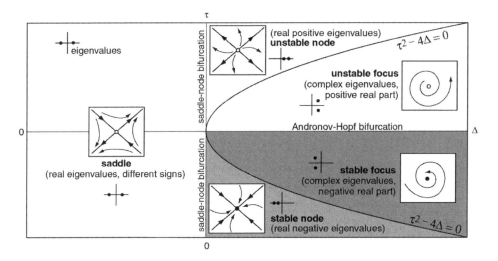

FIGURE 1.8: Classification of fixed points of dynamic systems, based on their eigenvalues. $\tau = \text{tr}(\mathbf{J})$ is the trace and $\Delta = \det(\mathbf{J})$, the determinant of the Jacobian, $\mathbf{J}$. The term Hopf bifurcation (or Andronov–Hopf bifurcation) refers to a periodic solution (i.e., a self-excited oscillation) from an equilibrium as a parameter crosses a critical value, whereas a saddle-node bifurcation refers to a collision and disappearance of two equilibria.

Furthermore, the real part (i.e., magnitude) of an eigenvalue $\mu_j > 0$, for any j, and $e^{\mu_j t}$ increases with time, t, so that trajectories tend to move away from stationary point. The complex part only contributes an oscillatory component to the solution. A *limit cycle* is a closed trajectory in a phase plot where at least one other trajectory spirals into it, in the limit of time approaching infinity or negative infinity. A system that experiences limit cycles has a dominant, complex eigenvalue (Figure 1.8). In higher-dimensional spaces, eigenvalues may be repeated several times (i.e., termed degeneracy).

Real-world systems are subject to known and unknown forces (i.e., environmental uncertainty or noise). Processes operate at different time scales (i.e., slow and fast), with different degrees of simultaneity. In addition, there are unresolved stochastic influences that can fundamentally alter how a system behaves and evolves (Wilks, 2008; Imkeller and Monahan, 2002). Such dynamics can lead to abrupt shifts between different regimes that can be catastrophic and irreversible (i.e., regime *shifts*, not to be confused with the term "regime *change*") (Biggs, 2009; Folke, 2004; Scheffer et al., 2001). In climate dynamics, *weather* is the fast scale on the order of several days, while *climate* is the slow scale, on the order of years or longer. Processes may be purely stochastic or involve both deterministic and stochastic contributions (i.e., mixed stochastic). A stochastic system (i.e., pure or mixed) involves having a sufficient degree of random variability whereby the state of a system involves probability—i.e., a situation where there are multiple plausible outcomes, each having varying degrees of certainty or uncertainty. Deterministic models can be extended by considering the statistical mean, variance and covariance contributions (i.e., main and interaction effects) on a modeled process. In this way, Ordinary Differential Equation (ODE) models can be reformulated as Stochastic Differential Equations (SDEs). One may be able to obtain analytical and/or numerical solutions of system observable behaviors and resolved variables, while at the same time offering insights involving underlying, hidden, or unresolved variables that influence system dynamics.

A general form for a stochastic differential equation (SDE) is given by (Moon and Wettlaufer, 2013; Hasselmann, 1976),

$$\frac{dX(t)}{dt} = f(t, X(t)) + u(t, X(t)) + h(t, X(t))\frac{dW(t)}{dt}, \quad t \geq 0 \tag{1.34}$$
$$X(t_0) = X_0$$
$$f(t, 0) = h(t, 0) = 0 \tag{1.35}$$
$$t_0 \geq 0$$

where f is a function of resolved (explicit) variables, and u is a function of unresolved (implicit) variables (both termed drift processes), and h is a diffusion process with diffusion matrix $\Sigma = h(t, X)h^{\mathsf{T}}(t, X)$. W is defined as a Weiner process—a continuous-time stochastic process where for $t \geq 0$ with $W(0) = 0$, the increments $(W(t) - W(s))$ follow a Gaussian distribution with a mean of 0 and variance of $(t - s)$ for any $0 \leq s < t$. Also, increments for non-overlapping time intervals are assumed independent. A common Weiner process is a Brownian process that is a random walk with random step sizes. The Brownian process is the limiting case of a random walk Weiner process, as its time increment goes to zero. Integrating Equation 1.34 yields,

$$dX(t) - X(0) = \int_0^T f(s, X(s))ds + \int_0^T u(s, X(s))ds + \int_0^T h(s, X(s))dW(s), \quad \forall t > 0 \tag{1.36}$$

If we consider that the functions in Equation 1.36 have both a term that depends on location $X(s)$, and one that does not, and consider that all system variables are resolved, then,

$$f(s, X(s)) = \alpha_1(t)X(s) + \alpha_2(t) \tag{1.37}$$
$$u(s, X(s)) = 0$$
$$h(s, X(s)) = \sigma_1(t)X(s) + \sigma_2(t)$$

When $\sigma_1(t) = 0$, then the stochastic integral does not depend on $X(s)$ and is termed "additive" noise. If $\alpha_2(t) = 0$ and $\sigma_2(t) = 0$, then it is termed "multiplicative" noise. For the case of multiplicative noise, the corresponding SDE is then,

$$dX(t) = \alpha_1 X(t)dt + \sigma_1 X(t)dW(t) \tag{1.38}$$

Hence, the stochastic term is a function of both t and $W(t)$, and one cannot apply the chain rule as per Riemann calculus, but needs to expand it to include higher-order terms involving second-order derivatives, such that,

$$df = \frac{\partial f}{\partial t}dt + \frac{\partial f}{\partial W}dW + \frac{1}{2}\frac{\partial^2 f}{\partial t^2}(dt)^2 + \frac{\partial^2 f}{\partial t \partial W}(dt dW) + \frac{1}{2}\frac{\partial^2 f}{\partial W^2}(dW)^2 + \dots \quad (1.39)$$

Neglecting all terms having dt to a power higher than 1, and assuming Brownian motion where $(dW)^2 = dt$ and $dt dW = (dt)^{3/2}$, we obtain,

$$df = \left(\frac{\partial f}{\partial t}dt + \frac{1}{2}\frac{\partial^2 f}{\partial W^2}\right)dt + \frac{\partial f}{\partial W}dW \quad (1.40)$$

This equation is known as "Itô's Lemma or Rule" for obtaining differentials of stochastic functions. Given a stochastic process $X(t)$ and a stochastic process, $Y(t) = g(t, X(t))$, then applying Itô's rule, $Y(t)$ is governed by the following SDE equation,

$$dY = \frac{\partial g}{\partial t}dt + \frac{\partial g^\intercal}{\partial X}dX + \frac{1}{2}dX dX^\intercal \frac{\partial^2 g}{\partial X^2} \quad (1.41)$$

We now reduce the general SDE equation (Equation 1.34) under the assumption that $X(t)$ has no unresolved variables, such that,

$$\frac{dX(t)}{dt} = f(t, X(t)) + h(t, X(t))\frac{dW(t)}{dt} \quad (1.42)$$

We then apply Itô's rule to this reduced equation governing the evolution of $X(t)$ over time, such that for any function $Y(t) = g(t, X(t))$, we obtain,

$$dY(t) = \frac{\partial g}{\partial t}dt + \frac{\partial g^\intercal f}{\partial X}dt + \frac{\partial f^\intercal h}{\partial X}dW + \frac{1}{2}\tau\left(hh^\intercal \frac{\partial^2 g}{\partial X^2}\right)dt \quad (1.43)$$

where $dt = 0$, $dX dt = 0$, and $dW_i dW_j = \delta_{i,j}dt$, and τ denotes the matrix trace and $(\cdot)^\intercal$ the vector/matrix transpose. It can be shown that (Oksendal, 2010),

$$\frac{\partial g^\intercal f}{\partial X} = \sum_i f_i(X, t)\frac{\partial g}{\partial X_i} \quad (1.44)$$

$$\tau\left(hh^\intercal \frac{\partial^2 g}{\partial X^2}\right) = \sum_i \sum_j (hh^\intercal)_{ij}\frac{\partial^2 g}{\partial X_i X_j} \quad (1.45)$$

An Itô differential generator or operator, $G(\cdot)$ can then be defined by taking expected values of the function Y,

$$G_t(\cdot) = \frac{\partial}{\partial t} + \sum_i f_i(X, t)\frac{\partial}{\partial X_i}(\cdot) + \sum_i \sum_j (h(X,t)h(X,t)^\intercal)_{ij}\frac{\partial^2}{\partial X_i X_j}(\cdot) \quad (1.46)$$

To solve the main evolution Equation 1.36, one must compute stochastic integrals, which differ from Riemann deterministic integrals. For the first two terms in Equation 1.36, the standard Riemann integral can be applied, for f and u that are assumed to be *smooth* functions with bounded derivatives across the definite limits $[0, T]$. A function is smooth if its partial derivatives to all orders exist and are continuous within a given domain. The integral range is partitioned into n intervals $0 = s_0 < s_1 < ... < s_{N-1} < s_{N=T}$, gives

$$\int_0^T f(s)ds = \lim_{N \to \infty} \sum_{j=0}^{N-1} f(s_j)(s_{j+1} - s_j) \tag{1.47}$$

The third term is a stochastic integral for which two different forms exist—the Itô and Stratonovich forms, depending on whether an interval's midpoint or endpoint is computed in relation to values of the Weiner process, $W(s)$. The Itô integral, based on the interval *endpoint*, is computed as,

$$\int_0^T f(s)ds = \lim_{N \to \infty} \sum_{j=0}^{N-1} f(s_j)(W(s_{j+1}) - W(s_j)) \tag{1.48}$$

whereas, the Stratonovich integral, based on the interval *midpoint*, is computed as,

$$\int_0^T f(s)ds = \lim_{N \to \infty} \sum_{j=0}^{N-1} f\left(\frac{s_j + s_{j+1}}{2}\right)(s_{j+1} - s_j) \tag{1.49}$$

In cases where analytical and/or numerical solutions of system stability are not possible due to strong nonlinearity, for example, a *Lyapunov function* may be obtained for assessing system stability (or instability) without actually solving a system of equations. A Lyapunov function has positive values everywhere except at a specified equilibrium point (i.e., for function V, $V(0) = 0$ and $V(X) > 0$, for all $X \neq 0$, termed "positive definite"), and decreases (or is non-increasing) along every possible system trajectory (i.e., has continuous first-order partial derivatives). In this way, the function must be continuous and have a unique minimum at the equilibrium point with respect to all other points within a domain of attraction, D, and V must never increase for any system trajectory within or bounded by D. For stochastic systems, trajectories must stay near an equilibrium point "on average" (i.e., mean or expected values of change in the function V satisfy the expectation, $E[dV(t)] \leq 0$ (Pukdeboon, 2011)).

In general, if there exists a positive-definite function V (i.e., $V(t, X(t))$), such that the first-order derivative satisfies,

$$\frac{d}{dt}(V(t, X(t))) = \frac{\partial V}{\partial t} + \sum_{i=1}^{d} \frac{\partial V}{\partial X_i} f_i(t, X(t)) \leq 0 \tag{1.50}$$

whereby system trajectories that are deterministic, are determined by the function, $f(t, X(t))$. According to Equation 1.25 the system is then stable and $V(t, X(t))$ is its Lyapunov function. There is generally no applicable method for finding Lyapunov functions in deterministic or stochastic systems, and so trial and error or stochastic optimization methods such as the variable gradient method or sum of squares decomposition are applied.

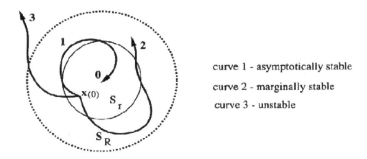

curve 1 - asymptotically stable

curve 2 - marginally stable

curve 3 - unstable

FIGURE 1.9: Alternative concepts of stability (Slotine and Li, 1991).

Applying the stochastic generator in Equation 1.46 for a Lyapunov function V, yields a *stochastic* Lyapunov function that can be applied in real-world problems to explore stochastic stability. This yields the equation,

$$G(V(t, X(t))) = \frac{\partial V}{\partial t} + \frac{\partial V}{\partial X} f(t, X(t)) + \tau \left(h(t, X(t)) h(t, X(t))^\intercal \frac{\partial^2 g(t, X(t))}{\partial X^2} \right).$$
(1.51)

For a fractional change (Zeng et al., 2014), the above equation becomes,

$$dV(t, X(t)) = G(V(t, X(t))) dt + \frac{\partial V}{\partial X} h(t, X(t)) dW(t)$$
(1.52)

For different proposed Lyapunov stability functions, stochastic stability can be assessed, such as stability of a given defined stochastic process, stability of the probability distribution of a stochastic process and observable events, or the stability of distribution moments (i.e., mean, variance, covariance) of a stochastic process or set of interacting processes. Ling et al. (2013) discuss the construction of Lyapunov functions for nonlinear stochastic dynamical systems. In particular, for real-world systems having large degrees of freedom/dimensionality and significant nonlinearity (so-called multi-degree-of-freedom systems or MDOFs), they outline how the integration of the Lyapunov function method, alongside statistical or stochastic averaging, could offer greater efficiency in evaluating the stochastic stability of stochastic systems than other individual approaches, such as the largest Lyapunov exponent method (refer to Ling et al. (2013) and references therein). The presence of time-delays, feedbacks, and the degree that slow and fast scales can be separated, all influence the reliability of a stability analysis or assessment method that is applied to spatial-temporal data of a given CAS.

A stochastic solution is termed *stochastically stable*, if, for every pair of $\epsilon \in (0, 1)$ and $S_r > 0$, there exists a $\delta = \delta(\epsilon, r, t_o) > 0$, and $x(t_o) = x_o$, such that the probability, P (Zeng et al., 2014; Mao, 1997),

$$P\left(|x(t, t_o.x_o)| < S_r, \forall t \geq t_o \right) \geq 1 - \epsilon$$
(1.53)

whenever $|x_o| < \delta$, Here, $| \cdot |$ denotes the absolute value of a measure. If this above condition does not hold, then a solution is termed, *stochastically unstable*. Furthermore, if a solution is stochastically stable it is also *asymptotically stochastically stable*, such that whenever $|x_o| < \delta$, and in the limit,

$$P\left(\lim_{t \to \infty} x(t, t_o.x_o)| = 0\right) \geq 1 - \epsilon. \tag{1.54}$$

The pth moment of the probability distribution, P, is termed *exponentially stable* if, given constants λ and C, the expected value, $E(\cdot)$, satisfies,

$$E\left(|x(t, t_o.x_o)|^p\right) \leq C|x_o|^p e^{-\lambda(t-t_o)}, \quad t \geq t_o. \tag{1.55}$$

Figure 1.9 depicts trajectories starting from the initial location, x_o, and a region S_r with inner radii δ and outer stability radii, S_R. It depicts the alternative definitions of stochastic stability that relate to the existence of a solution, its uniqueness, and strength of convergence (e.g., strong, weak). Strong convergence is when the *mean of the error ϵ converges* to the exact solution and so we try to approximate the sample paths. Weak convergence is when *only the expectation, E, or error in the mean converges*.

Another approach applied to compute distribution moments for stochastic processes is to derive the corresponding Fokker–Planck equations for a model system. This provides a fundamental, theoretical link between individual processes and their behavior and a universally representative or "governing" statistical distribution. A general representation of the Fokker–Planck equation will be derived next. Only the so-called transition probability is needed to describe the joint probability density function of a stochastic process such as the Weiner process that has the Markov property. Its evolution or trajectory is governed by a global or universal transition rule that only depends on the current state. Under this condition, the distribution function, $F_X(\mathbf{X})$, for a vector $\mathbf{X}$ of random variables, i.e. $\mathbf{X} = (X_1, \ldots, X_n)$, is given by,

$$F_X(\mathbf{X}) = P(X_1 \leq x_1, \ldots, X_n \leq x_n) = \int_{-\infty}^{x_1} \cdots \int_{-\infty}^{x_n} f_X(y_1, \ldots, y_n) dy_1 \ldots dy_n, \tag{1.56}$$

where f_X is the *joint probability density function*, $f_X = p(x_1, t_1 \ldots x_n, t_n)$, of a Markov process comprising the multiplication of individual probabilities associated with each time step, or the so-called *transition probabilities*, $p(x_{i-1}, t_{i-1}|x_i, t_i)$.

Starting with Itô's Lemma (Equation 1.40), where under general assumptions, α and σ are dependent on location x and time t, and then integrating the resulting expression, yields,

$$f(X(t)) - f(X(0)) = \int_0^t \left[\alpha(x,t)\frac{\partial f}{\partial x} + \frac{\sigma(x,t)^2}{2}\frac{\partial^2 f}{\partial x^2}\right] ds + \int_0^t \sigma \frac{\partial f}{\partial x} dW. \tag{1.57}$$

Given the following statistical relations for the expected value, $E(f(x))$, and its time derivative, $d/dt(E(f(x))$ in relation to the joint probability distribution, as,

$$E(f(X(t))) = \int_{-\infty}^{\infty} f(x)p(x,t;x_0,t_0)dx, \qquad (1.58)$$

$$\frac{d}{dt}E(f(X(t))) = \int_{-\infty}^{\infty} f(x)\frac{\partial p}{\partial t}(x,t;x_0,t_0)dx. \qquad (1.59)$$

Obtaining the expectation of Equation 1.57 and differentiating with respect to time, t, yields,

$$\frac{d}{dt}E(f(X(t))) = \int_{-\infty}^{\infty} \left[\alpha(x,t)\frac{\partial f}{\partial x} + \frac{1}{2}\sigma^2\frac{\partial^2 f}{\partial x^2}\right] p(x,t|x_0,t_0)dx \qquad (1.60)$$

Combining Equations 1.60 and the time derivative of the expected value in Equation 1.58, gives,

$$\int_{-\infty}^{\infty} \left[\alpha(x,t)\frac{\partial f}{\partial x} + \frac{1}{2}\sigma^2(x,t)\frac{\partial^2 f}{\partial x^2}p(x,t|x_0,t_0) - f(x)\frac{\partial p}{\partial t}(x,t;x_0,t_0)\right] dx = 0 \quad (1.61)$$

In the limit $x \to \pm\infty$, the probability p and $\partial p/\partial x \to 0$, so that for an arbitrary function, f,

$$\int_{-\infty}^{\infty} f(x)\left[\frac{\partial p}{\partial t} + \frac{\partial(\alpha p)}{\partial x} - \frac{1}{2}\frac{\partial^2(p\sigma^2)}{\partial x^2}\right] dx = 0, \qquad (1.62)$$

and this specifies the full statistical distribution of a stochastic process as the *Fokker–Planck equation*, given by,

$$\int_{-\infty}^{\infty} f(x)\left[\frac{\partial p}{\partial t} + \frac{\partial(\alpha(x,t)p)}{\partial x} - \frac{1}{2}\frac{\partial^2(p\sigma^2(x,t))}{\partial x^2}\right] dx = 0 \qquad (1.63)$$

$$\left(\frac{\partial p}{\partial t} + \frac{\partial(\alpha(x,t)p)}{\partial x} - \frac{1}{2}\frac{\partial^2(p\sigma^2(x,t))}{\partial x^2}\right) = 0 \qquad (1.64)$$

and by setting the time-derivative to zero in the above equation, the *stationary* probability distribution is obtained.

The Fokker–Planck equation has been applied in modeling financial market behavior, risk and system response (i.e., Black–Scholes equation) (Wei, 2000). An integrated approach to solving the Fokker–Planck equation, applied to the Black–Scholes futures market economic risk problem, has also been developed (Ewald et al., 2004). The Fokker–Planck equation has also been applied in modeling environmental/climate dynamics (Majda, 2012). In particular, deriving conceptual based and more reduced (i.e., less complex) stochastic models for climate and extended-range weather prediction is of great importance. This is because low-frequency (i.e., long duration) variability linked with a few large-scale teleconnection patterns (e.g., El Niño Southern Oscillation or ENSO) can exert a huge impact on regional climate and seasonal predictability. In each scientific domain, the resultant equations derived from the Fokker–Planck equation serve as important benchmarks for more complex and higher dimensional statistical-based problem solving.

Moon and Wettlaufer (2013) developed a perturbation theory for first-order nonlinear, non-autonomous SDEs. Various numerical solvers/schemes have also been developed and tested—Euler, Mil'shstein, Adams-Bashforth, Runge-Kutta, and other higher-order schemes are discussed in detail in Kloeden and Platen (1999). As highlighted by Ewald et al. (2004), the choice of integration technique can greatly affect the outcome of a stochastically forced numerical model, and a large variety of different numerical results can be generated by different implementations, which they show is the case in solving the dynamics of a stochastic model of sea surface temperature (SST) and its irregular inter-annual variability, within the El Niño region of the Pacific ocean (Ewald et al., 2004). In this study, they find that there is no significant increase in accuracy gained by assuming multiplicative (versus additive) noise, nor is there a significant asymmetry between warm and cold El Niño events. An observed historical trend in the leading climate signal (i.e., specifically the principal component of tropical Indo-Pacific SST anomalies over the past 60 years), was also determined to be unlikely attributable to sampling or measurement uncertainty. This finding suggests, therefore, that the stochastic variability in El Niño is mediated by another deterministic external forcing variable over an extended period of time that may help to better explain the SST variability and increase model prediction power.

Linear and nonlinear climate models focusing on separating slow and fast variables and the influence of high-frequency (i.e., short duration) stochastic processes have been studied that include variability in: 1) the thermohaline circulation (THC) in the world's oceans involving slow currents driven by gradients in the density of seawater dominating the deep circulation, believed to play a significant role in climate variability on timescales of decades and longer, 2) modeling of glacial/interglacial oscillations and millennial climate variability (i.e., "the bipolar see-saw" interaction of the northern and southern hemisphere during abrupt climate transitions) and evidence in the paleoclimate proxy data obtained from ice cores and ocean sediments, and, 3) the atmospheric-ocean coupling interaction of ENSO (El Niño Southern Oscillation) in the central and eastern equatorial Pacific (Stocker, 2011; Wilks, 2008; Stocker and Johnsen, 2003; Imkeller and Monahan, 2002; Hasselmann, 1976). These stochastic models capture, test, and explore essential and critical aspects and processes within the Earth system and is an approach that can be applied in exploring other CAS systems (Majda, 2012).

Ecosystems comprise dynamic processes that interact over multiple spatial and temporal scales. As a consequence, there are significant changes in the leading predictors that control the underlying state for an observed response at any point in time and space. As leading predictors change, so to does the "signal to noise ratio (SNR)" of the underlying processes, introducing loss and gain of signal information. The SNR can be estimated as a ratio of total variance output from a model to the total variance of model inputs. Deterministic or mechanistic ecosystem models of inter-connected systems of plant, soil, atmospheric and other underlying, interacting processes neglect measurement or process noise. Statistical models, on the other hand, typically focus on representing and understanding individual components, specific processes and/or scales of interactions within an ecosystem. Statistical models are being increasingly developed and applied to understand whole system dynamics and behavior and "big data" problems. Traditionally, ecological forecasting has typically been based on process-oriented models, informed by data in arbitrary ways. Although most ecological models incorporate some representation of mechanistic processes, many models are generally not adequate to quantify real-world dynamics and provide reliable forecasts with accompanying estimates of uncertainty (Littell et al., 2011). This is because such mechanistic models can be easily over-parameterized and require tens to hundreds of parameters and variables to explain an observed pattern or response "signal", without substantially increasing the SNR by explaining a greater fraction of unexplained "noise." In contrast, probabilistic approaches seek to optimize the smallest number of parameters and variables required to explain a signal and minimize the unexplained noise concurrently. This ensures that the SNRs of model predictions and forecasts decrease and are more robust to any changes in input parameters and variables. Moreover, it may be the case that there are "limits to predictability," measurable as limits in the achievable SNR. Such limits may exist, independent of any improvements in predictive accuracy and model identification, in terms of whether a deterministic or stochastic model is used or the complexity of a predictive model in terms of the best selection of predictors.

Let's examine noise in greater detail. If one considers a stochastic process following Brownian motion under the influence of friction (i.e., a noisy relaxation process), then an additional drift term is added to a Wiener process such that there is a tendency for reversal motion toward an attraction location, whereby attraction is greater at a greater distance away from the attraction area. This is a stationary, Gaussian and Markovian process called the Ornstein–Uhlenbeck stochastic process and is the continuous-time analog of an autoregressive AR(1) (i.e., order 1) discrete-time process, such as,

$$dX(t) = \gamma(\mu - X(t))dt + \sigma dW(t), \tag{1.65}$$

where $W(t)$ follows a Wiener process. For the Ornstein–Uhlenbeck process, it can be shown that the solution is,

$$X(t) = e^{-\gamma t}\left[X(0) + \int_0^t e^{\gamma s}\sigma dW_s\right], \tag{1.66}$$

with covariance $\Sigma_{i,j}$ given by $E[(X(i) - \mu_{X(i)})(X(j) - \mu_{X(j)})]$ or $cov(X(t), X(s))$, with respect to means denoted as μ, is given by,

$$\Sigma_{i,j} = \frac{\sigma^2}{2\gamma}\left(e^{-(t-s)} - e^{-\gamma(t+s)}\right), \quad s < t. \tag{1.67}$$

Substituting $\tau = (t - s) > 0$ and $t, s \to \infty$, yields the autocorrelation function $C(\tau)$ and its corresponding spectral power or density function, $S(\omega)$ (i.e., as the Fourier transform of $C(\tau)$), as,

$$C(\tau) = \frac{\sigma_\omega^2}{2\gamma}e^{-\gamma t}, \tag{1.68}$$

$$S(\omega) = \frac{\sigma_\omega^2}{\gamma^2 + \omega^2} \tag{1.69}$$

Moving from continuous-time to discrete-time, given a general environmental autoregressive moving-average or ARMA process (i.e., of order AR(p) and MA(q)) is given by,

$$X(t) = \mu + \epsilon(t) + \sum_{i=1}^{p} \phi_i X(t-i) + \sum_{i=1}^{q} \theta_i \epsilon(t-1), \qquad (1.70)$$

having the spectral density function, $S(\omega)$, where ω is the angular frequency (i.e., $\omega = 2\pi f$, for linear frequency f), given by,

$$S(\omega) = \frac{\sigma_\omega^2}{1 - \phi_1^2 - 2\phi_1 cos(2\pi f)}. \qquad (1.71)$$

When $\phi_1 > 1$, then red/Brownian noise as a slow, low frequency (so-called low-pass filter) is statistically coupled to an environmental signal. When $\phi_1 < 1$, blue noise results, as fast, high frequency (high-pass filter), and when $\phi_1 = 0$, no frequencies dominate and there is only white noise in a stochastic process. Many studies have explored stochastic models with "colored noise" that involves correlation in time between a given signal of interest and uncertainty or noise. Such correlation in time is termed autocorrelation.

In the environmental context, Fowler and Ruokolainen (2013) have examined the frequency distribution of AR(1) red and blue noise, and $S(f) \sim 1/f^\alpha, 0 < \alpha < 2$ pink-noise colored stochastic processes. They compare colored series with Gaussian- or normally-distributed times-series with white noise (Wiener process), and shows that colored series tend to deviate from a normal distribution, which may have a confounding effect. Recall that the Fokker–Planck equation assumes stochastic processes are Markovian and Gaussian-distributed with independent "white noise" behavior (i.e., serially uncorrelated noise). They show how noise assumptions can change a signal's distribution shape and lead to under- and over-estimation in model predictions, leading to underestimation of population extinction risk. For example, environmental *reddening*, where red noise dominates in a stochastic process, may reduce the probability of extreme climatic or weather events, but it requires sufficient length of record in time-series to determine (Schwager et al., 2006). Obtaining reliable predictions from models following the effects of signal or cumulative disturbances and shocks under Markovian and Gaussian white noise assumptions is limited in its real-world application (Hänggi and Jung, 1995). Moreover, depending on the stability of a system, shocks may lead to persistent, non-equilibrium and non-stationary conditions. The fundamental science of stochastic processes is still largely unexplored, and there is a great need for numerical methods for treating a highly complex system consisting of important interactions under a wide range of spatial and temporal scales. The needs of ecosystem and climate science likely will play a guiding role in the further development of the theory and application of stochastic analysis and modeling (Ewald et al., 2004)

1.1.6   Earth system carrying capacity and population stability

Uncertainty in the Earth system poses huge challenges and an enormous debate on the understanding and defining of the best long-term relationship between economic growth and environmental sustainability. Early misconceptions, misinterpretations, and misunderstandings that growth would come to an abrupt end, have fueled scientific criticism and social controversy and divide[14]. Such predictions assume that our planet has a finite level of resources with an accelerated human ecological footprint and the use of such finite resources. At the core of this debate have been questions regarding the level of complexity and realism in model-based assumptions, and in turn, limits to economic and environmental growth.

In 1972, three scientists from the Massachusetts of Technology (MIT) created a computer simulation of interactions between population, industrial growth, food production and limits in the ecosystems of the Earth called the World model[15]. This model was an extension of Jay Wright Forrester's model[16]. A report stemming from this modeling study, called "Limits to Growth," generated new insights and contentious debate on the possible future of our planet Earth, and was commissioned by the Club of Rome[17] (Meadows et al., 2004, 1974, 1972). The World model consists of several interacting components: the food system, dealing with agriculture and food production, the industrial system, the population system, the non-renewable resources system, and the pollution system. The model was simulated to explore different assumptions and 12 future scenarios out to 2100 of global resource consumption and production, showing that resource use was heading beyond the "carrying capacity" of the planet. *Carrying capacity* is the level of population size which the resources of the environment can just maintain or carry without a tendency to either increase or decrease (De Vries, 2013). Alternatively, it is the maximum population size of a biological species that the environment can sustain indefinitely, given the food, habitat, water, and other necessities available in the environment. What the environment can sustain is prescribed by biogeochemical cycles, biophysical limits and thresholds. Carrying capacity can also be assessed by ecological footprinting that can be defined as how much biologically productive area is required to produce the resources required by the human population and to absorb humanity's CO_2 emissions.

The World model assumed that population and industrial capital grows exponentially, leading to a similar growth in demand for food, non-renewables, and air pollution. The supply of food and non-renewable resources was, however, taken to be absolutely finite. In this way, exponential growth within finite limits resulted in overshooting the Earth's carrying capacity. The three main conclusions of the Limits to Growth study were:

1. If the present growth trends in world population, industrialization, pollution, food production, and resource depletion continues unchanged, the limits to growth on this planet will be reached sometime within the next one hundred

---

[14]Club of Rome, www.clubofrome.org/flash/limits_to_growth.html

[15]World3 model, insightmaker.com/insight/1954

[16]Forrester is considered the founder of the scientific study of systems dynamics that explores the dynamics of interactions between objects and components in dynamic systems, and who authored, Industrial Dynamics (1961), Urban Dynamics (1969) and World Dynamics (1971).

[17]Club of Rome, www.clubofrome.org/?p=326

years. The most probable result will be a rather sudden and uncontrollable decline in both population and industrial capacity.

2. It is possible to alter these growth trends and to establish a condition of ecological and economic stability that is sustainable far into the future. The state of global sustainability could be designed so that the basic material needs of each person on Earth are satisfied and each person has an equal opportunity to realize his or her individual human potential.

3. If the world's people decide to strive for the second outcome rather than the first, the sooner they begin working to attain it, the greater will be their chances of success.

These three conclusions are further summarized as follows: Unless special action is taken, human resource use and emissions will continue to increase as a consequence of growth in population, and human activity, if unchecked, will grow beyond the carrying capacity of the Earth, that is, beyond what the Earth can provide on a sustainable basis. If such expansion into unsustainable territory is allowed to happen, the decline, or collapse of human resource use and emissions become unavoidable (Meadows et al., 2004)[18].

Ekins (1999) details various critiques of the assumptions and associated findings of the original World model. Cole et al. (1973) and Nordhaus (1973), for example, re-ran the model under different assumptions, introducing successive (i.e., step-wise) exponential increases in available resources involving discovery and recycling, introducing pollution control, changing population growth, technological change, and substitutions. The result of all these changes was either continual consumption or a postponement of an inevitable overshoot and collapse. The foreseeable, cumulative effect of simultaneously changing the composition of outputs, substitution between input factors, and technological change, was to shift the development path whereby limited resources were consumed far slower than the rate of growth. The so-called "IPAT equation" incorporates two of these aspects—change in input consumption and technology in gauging resultant impacts on the environment, I, in terms of, functions of population size (P), per-capita consumption or affluence (A) and a technology efficiency factor (T) (Ehrlich and Holdren, 1971), as,

$$I = f(P) \cdot f(A) \cdot f(T) \qquad (1.72)$$

However, a recent evaluation of the benefit of technological efficiency on system dynamics of SES demonstrates that optimizing material and energy flows does not necessarily change an entire system's vulnerability nor its responsiveness to disruptions (Zhu and Ruth, 2013). Chertow (2001) discusses the IPAT equation and its variants in connection with the industrial ecology perspective. This includes a variant, whereby technology is not expressed as a separate variable but is instead related to per capita impact, F, and Gross Domestic Product (GDP) as a measure of affluence.

---

[18]40 years - Limits to Growth, `www.clubofrome.org/flash/limits_to_growth.html`

In *Beyond the Limits* (1992), an update to *Limits to Growth, Simulations of the World3* (version World3.91) dynamics model were explored for a wider range of possibilities, from collapse to sustainability, relying on basic global policy assumptions to infer that the human use of essential resources and the generation of pollutants have both surpassed their sustainable rates (Meadows et al., 1992). Scenario simulations cover the period from 1900 to 2100. Reported findings indicate that unless there are significant reductions in material and energy flows, the world faces an uncontrolled decline in per capita food output, energy use, and industrial production. Most updated scenarios with the World3/2004 model version similarly result in ongoing growth of population and of the economy until an abrupt shift or turning point around 2030. Only drastic measures for environmental protection can favorably change system behavior, and only under these circumstances, scenarios could be calculated in which both world population and wealth could remain at a constant level. However, many necessary political measures, up to 2004, have not been adopted (Meadows et al., 2004).

Brian Hayes provides an examination of the 150 equations that govern the evolution of the simulated World3 model, its structure and predictive power (Hayes, 2012). A JavaScript simulation of the model is available[19]. As with Nordhaus' earlier critique that the model simulated world dynamics without sufficient empirical data, Hayes critically examines the model's structure, its high degrees of freedom, and its reliance of its equations and dynamics on more than 400 assumed constants, coefficients, look-up-tables, and initial conditions that required empirical knowledge of the real-world. Furthermore, the model includes no assumptions or description of interactions between auxiliary variables. So, while it is generally recognized that the World3 model illustrates "basic dynamical tendencies" of the Earth system, it lacks predictive power. Global climate models (GCMs), while also complex, in contrast to the World model, combine data and equations on biogeochemical processes to gain sufficient predictive power. Such models have generated precise predictions on the order of 1% change in global average temperature. Nonetheless, the more complex the model in terms of its inputs, assumptions and internal logic, the more difficult it is to validate its output and understand how it would change under different parameters and model structural assumptions (Figure 1.10).

Let's turn our attention now away from Earth dynamics to human population dynamics. Historical global data on the human population reveal that it can be modelled over time as a hyperbolic function, given by Aral (2014b,a); Akaeva and Sadovnichiib (2010); Foerster et al. (1960).

$$P = \frac{c}{(t - t_0)} \tag{1.73}$$

where P is the world's human population, c is a calibration constant, t is time (years, yr), and t_0 is a reference time. Foerster et al. (1960) use reference values of c of 200 billion people-yr, and t_0 as the year (yr) 2025 (Foerster et al., 1960). This

---

[19]World3 interactive model online: `http://bit.player.org/limits`

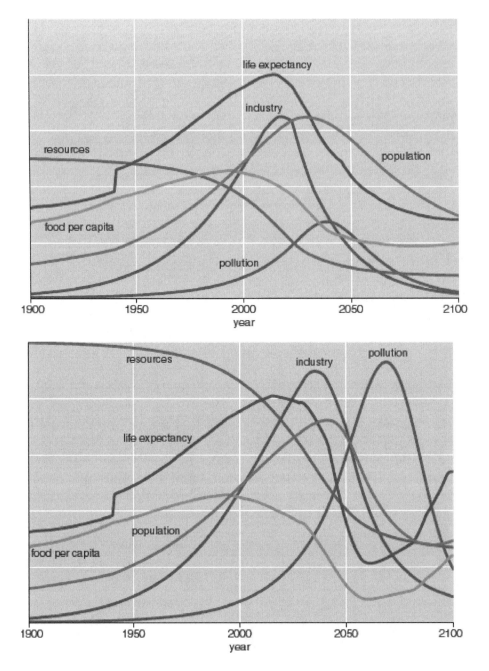

FIGURE 1.10: Output of the World3 model traces the state of the key variables over the period 1900–2100. The upper panel shows the "standard run," based on best estimates of initial conditions. Population and economic activity grow throughout the 20th century but collapse before the middle of the 21st century, because nonrenewable resources are severely depleted. The outcome is no more appealing in the lower panel, where the initial stock of resources has been doubled. A pollution crisis brings an even harsher collapse (Hayes, 2012).

relationship can be modeled by the following differential equation,

$$\frac{dP}{dt} = \frac{P^2}{c}.$$ (1.74)

However, this assumes that population P increases exponentially forever, which is not sustainable, and when $t = t_0$, $P \to \infty$, which is nonsensical, so it requires adjustment with respect to a finite human lifetime adjustment factor (τ) that compensates for finite population growth, giving,

$$\frac{dP}{dt} = \frac{c}{(t_1 - t)^2 + \tau^2}.$$ (1.75)

The analytical solution to this equation is,

$$P = \gamma^2 \cot^{-1}\left(\frac{t_1 - t}{\tau}\right), \quad \gamma^2 = \frac{c}{\tau}.$$ (1.76)

This assumes that the human population lifetime is governed by population level only such that it approaches a stable limit. For $c = 1.63 \cdot 10^{11}$ people-yr, $\tau = 45$ yr and t_1 the year 1995, assumed constant over time, the human population is predicted to stabilize at 11.36 billion people. This assumes that the Earth's biosphere can sustain it (Kapitza, 1996). Aral (2014a) has recently fit this model to updated data that is more recent, with a τ of 60 years, $c = 2.58 \cdot 10^{11}$ people-yr and a t_1 of 2030 yr, yielding a stable world population of 13.51 billion past 2030 to 2800, assuming the Earth can sustain such a human population.

Such a simple (so-called "first order" model) assumes a homogeneous human population with no consideration of internal dynamics influencing population growth rate, such as country-specific reproductive rates, environmental health factors, the adoption of new technologies, social learning, epidemics, and a wide array of environmental health factors. It also includes no consideration of dynamic factors affecting both human population levels and the Earth's carrying capacity. Such models based on fixed resource limits and a single, fixed carrying capacity are unrealistic (Meyer and Ausubel, 1999).

Following Meyer and Ausubel (1999), one can build more complex (and realistic) models of carrying capacity by introducing a dynamic carrying capacity for both the human population system and Earth system, whereby,

$$\frac{dP(t)}{dt} = c_1 P^2(t)(\kappa(t) - P(t)),$$ (1.77)

$$\frac{d\kappa(t)}{dt} = c_2 \frac{dP(t)}{dt}.$$ (1.78)

These equations represent a system that comprises a dynamic human population and a dynamic Earth carrying capacity. This system model assumes parabolic population growth (i.e., P^2 factor) and is a "third-order" model having three variables, time t, the human population, $P(t)$, and carrying capacity, $\kappa(t)$ that are dynamic (i.e., time-dependent). Here, population growth is assumed to be exponential at a constant rate (i.e., Malthus' principle) specified by the constant c_1, and carrying capacity is also directly proportional to population growth (i.e., the Condorcet principle) (Avery, 2014). This model assumes carrying capacity starts from zero.

If one further assumes that carrying capacity, $\kappa(t)$ is modeled as a logistic, sigmoidal "S" shape function, inside of a logistic, and an initial, non-zero carrying capacity, then (Meyer and Ausubel, 1999),

$$\frac{dP(t)}{dt} = \alpha_P P(t)\left(1 - \frac{P(t)}{\kappa(t)}\right) \tag{1.79}$$

$$\frac{d\kappa(t)}{dt} = \alpha_\kappa(\kappa(t) - \kappa_1)\left(1 - \frac{(\kappa(t) - \kappa_1)}{\kappa_2}\right) \tag{1.80}$$

where α_P is the growth rate of human population, α_κ is growth rate in carrying capacity between an initial κ_1 and a final, κ_2 value. Note that this model reparameterizes carrying capacity and decouples it from being directly dependent on population, P. Solving for carrying capacity, $\kappa(t)$, yields,

$$\kappa(t) = \left(\kappa_1 + \frac{\kappa_2}{1 + e^{-\alpha_\kappa(t - t_m)}}\right) \tag{1.81}$$

where t_m is an inflection point. This equation has a sigmoidal form with bi-logistic growth whereby carrying capacity has a growth trajectory approaching the initial carrying capacity level, κ_1, and then grows further again to the second level, κ_2. The above coupled population-carrying capacity model has a total of six parameters, $P(0), \alpha_P, \alpha_\kappa, \kappa_1, \kappa_2$ and t_m. Aral (2014b,a) has extended this model, formulating a joint homogeneous-heterogeneous model for human population growth and the Earth's biosphere carrying capacity (termed "biocapacity"). A homogeneous model is first proposed as,

$$\frac{dP(t)}{dt} = \alpha_P P^2(t - \tau_1)\left(1 - \frac{P(t)}{\kappa(P, \tau_2, \tau_3)}\right), \tag{1.82}$$

$$\kappa(P, \tau_2, \tau_3) = P_c + \gamma(P(t - \tau_2) - P_0)exp(-\kappa(P(t - \tau_3) - P_0))$$

where P_0 and P_c are the static biocapacity and Earth's dynamic carrying capacity, α_P is the human population growth rate, and τ_1, τ_2, τ_3 are the human mean reproductive age, diffusion time of basic technologies within the human population, and delay in time for the biosphere to respond to anthropogenic impacts/loads, respectively. The parameters γ and κ are rate constants. This model has been extended to a joint homogeneous-heterogeneous to include external variables within the Earth system that affect carrying capacity, such as the Earth's global temperature given by,

$$\frac{dP(t)}{dt} = \alpha_P P^2(t - \tau_1)\left(1 - \frac{P(t)}{\kappa(P, \tau_2, \tau_3)}\right) - \left(\frac{P(t - \tau_3)}{P_c + P(t - \tau_3)}\right)\left(\frac{\beta T(t - \tau_3)}{T_0}\right), \tag{1.83}$$

$$\kappa(P, \tau_2, \tau_3) = P_c + \gamma(P(t - \tau_2) - P_0)exp(-\kappa(P(t - \tau_3) - P_0)).$$

The variable T is the Earth's global temperature and the two additional parameters, namely, β and T_0, are the increase/decrease in temperature (°C), relative to a defined baseline (e.g., 2°C (Guan et al., 2013; IPCC, 2013)), and the calibration temperature at the start of historical temperature records, respectively.

This model predicts that a stable limit for the world's population cannot be maintained beyond a global temperature of 5.4°C, corresponding to values for P_o of 1 billion (exceeded in the early 1980's), P_c of 7 billion, α_P: (0.05-0.07) yr^{-1}, γ:(0.4-0.85) and τ:(1.31-0.51) yr^{-1}, respectively, and $(\tau_1, \tau_2, \tau_3) = (25, 25 - 30, > 100)$ yr (Aral, 2014b). Further regional correction and improvement in this model to make it more realistic is possible. This could involve adjusting P_c for different countries and regions, including other parameters for spatially heterogeneous population growth, technology adoption, and regional temperature warming.

Given the importance of demographic assumptions in human population growth, new alternative socioeconomic scenarios (so-called "shared socioeconomic pathways (SSPs)") now include a comprehensive set of human demographic conditions on population, urbanization, education and other societal aspects. While there are still open questions related to the internal consistency of demographic assumptions in relation to economic trends between alternative SSPs, they are generally consistent with current knowledge and evidence reported in the literature. Jiang (2014) discusses the need for more consistency in definitions of regions, and limitations of our understanding of future changes (i.e., via forecasts or projections) involving observed regional patterns and assumed interactions between different socioeconomic development stages. Grasman et al. (2005) also discuss resilience and persistence in the context of stochastic population models. It is important to remember that the above models are inherently deterministic. In such models, while parameter sensitivity analysis linked with predictions and uncertainty around scenario projections can be performed, there is still a critical need to model uncertainties and for improved forecasting via stochastic models of carrying capacity and human population growth that incorporates regional variability and uncertainty.

Addressing the need for stochastic models, Raftery et al. (2012) have applied a probabilistic (i.e., Bayesian) model for generating population projections using United Nations data for total fertility rate (TFR) (assumed to converge and stabilize at 2.1 set by expert judgment), life expectancy at birth, and potential support ratio (PSR) (i.e., defined here as the number of people aged 20–64 for each person 65 or over) around the globe. They illustrate their approach by applying this method for five countries of different demographic stages, continents and population size, namely: Brazil, the Netherlands, Madagascar, China, and India. The model was formulated by Alkema et al. (2011) as a Bayesian hierarchical model having random walk with a drift that is a double logistic function of TFR during a declining, or demographic transition, phase and then an autoregressive AR(1) function during a slower recovery phase. This is similar to the bi-logistic model presented previously. They validated their model with historical time-series data (1950–1990) and generated predictions for the time-period 1990–2010. PSR is projected to decline dramatically for all scenarios. While this stochastic model takes into account uncertainty about the future overall levels of fertility and mortality, it does not consider

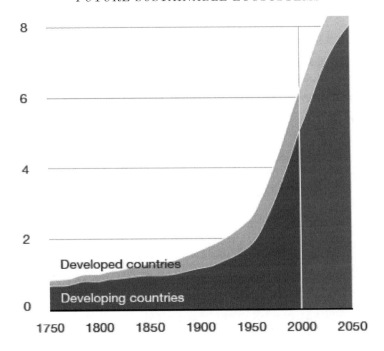

FIGURE 1.11: Human population growth (billions of people) in developed and developing countries (mid-range projection). Continued population growth remains
one of the biggest challenges to world food security and sustainability (UNEP,
2009) (©2007 UN Population Division) (Cartographer/designer: Hugo Ahlenius,
UNEP/GRID-Arendal).

international migration as the largest component of forecast error (Raftery et al.,
2012).

Population growth between developing and developed countries according to the
UN Population Division's mid-range projection is shown in Figure 1.11. Applying
the stochastic model of Alkema et al. (2011) to updated United Nations' data up
to 2012, Gerland et al. (2014) find that, contrary to previous reports, the world
population is unlikely to stop growing this century, with an 80% probability that
the world population of now 7.2 billion people will increase to between 9.6 billion
and 12.3 billion in 2100. This stochastic-based projection incorporates uncertainty
contrasts with deterministic ones that indicate a rapid increase and then rapid decrease in population, approaching a stable population of 7 billion people through
to 2300 (Aral, 2014a). This latest analysis has smaller uncertainty in its projections (Gerland et al., 2014). Clearly, stochastic assumptions dramatically change
the predictions and its reliability. The largest increase projected by this latest
stochastic model is expected in Africa, due to higher fertility rates. The ratio of
working-age people to older people is also predicted to decline substantially in all
countries, even those that currently predominantly comprise younger populations.
The growth of our the global human population and per-capita consumption rate
underlie all of the other present leading variables (i.e., "drivers") of global changes

in our Earth system (Bradshaw and Brook, 2014). As Barnosky et al. (2012) highlight, projected growth in the human population (now approximately increasing at a rate of 77 million people per year) is three orders of magnitude higher than the average annual growth from 10,000 to 400 years ago (i.e., a rate of 67,000 people per year). In just this past century, the human population has quadrupled, with UN projections (considered to be more conservative) indicating it will grow to 9 billion people by 2045, and then to 9.5 billion by 2050 [20,21] (Barnosky et al., 2012). Such estimates are within the current projected uncertainty of stochastic population and carrying capacity models.

1.1.7 Earth system tipping points and planetary boundary thresholds

All of Earth's ecosystems may, in the future, undergo a critical transition. A *critical transition* is an abrupt change in a dynamical system that occurs rapidly in comparison to past system dynamics whereby the system crosses a threshold near a transition, and the new state of the system is far away from its previous state. Such a transition could emerge from synergistic, cumulative or "integrated" global forcings (e.g., population, food, water, and energy ecosystem services with respect to Earth's carrying capacity, climate variability and warming, pollution and land use and land degradation). These drivers or forcings are scheduled to reach their threshold values by 2045, according to Barnosky et al. (2012). They provide scientific evidence and plausible arguments for anticipating a loss of stability in the Earth system, such that it will undergo a rapid transition following a fold bifurcation and hysteresis when the ratio of the percentage of transformed lands and global human population level reach a threshold of between 50–90% (Figure 1.12). Hysteresis occurs when a single variable drives a system after a delay or time-lag i.e., "memory"), whereby its influence persists, or similarly a cause and effect or "causal" lag in a given variable or property of a system with respect to its effect, such that its effect is generated as it magnitude varies. Such changes are typically observed to be nonlinear. Given such nonlinearity, a slowing down of intrinsic transient response within available time-series data is predicted to occur before instabilities due to bifurcation dynamics. Such slowing down is associated with an increase in the autocorrelation and variance of the fluctuations in a stochastically forced system approaching a bifurcation (Scheffer et al., 2009; Levina and Lenton, 2007).

Let's look at possible, future Earth transitions in greater detail. When ecosystems increase in complexity (i.e., size, connectivity, and/or interaction strength between components), they decrease in stability. This assumes that ecological interactions (e.g., species trophic, food-web, or predator-prey interactions) are assumed to occur randomly (May, 1972). Further details and discussion on ecosystem stability linked with time-delays is provided in *Stability and Complexity in Model Ecosystems* by Robert May (May, 2001). Golinksi et al. (2008) have determined that popula-

---

[20] United Nations, Department of Economic and Social Affairs. World Population Prospects, the 2010 Revision, esa.un.org/unpd/wpp/Analytical-Figures/htm/fig_1.htm

[21] United Nations.World Population to 2300 1–254 (United Nations, Department of Economic and Social Affairs Population Division, 2004).

tions within a stochastic environment are less stable (i.e., based on the deviation of population density with respect to carrying capacity) if they possess an extended or longer-term "ecological memory" of the previous changes or disturbances they have experienced. Allesina and Tang (2012) have also recently shown that community dynamics can be stabilizing (predator-prey interactions), with mutualistic and competitive interactions being destabilizing. Recent investigations by Sinha et al., (2005; 2006) on static and dynamic real-world non-random and dynamic trophic networks, support the universality of May's findings on static ecosystem networks, and suggest that Nature works to continually increase the robustness of ecosystems/ecological networks slowly over time, by eliminating species and links that are destabilizing (Wilmers, 2007; Sinha and Sinha, 2006; Sinha, 2005). Additional findings indicate that the number of network interactions in real ecosystems may be a fundamental intrinsic property, and is less dependent on extrinsic biotic and abiotic conditions. Too many interactions within an ecosystem is destabilizing and affects ecosystem persistence and resilience. Thrush et al. (2009) caution that empirical research still needs to be better integrated with ecological theory to understand how to anticipate abrupt ecological shifts (termed "regime shifts") and requires both theoretical and field ecologists to shift their scientific perspective from one that is strongly based on hindsight to one that is far more predictive and forecast based.

Critical transitions may also, though, be strongly associated with just a single, dominant driver (Thompson and Sieber, 2011) interacting as a "control" variable that propagates slower across a system experiencing other faster changes. As discussed by Thompson and Sieber (2011), "tipping elements" can be well-defined or distinguishable sub-systems of the climate that function quasi-independently and are susceptible to rapid changes. Such elements may enact a broad, dominant influence over Earth system dynamics. A list of such tipping elements has been proposed by Lenton et al. (2012); Lenton (2011); Lenton et al. (2008) as leading candidates that could generate future tipping due to human activities, with high relevance for political decision making. These tipping elements are listed below, and have a varying likelihood of exhibiting critical, bifurcation dynamics:

- Loss of Arctic summer sea ice
- Collapse and loss of the Greenland and West Antarctic ice sheets
- Shut-down of the Atlantic thermohaline circulation (THC)
- Increased amplitude or frequency of the El Niño Southern Oscillation (ENSO)
- Switch-off of the Indian summer monsoon (possible bifurcation)
- Changes to the Sahara/Sahel and West African monsoon, greening the desert
- Substantial loss of the Amazon rainforest and northern boreal forest

In addition, in the future there may be multi-stability and hysteresis in the climate-cryosphere system under orbital forcing (Calov and Ganopolski, 2005), accompanying changes in ocean heat storage between surface and deeper layers that infer a recent slowdown or hiatus in the rise of global temperature due to global warming (Chen and Tung, 2014).

Oceans act as a planetary sink for excess heat from higher atmospheric greenhouse gas emissions from anthropogenic sources, and therefore they will inevitably

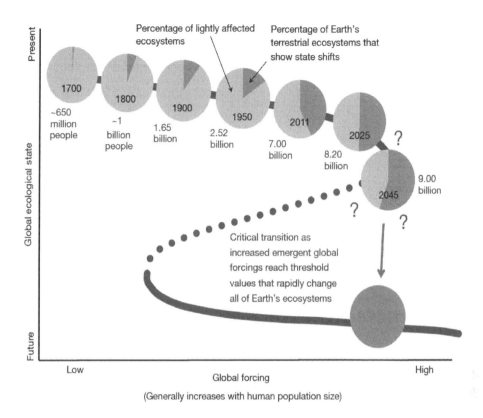

FIGURE 1.12: Quantifying land use as one method of anticipating a planetary state shift. The trajectory of the grey line represents a fold bifurcation with hysteresis. At each time-point, light gray represents the fraction of Earth's land within stability limits of the past 11,000 yrs. Dark grey indicates the fraction of terrestrial ecosystems that have lost stability and undergone rapid changes in ecosystem state; these are minimum values because they count only agricultural and urban lands. The percentages of such transformed lands in 2011 when divided by 7,000,000,000 (the present global human population) yield a value of approximately 2.27 acres (0.92 ha) of transformed land for each person. That value was used to estimate the amount of transformed land that probably existed in the years 1800, 1900 and 1950, and which would exist in 2025 and 2045 assuming conservative population growth and that resource use does not become any more efficient. An estimate of 0.68 transformed acres (0.28 ha) per capita (approximately that for India today) was used for the year 1700, assuming a lesser effect on the global landscape before the industrial revolution. Question marks emphasize that at present we still do not know how much land would have to be directly transformed by humans before a planetary state shift was imminent, but landscape-scale studies and theory suggest that the critical threshold may lie between 50 and 90% (although it could be even lower, owing to synergies between emergent global forcings) (Billion, 10^9) (See Barnosky et al. (2012) and references therein).

switch from being a sink, becoming a source of emissions that may induce an episode of accelerated global warming. An updated global surface temperature analysis now reveals that the recent global warming trend (since 2000) is actually higher than previously reported by the United Nations Intergovernmental Panel on Climate Change (IPCC) and does not support any warming slowdown or hiatus. This analysis indicates that the earlier trend of $0.113°C$ decade^{-1} (1950–1999) is virtually indistinguishable from the recent trend $0.116°C$ decade^{-1} (2000–2014) (Karl et al., 2015; IPCC, 2013). This updated analysis has corrected time-dependent biases between buoy- and ship-based sea-surface temperature (SST) historical (1980–present) data, whereby the buoy automated sampling data was determined to be more accurate and reliable. In addition, this latest analysis includes better estimates of land surface air temperature, correcting for operational changes or data inhomogeneities in monitoring networks involving changes in station locations, instrumentation, observer bias, urbanization, and other considerations. There is a continued need to better understand emerging differences in the degree of warming with latitude, and how it may be mediated by regional-scale impacts of atmospheric teleconnections.

Barrett and Dannenberg (2013) show how "good" collective/societal behavioral shifts to avert catastrophes, require strong incentives and robust trust relationships that reduce the threshold uncertainty associated with tipping points. The tipping elements listed above need to be extended with additional social (i.e., human) variables that very likely characterize tipping elements that drive criticality in social dynamics leading to possible bifurcation and tipping (Bentley et al., 2014). Kriegler et al. (2009) have determined conservative lower bounds for the probability of triggering at least one of the above events as 16% for medium (2–4°C), and 56% for higher global mean temperature change (above 4°C) relative to the year 2000 levels. Such estimates were determined by expert judgment and there is large uncertainty associated with them, in part due to their main interactions. Nonetheless, they are considerably higher than the probability allocated to catastrophic events in current climate damage assessments (Kriegler et al., 2009).

Scheffer et al. (2012, 2009, 2001) detail how humans can anticipate and identify early-warning signals of critical Earth system transitions, and catastrophic shifts between multiple stable steady states within different ecosystems. They identify leading statistical early-warning indicators for CAS integrated across different fields of science, such as recovery rate, return time, rate of change, autocorrelation at a given time lag, variance, skewness, noise reddening, and spatial correlation. Interestingly, Brook et al. (2013), in a study of tipping points, identify that large spatial heterogeneity in drivers and responses, and the lack of strong continental interconnectivity likely induce relatively smooth changes at the global scale, without any expectation of marked tipping patterns. They conclude that because of high spatial uncertainty, and scaling effects, identifying critical points may be unfeasible (i.e., involve a broad domain of attraction involving many different interacting driver variables) (Brook et al., 2013).

In addition to the tipping points, "planetary boundary thresholds" have also been identified. Rockström et al. (2009a,b) have proposed nine planetary boundary thresholds that humanity can operate safely within, seven of which can be quantified at the current time, and two that could not be quantified (i.e., chemical pollution and atmospheric aerosol loading). These seven quantifiable thresholds are: 1) climate change (CO_2 concentration) in the atmosphere < 350 ppm and/or a maximum change of $+1$ W m^{-2} in radiative forcing, 2) ocean acidification (mean surface seawater saturation state with respect to aragonite $> 80\%$ of pre-industrial levels), 3) stratospheric ozone ($< 5\%$ reduction in O_3 concentration from pre-industrial level of 290 Dobson Units), 4) biogeochemical nitrogen (N) cycle (limit industrial and agricultural fixation of N_2 to 35 Tg N yr^{-1}) and phosphorus (P) cycle (annual P inflow to oceans not to exceed 10 times the natural background weathering of P), 5) global freshwater use (< 4000 km^3 yr^{-1} of consumptive use of runoff resources), 6) land system change ($< 15\%$ of the ice-free land surface under cropland), and 7) the rate of loss of biological diversity (annual rate of <10 extinctions per million species). Mace et al. (2014) have further reviewed evidence in support of a planetary boundary, beyond which anthropogenic change will put the Earth system outside a safe operating space for humanity, for biodiversity. They identify weaknesses with the use of both an extinction rate and species richness as metrics for assessing a planetary biodiversity threshold or "biodiversity loss boundary," as they do not scale well from local to regional to the global scale. Instead, they suggest metrics based on three facets of biodiversity, namely: the genetic library of life, functional type diversity; and biome condition and extent (Mace et al., 2014). These suggested improvements in metrics for biodiversity loss, complement previous improvements that have been made in the Species Red List Index (RLI) for non-avian taxa that account for statistical artifacts introduced when value of zero is reached, bias in the relative influence of assessment frequency on RLI values and the introduction of newly evaluated species (Butchart et al., 2007).

Tipping elements and planetary boundary thresholds may lead to a combination of both predictable and unpredictable future ecosystem dynamics. For example, Kwuimy et al. (2014) have explored the complexity of bifurcation dynamics of a nonlinear pendulum between regular to chaotic motion, as a system that mimics more complex systems whose dynamics are driven by forces with damping or drift effects. They compare so-called recurrence quantification analysis (RQA) with the use of statistical parameters such as skewness, amplitude crest factor, and kurtosis to measure the orientation and degree of asymmetry within the system. They find that both regular (i.e., bi-periodic) as well as more chaotic, transient dynamics of the system as the forcing frequency is varied can be more reliably characterized and distinguished with such statistical parameters. Kuehn (2011) has formulated an integrated mathematical framework for critical transitions involving bifurcations, fast–slow systems, and stochastic dynamics. Kwasniok (2013) has also shown how the critical transitions in dynamic systems can be predicted from multivariate time-series data using non-stationary stochastic models (i.e., involving probability density). In a stochastic sense, an estimate of the probability of an early escape as a noise-induced escape from a steady-state potential well is also possible (Thompson and Sieber, 2011).

In summary, there is accumulating evidence of the existence of a set of underlying statistical "rules" that may govern the critical dynamics of all CAS. This lends additional support for governing statistical elements of pattern and process and their interaction within CAS systems (refer to Figure 1.6). Statistical investigation across a broad range of spatial and temporal scales must be undertaken to better understand what drives CAS dynamics and ways to identify and track alternative critical transitions from more stationary and non-stationary changes of state.

1.1.8   The way forward with an integrated approach

Identifying simpler underlying, 'hidden' rules from complex, observable dynamics raises philosophical questions about how one approaches science and its integration. The importance placed on simple approaches and solutions over more complex ones has very large ramifications in terms of our ability to find the necessary innovative, creative and "outside the box" solutions to our environmental problems. Occam's Razor, the principle of parsimony, which asserts the simplest solution having the least assumptions as possible to a given problem tends to be the best one, is widely applied, but its validity has never been proved in full generality (Wolpert, 1990). Occam's Razor essentially supports eliminating assumptions that are considered to make no difference in the observable predictions of an explanatory hypothesis or theory. This approach, however, not only is inherently subjective, but ignores the consequences of extended sets of assumptions. It also reduces the range of possibility, creating knowledge gaps and blind spots. As highlighted by Sir Francis Crick regarding the limitations of applying Occam's Razor in biology and ecology, "While Occam's Razor is a useful tool in the physical sciences, it can be a very dangerous implement in biology. It is thus very rash to use simplicity and elegance as a guide in [such] research." Evolution has led to complex designs, not simple ones (Northrop, 2011). Moreover, as the world-renowned ecologist, Simon Levin in, *Fragile Dominion* explains, "The structure and functioning of ecological communities, and of the entire biosphere, reflect the complex interplay between developmental (ontogenetic) and evolutionary processes occurring at a wide spectrum of scales. A perspective drawn from one scale alone is bound to mislead; to understand and explain observed trends, we must be able to integrate the effects of a rich panoply of influences." (Levin, 2000). Many other distinguished scholars and writers have highlighted the fragility of our natural resources and the benefits and challenges of the integrated, systems approach at the core of sustainability science for the 21st century (Elton, 2013; De Vries, 2013; Meadows, 2008; Walker and D., 2006; Schumacher, 1999; Walker, 1999).

Scientists, technologists, funding agencies, governments, institutions, and broader society are increasingly identifying that complex problems involving complex systems (i.e., ecological and socio-economic) require not simple, but more complex, systems level solutions that account for and manage uncertainty. Reduced or compartmental solutions may increase uncertainty by not explaining what is ignored and thus hinder knowledge and discovery progress (Folke et al., 2002). It is important to identify here that we define "complex" not to mean "complicated," despite it being derived from the Latin word, *complexus.* Such a definition would imply a subjective attribute and infer what is complex to one observer may not be complex to another, and is therefore defined in the eyes of the beholder, based on knowledge, skills, and insight (Northrop, 2011). Instead, we use an objective definition referring to complexity in natural ecological systems arising from their innate properties of being strongly interdependent, highly interacting, having open boundaries, and exhibiting nonlinear behavior with memory and adaptive learning. Complexity science—describing and modeling complex ecological systems—is in its infancy, but is set to introduce a paradigm shift in our understanding of natural, physical and social environment. Uncertainty within complex systems is both reducible and irreducible, increasing the knowledge gap that exists on how we can predict unanticipated behavior and minimizing unintended consequences—making predicting and manipulating their current and future dynamics very challenging. For this reason, interdisciplinary and integrated learning and actions are needed, in addition to, those that are more disciplinary and reductionist oriented. Falloon et al. (2014) have recently identified major issues with knowledge integration and model complexity, and discusses improved ways in which uncertainty frameworks can be used to inform adaptation decisions with case studies focused on agriculture.

Even though real-world conditions are never constant, not having a full or complete knowledge of dynamics of the world's system, is not, in itself, a reason for not furthering a sustainable developmental path (Scheffer et al., 2001). Instead, taking action relies on shared risk, whereby individuals, organizations or the world's economy as a whole work congruently to adapt and transition to a sustainable developmental path (Arnell, 2015). Such collective action requires reliable knowledge frameworks that are well-supported by the best available knowledge on theories, principles, and definitions. It relies on such knowledge frameworks being structurally sound to effectively facilitate meaningful action and provide practical tools, indicators and metrics for assessing sustainability progress. Applying an integrated approach enables one to make use of a wide array of data and types of conceptual, mathematical and computational models for exploring interrelationships within functioning CAS, and to systematically reduce the known risks associated with how we interact with them.

An integrated approach may offer spectacular and unforeseen discovery and advances, not just within one discipline, but in many areas of science and its application in the context of solving urgent real-world resource and environmental-based problems (Homer-Dixon, 2006; NRC, 1997b). Determining whether our current path of development is sustainable relies on a fundamental understanding of how our ecological and socio-economic systems operate and the consequences of our strong level of interactions within them. Achieving sustainability requires new thinking across the spectrum of human endeavor, not merely among scientists and technologists. Economic, social, and institutional innovations must keep pace with technological innovations. Multi-disciplinary and multi-dimensionality knowledge is required, spanning science, innovation, technology, information technology and e-commerce, economic development, health, foreign direct investment and multinational companies, international debt and aid, trade, politics, war, natural disasters, population growth, terrorism and related issues (Ahmed and Stein, 2004). Ian Hanington and David Suzuki in their book, *Everything Under the Sun: Toward a Brighter Future on a Small Blue Planet* emphasize the importance and need for the integrated perspective in developing viable long-term sustainable solutions and actions to a myriad of our interconnected environmental, economic, and social challenges (Hanington and Suzuki, 2012).

1.2 Building Adaptive Capacity and Opportunity

Adaptive capacity is the capacity of a system to adapt to its changing external environment. The adaptive capacity of a SES encompasses the ability of institutions and networks to learn and store knowledge and experience, and to provide creative flexibility in multifunctional decision-making and problem solving. It also involves a consideration of the responsiveness of system components and system structure to its various functions. This definition of adaptive capacity involves the ability to prevent and mitigate economic, social, and environmental threats. Often, adaptation to change and mitigation of the threats or harmful impacts of change (i.e. of human-induced climate change) are distinguished as separable aspects, while in reality they are inextricably linked. We survey next our current adaptive capacity for addressing sustainability and climate change, in terms of international standards and consensus, sustainable economic growth, identifying mitigation and adaptive actions and challenges for public policy.

1.2.1 International consensus, agreements and standards

The international community, through the United Nations, formally recognized sustainability and environmental awareness as a global issue at the United Nations Conference on Human Environment, held in Stockholm, Sweden in 1972. This provided a foundation for building capacity and discussion on sustainability and sustainable development linked with collective responsibility, rights, and the extent of environmental harm within various jurisdictions (Tuazon et al., 2013). It was 15 years later, however, in a 1987 report of the UN World Commission on Environment and Development (UNWCED) entitled, "Our Common Future," also known as the Brundtland report, that the first definition of sustainability was formalized. "Sustainable development" was defined as balancing the fulfillment of human needs with the protection of the natural environment so that these needs can be met not only in the present, but in the indefinite future. This later definition of sustainable development was adopted and used by the Brundtland Commission who further defined sustainable development as development that "meets the needs of the present without compromising the ability of future generations to meet their own needs."

This United Nations (UN) report established two key concepts, namely: 1) the concept of needs, in particular the essential needs of the world's poor, to which overriding priority should be given and, 2) the idea of limitations imposed by the state of technology and social organization on the environment's ability to meet present and future needs. It also identified seven critical actions that are necessary to ensure a good quality of life for people around the world (Ahmed and Stein, 2004; Brundtland, 1987):

- Revive growth
- Change the quality of growth
- Meet essential needs/aspirations for jobs, food, energy, water and sanitation
- Ensure a sustainable level of population
- Conserve and enhance the resource base
- Re-orient technology and manage risk
- Integrate environmental and economic considerations in decision-making

This led further to the UN Convention on Education and Development (UNCED) at the Earth Summit in Rio de Janeiro, Brazil in 1992, where the United Nations Convention on Biological Diversity (CBD), the Framework on Climate Change, the Rio Declaration, and the Agenda 21 Agreement were signed. Agenda 21 is a non-binding program of action adopted by more than 178 governments. While this agreement lacks the support of international law, it does establish a strong moral obligation to ensure strategy implementation. Agenda 21 forms the basis of our global partnership that encourages cooperation among nations to support a transition to sustaining life on Earth. The UN Commission on Sustainable Development (UNCSD) was formed specifically to review progress on the implementation of Agenda 21.

Soon after this agreement, in 1996, the International Standards Organization (ISO) established the ISO 14000 series of standards to address various aspects of environmental management, so as to provide practical tools for companies and organizations looking to identify and control their environmental impact and constantly improve their environmental performance. This helped to build capacity and opportunity for industry, corporations, and Small to Medium Business Enterprises (SME) to establish consistent requirements and stronger compliance measures for processes, practices and reporting linked with sound governance and environmental principles. ISO 14001:2004 provides requirements with guidance for use, and ISO 14004:2004 provides general guidelines on principles, systems and support techniques. ISO 14064-1:2006 provides specifications and guidance for organizations in quantifying and reporting GHG and removals, and ISO 14006:2011 provides guidelines for incorporating ecodesign for Environment Management Systems (EMS). The other standards in the ISO 14000 series focus on specific environmental aspects such as life-cycle analysis, communication, and auditing[22]. ISO standards can be used by any organization regardless of its activity or sector, and provides assurance to company management and employees, including external stakeholders, that environmental impact is being measured and improved. While it does not go as far as stipulating specific requirements, goals, or targets for environmental performance and sustainability, it institutes a framework that a company or organization can initiate, adhere to, refine, and improve in the areas of waste management, energy

---

[22]ISO 14000 series of standards,www.iso.org/iso/home/standards/management-standards/iso14000.htm

and material consumption, and distribution costs. It also establishes a procedural protocol for ensuring open lines of communication with regulators, customers and the public.

1.2.2 Green growth

It is now broadly recognized that environmentally sustainable economic growth or "green growth" is not a contradiction in terms. Instead, it *is* considered to not only be theoretically possible, but also economically *achievable*, under the right set of environmental and economic policies (Ekins, 1999). Green growth, as agreed at the 5th Ministerial Conference on Environment and Development in Asia and the Pacific, is a strategy for achieving sustainable development. It focuses on adapting the economy in a way that links economic growth and environmental protection. It emphasizes the importance of building a green economy in which investments in resource savings as well as sustainable management of natural capital are drivers of growth. An economy which is in closer alignment with sustainable development objectives yields greater opportunities in using financial resources more efficiently and effectively in meeting growth and development needs. At the same time, there are value-added benefits in terms of reducing vulnerability and anticipating cycles or adapting to longer-term changes in resource constraints. Green growth strategies can help economies and societies become more resilient as they work to meet demands for food production, transport, housing, energy, and water. Strategies can help mitigate the impacts of adverse shocks by reducing the intensity of resource consumption and environmental impacts, while alleviating pressure on commodity prices. Green growth also offers competitive advantages to those countries that commit to policy innovations.

The global market for green goods and services is vast and growing fast, offering countries the dual benefit of prosperity and job creation. However, population demographics around the world are also rapidly changing. According to Harper (2014), aging population is a global phenomenon that will continue to affect all regions of the world, and by 2050 there will be the same number of old as young in the world (i.e., 21%), with 2 billion people aged 60 or over and 2 billion under age 15. This has critical economic and social implications. Here there is a disconnect between where the market is going in the future and what governments can afford to invest in, guide, and influence through policy. This is because governments continue to make critical trade-offs between growing their national debt (associated with increased social pressure for increasing public spending on pensions, health care, new green infrastructure and environmental sustainability) and fiscal restraints to reduce their national budget deficits. While the latter helps to create a better environment for investment, different types of investment are likely required to transform policy, and direct spending to areas that will translate into the growth of green goods and services.

There have been many major international initiatives that have helped to spur and support green growth. In 2008, the UN Environment Programme (UNEP) led an international Green Economy Initiative. In 2011, the OECD published a strategy toward green growth, and in 2012, The World Bank published a report "Inclusive Green Growth: The Pathway to Sustainable Development" (WBG, 2012). This report provides a real-world framework for inclusive green growth based on input, efficiency, stimulus, and innovation, the main effects in the production function. Trade-offs, synergies and benefits of green growth are detailed. Price-based incentives and instruments (i.e., taxes and pollution offset permits) are considered preferable to standards under simple production assumptions; however, standard-based enforcement rules and regulations may be more effective yet require accurate valuation, to ensure they are not underestimated (WBG, 2012). As well, it is noted that enforcement approaches introduce a risk of unintentionally creating new barriers by favoring incumbent organizations over the entry of new businesses and their entry into markets which are often most innovative and create the most jobs. Then, in 2012, the Global Green Growth Institute (GGGI), the UNEP, and the World Bank signed a Memorandum of Understanding (MOU) to formally launch the Green Growth Knowledge Platform (GGKP). The GGKP's mission is to enhance and expand efforts to identify and address major knowledge gaps in green growth theory and practice, and to help countries design and implement policies to move toward a green economy. Green infrastructure will require *integrated* design and planning, risk assessment, upgrading, and new construction as different subsystems are more and more dependent on each other (e.g., urban plans, transport, land-use and reserve planning, water reservoirs, dams and use, coastal and flood defenses, housing and buildings, and energy production). The OECD has subsequently explored how to integrate green growth within current development pathways covering 74 policies and measures from 37 countries and 5 regional initiatives as an action-oriented guide for policy setting at the national and international level and for implementing sustainable practices with a higher chance of success (OECD, 2013). Recent findings indicate that rapidly accelerating growth in developing countries raises the stakes for investments in development, but also increases opportunities in building capacity in having a range of pathways and options to develop. In this way, green development offers an alternative that relies on and values natural assets which are essential to the well-being and livelihoods of people in developing countries, so vital to their sustainable future. These recent initiatives and actions relate, in part, to the need for resource-based yields, production, consumption, technology, and affluence gaps to be more comprehensively assessed, modeled and forecasted. The prototype, Global Sustainable Development Report, "Building the Common Future We Want," released by the UN Department of Economic and Social Affairs (UNDESA)'s Sustainable Development Division in 2013 [23], illustrates a range of perspectives, knowledge, alternative approaches, and various ways of engaging the scientific community with policy makers in Green Growth (UNDESA, 2013). A 2014 report from the World Bank on shifting priorities for increasing future global economic prospects, highlights that "future growth must increasingly be driven by domestic efforts to boost productivity and competitiveness" linked with structural reforms that increase the potential for growth (WBG, 2014).

---

[23] sustainabledevelopment.un.org/globalsdreport/

1.2.3   Big challenges in translating knowledge into action

While there are many general approaches to solving global sustainability problems and challenges, there are huge gaps in translating them into meaningful action. Action shouldn't involve a mass rush and panic, even though there is a clear urgency, but should involve adequate time for the confluence of ideas and reflection. Scientific-based sustainability frameworks that go beyond broad concepts and generalities, therefore, need to be devised, tested, and put into action. This acknowledges that such frameworks will evolve, being continually refined and improved, but that the stage must be set for sustainability and how meaningful actions can be identified, taken, and assessed. International and national institutions need to establish new sustainable metrics capable of integrating theoretical and applied knowledge, while providing adaptive frameworks for their improvement and refinement. These frameworks must be well-structured, salient, credible, and legitimate to adequately support and instill urgent action by better engaging organizations and citizens and helping them individually or collectively respond—increasing societal adapting capacity (Meinke et al., 2009). We need to fundamentally restructure our current developmental perspective to help maintain *financial* (sound macro-economic planning and prudent fiscal management), *physical* (infrastructure assets, such as buildings, machines, roads, power plants, and ports), *human* (good health and education to maintain labor markets), *social* (people's skills and abilities as well as the institutions, relationships, and norms that shape the quality and quantity of community social interactions) and *natural capital* (natural resources, both commercial and non-commercial, and ecological services which provide the basic requirements, including food, water, energy, fibers, waste assimilation, climate stabilization, and other life-support services). During long-term re-orientation, sudden resource supply shortages, disruptions and climate/weather-related extreme events may still occur. Simply put, our ability to foresee and forecast the occurrence, vulnerability and associated impacts of sudden shocks and tipping-points may not extend as far as needed beyond our current capability. Ecosystem dynamics and human behavior will likely always possess some inherent level of irreducible uncertainty or indirect costs that result from adapting. Nonetheless, an integrated perspective and approach to learning and developing scientific knowledge is crucial in finding and implementing realistic, feasible solutions to solving real-world problems linked with ecosystems and natural resources.

Meaningful, collective actions must be taken to help avoid exacerbating our risk and vulnerability to known and unknown impacts. Uncertainty has been widely blamed for inaction in climate policy (Lorenzoni et al., 2007). A large contribution to such uncertainty is due to the fact that knowledge to support policy comes from an ensemble of many different forms of evidence, including compartmental models with varying assumptions and prediction skill. Instead, a systems-level or systems-wide modeling perspective is needed. Huber et al. (2014) identify that integrated modeling frameworks need to be devised that make the most of the available model-based tools such that climate impact and adaptation proceeds in an integrated way, and moves beyond patchwork. For example, using fuzzy decision rules and clustering with the AHEAD (Adequate Human livelihood conditions for wEll-being And Development) methodology, Lissner et al. (2014) demonstrate how available/reported uncertainty ranges on water availability actually lie outside of model thresholds for 65 out of 111 countries. Better, robust policies can be developed when our different perspectives, insights, and knowledge-sets are well-integrated. Furthermore, by coordinating our activities and actions, more reliable solutions may be found to ensure the sustainability of ecosystems and green goods and services.

Transforming our current SES system structures and functions against the status-quo is neither simple nor easy, but is best accomplished with a fully integrative approach, rather than in various disconnected incremental steps. An integrated perspective that seeks to combine environmental, economic, and social perspectives may provide a solid foundation for finding and implementing workable, long-term solutions to provide for our rapidly increasing resource demands, minimizing waste, maximizing production and use-efficiencies, and recycling involving renewable sources. Viable, long-term solutions cannot be built on non-renewable resources that are not only more and more scarce, costly to use, but also offer diminishing returns in being depleted far past known ecological thresholds and tipping points (Westley et al., 2011). Implementing integrated approaches, methods and plans within (or to substitute for) existing, unsustainable approaches and practices requires a comprehensive interdisciplinary knowledge and understanding of managed ecosystems. For a sustainable solution to be workable (and ultimately associated with a future developmental path that we can collectively rely on) it must be sufficiently well tested, debated, and validated against other competing theories and existing methodologies. Sustainable solutions must, however, also be supported collectively, transferred, adapted, improved, and refined with institutional arrangements and agreement, industrial participation and assistance, and social consensus. Let's look next at how such multi-sector structures and functionality pose a challenge for future public policy.

1.2.4 Sector multifunctionality and public policy challenges

Harold Hotelling in 1931 wrote a seminal article that spurred exploration on the "Economics of Exhaustible Resources" and the change in resource price under competitive and monopoly market assumptions (Devarajan and Fisher, 1981). From an economic perspective, with the availability and demand for non-renewable resources (in terms of future price) being more volatile and uncertain over the longer-term, current risk-adverse strategies may actually seek to accelerate short-term resource use and depletion. Moreover, without reliable information, competition and monopoly influences can result in significant policy distortions and make it difficult to provide any reliability or accountability. This was exemplified in agricultural rice production and trade policy across 33 rice-producing countries recently (Rakotoarisoa, 2011). Critics also argue that decades of skewed agricultural policies, inequitable trade, and unsustainable development have thrown the world's food systems into a volatile boom and bust cycle, and widened the gap between affluence and poverty. They argue that the world's recurrent food crises are making a handful of investors and multi-national corporations very rich, such as in the case of oil (Rubin, 2012), even as they devastate the poor and put the rest of the planet at severe environmental and economic risk. Recurrent famines, food and water shortages, and riots occur in both resource-poor and resource-rich countries and developing and developed economies, reflecting possible sector manipulation, inconsistencies and problematic fundamental assumptions supporting our current global agricultural system. Even referring to resource sectors, like agriculture, as multifunctional has proved both controversial and contested within many global trade negotiations, with debates taking place on whether subsidies are needed. This debate on one end considers that subsidies promote market equity, while on the other that they instead distort trade and commodity prices. Critics argue that agricultural subsidies, international trade and related policy frameworks do not stimulate markets to transition toward equitable agricultural and food trade relations or sustainable food and farming systems. Critics instead favor other ways to address issues impacting sustainability such as poverty, human health, and nutrition, that don't involve intervention in free markets (IAASTD, 2009a).

A striking example of contentious debate on the usefulness of subsidies and the need to recognize modern agriculture's multifunctionality is in the use and trade-off of agricultural crops for food versus bioenergy (i.e., bioethanol and biodiesel) in the world's bioeconomy. Real-world inter-sectoral decisions are more often than not trade-offs between different response options (Schellnhuber et al., 2014; Schlicken-rieder et al., 2011). For example, bioenergy developments have also been a cause for deep concern regarding their economic, social, and environmental viability, because of their potential negative impacts on food security through crowding out of staple food production, and on the environment due to natural resource scarcity and intensive agriculture production. While there has been a rush by many governments to develop bioenergy alternatives to fossil fuels, this has often been done in the absence of a wider understanding of the full costs and benefits of bioenergy (FAO, 2010). The Global Subsidies Initiative (GSI) of the International Institute for Sustainable Development (IISD), in a synthesis report has addressed the use of government subsidies in promoting biofuels in Australia, Canada, the European Union (EU), Switzerland, and the United States of America (US). This report outlines the many environmental and developmental pressures that subsidies have induced (GSI-IISD, 2007). Such changes include changes in crop patterns in the EU and US to two row crops of canola (oilseed) and corn (maize) that require high amounts of fertilizer inputs. The report concludes that such changes, "have also had knock-on effects on commodity markets, helping to push up prices not only of the biofuel feedstocks themselves, but also of close substitutes, putting pressure on an already strained world food system." Farmers in developing countries will no doubt benefit from these higher prices, but the urban poor, especially in net food-importing countries, will not.

Given the above problems and concerns with direct government policy and subsidy-based market *interventions*, the use of market *incentives*, as an alternative approach, places far higher societal responsibility to lead and transition to sustainability on individuals and business organizations. As evidence of this, the World

Business Council for Sustainable Development (WBCSD)'s Vision 2050, released in 2010, a new vision and agenda for the future of business, led by Pricewater-houseCoopers, Storebrand, Syngenta International, and Alcoa, along with 29 other large and influential member-companies (WBCSD, 2010). This new agenda involves shifting from climate change and resource constraints as environmental problems to economic ones related to the sharing of opportunity and costs, involving a balanced use of renewable resources and recycling non-renewables. Vision 2050 is a call for further dialogue and a call for action. It signals a huge transformation of regulation, consumer preferences, pricing of inputs, and measurement of profits and losses. Trust and long-term thinking are considered essential ingredients for enabling businesses to better address issues and for building capacity for greater knowledge sharing and inclusive decision-making processes. Figure 1.13 highlights business domains for the next decade and their sector-oriented opportunities as well as multi-sectoral ones identified in Vision 2050.

Many food retailers and manufacturers are themselves taking the lead looking beyond their own operations to realize improvements in environmental performance as an estimated 90% of the food industry's environmental footprint occurs in *commodity* production, an area outside their direct control. For example, IRIS is an initiative of the Global Impact Investing Network (GIIN). The IRIS catalog includes various metrics that can be used to assess the performance of organizations in the agriculture sector. The agriculture metrics have been designed to capture many of the environmental aspects of agriculture practices and productivity measures. These metrics may be pertinent to organizations operating throughout the agricultural value chain. There are also additional cross-sector metrics within the catalog that can be applied to organizations in the agriculture sector that can be viewed by browsing the full catalog[24]. The Finance Alliance for Sustainable Trade (FAST) is also an organization representing a global community of financial institutions and small and medium enterprises (SMEs), including producer organizations responsible for developing sustainable products and sustainable finance to market.

On April 15, 2014, Europe (i.e., the European Parliament) passed a historic law requiring big corporations to report on their sustainability. It was first proposed in 1999 by Richard Howitt, European Parliament Rapporteur on Corporate Social Responsibility. As he indicates, "All the evidence suggests that transparency is the best way to change business behavior. This European law will prevent corporate scandals and make a leap in the transition toward a sustainable, low-carbon economy for the future." The law requires the biggest companies to include sustainability factors as part of their annual financial reports, applicable to publicly-traded companies with more than 500 employees. They must now address "policies, risks and results" in relation to "social, environmental and human rights impact, diversity and anti-corruption policies" in their annual reports. While, reportedly, 2500 companies voluntarily produce such sustainability reports, it is expected that this number will rise to 7000 by 2017 due to this new legal requirement. Companies are encouraged to use standardized, recognized frameworks, such as the Global Reporting Initia-

---

[24]IRIS Catalogue, `iris.thegiin.org`

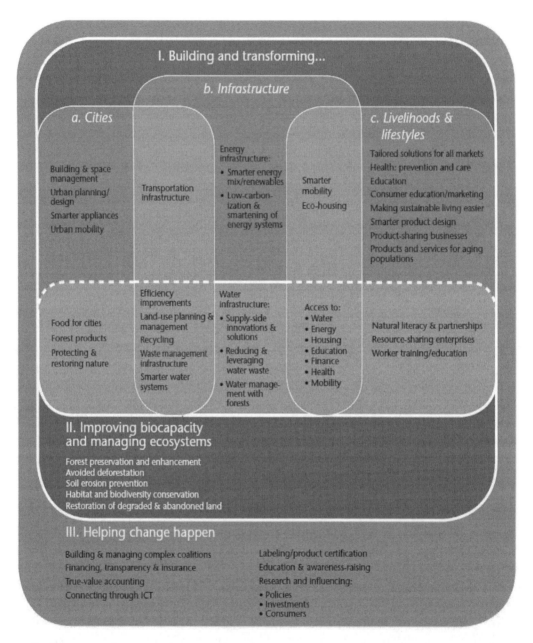

FIGURE 1.13: Business domains for the next decade: Opportunities and overlaps. This figure provides a relative mapping of opportunity spaces. Overlaps in this diagram indicate areas that are ripe for multi-sectoral collaboration, as well as where there might be demand for specified products and services (WBCSD, 2010).

tive (GRI) Sustainability Reporting Guidelines and the UN Guiding Principles on Business and Human Rights. This is regarded by many as a major step toward integrated reporting by businesses worldwide. According to Teresa Fogelberg, Deputy Chief Executive of GRI, the new legal directive is a "...vital catalyst needed to usher in a new era of transparency in the largest economic region in the world." As of 2011, 95% of the world's largest 250 corporations produce annual sustainability reports, up from 80% in 2008, according to KPMG, and 86% of the largest US companies do the same. That year, SunPower led with the first sustainability report in the solar industry. Addressing their supply chains is the next big task that companies like Ford and Microsoft are leading the way on[25].

1.2.5 Science for adaptation

Having briefly surveyed various contexts and instruments that are strengthening the adaptive capacity requried to transform our current SES systems so as to ensure future sustainable ecosystems, we next ask what adaptive capacity our current and future science and technology offers.

To build future adaptive capacity, science must become more cooperative and integrative. Traditionally, science and technology learners have been largely encouraged to work within only one discipline and focus at increasing depth on a problem. This has the unfortunate tendency to convince a learner over time that their chosen question and approach are better and more interesting, valuable, compelling, and more likely to succeed than others. At the same time, individual and/or collective success are not always accompanied by real-world shared, beneficial impacts, consequences, and outcomes. It is, therefore, impossible to judge or know ahead of time what focus or innovation will yield the greatest knowledge gains, benefits, and impacts for society. This is especially true if a broader, integrated perspective and approach to connect fragmented and distributed blocks of disciplinary knowledge is not available. This is analogous to the way a child plays with Lego^{TM} blocks (without a grand design or plan in mind), innovating continuously to create, while also continuously learning and responding as new integrated structures and functional capacities emerge. Science and technology and how they innovate is about to change dramatically in this respect, tantamount to approaching problem solving almost as children naturally learn when faced with complexity, risk, and uncertainty all at once. Integrating knowledge is now widely viewed as an educational imperative for the 21st century (NRC, 1997a). Integrated scientific domains are also emerging, such as "Industrial Ecology" and "Data Science" linked with sustainability science that seeks to bridge industry standards, processes, and practices with sustainable development. Clark and Dickson (2003) discuss the emerging research program for Sustainability Science and its focus on "the dynamic interactions between nature and society." They indicate that, "Sustainability science is not yet an autonomous field or discipline, but rather a vibrant arena that is bringing together scholarship and practice, global and local perspectives from north and south, and disciplines across

---

[25]Global Reporting Initiative (GRI),
  www.sustainablebusiness.com/index.cfm/go/news.display/id/25651

the natural and social sciences, engineering, and medicine. Its scope of core questions, criteria for quality control, and membership are consequently in substantial
flux and may be expected to remain so for some time." Lines of scientific inquiry in
Data Science and Industrial Ecology specifically address the need to further develop
and advance industry or sector-specific requirements, goals, or targets for environmental performance and sustainability beyond current international standards. For
instance, Industrial Ecology investigates water, energy, land resource use patterns,
efficiencies, and environment risks and uncertainties associated with human activity
and industrial production systems. It aims to develop and optimize new models of
highly efficient resource use and production to guide and improve the environmental
performance of different industries, with strong collaboration and communication
with a broad set of stakeholders within the agriculture, forestry, fisheries, mining,
bioeconomic and other natural resource industries. It has been argued that industrial ecology has, to date, largely engaged with the ecological sciences at a superficial
level. This has both attracted criticism of the field and limited its practical application for sustainable industrial development. However, a new approach has been
proposed—that context-specific observation and analysis of industry be advanced
before theoretical and practical advancement of the field can be achieved (Jensen
et al., 2011). Simply put, concepts without sufficient context have little adaptive
capacity, at least as far as achieving sustainability and sustainable development is
concerned.

Swart et al. (2014) distinguish between the "science of adaptation" to climate
change" and a "science for adaptation." They argue that fundamental inquiry and
concept development needs to be better integrated with participatory, practice-
oriented research, and that both must better integrate the diverse knowledge that
spans scientific disciplines, but at the same time, some separation needs to be recognized to enable reflection, synthesis, and learning within these two domains (Swart
et al., 2014). They highlight global (e.g., Future Earth[26]), continental (e.g., the
European Joint Programming Initiatives, JPI) in water, climate, food security, and
environmental health[27] initiatives. They also identify many other major, national-
based adaptation research and development initiatives, like the North American Climate Change and Human Health (NACCHH) Working Group, established in 2014
by Health Canada, the Public Health Agency of Canada, and the United States Centers for Disease Control and Prevention. Despite such collaborative partnerships,
current programs, they argue, typically "pay less attention to more fundamental research leading to appropriate theoretical frameworks and associated methodologies
for adaptation to climate change," and identify major "pitfalls of too much emphasis
on an imprecisely defined, interdisciplinary, practice-oriented forms of science." Put
simply, focusing on "a science for adaptation without a substantive science of adaptation." They identify several major pitfalls as barriers to innovation and learning,
that are typically encountered and need to be better addressed to enable stronger
adaptation through integration:

---

[26]Future Earth, `www.icsu.org/future-earth`
[27]JPI, `ec.europa.eu/research/era/joint-programming-initiatives_en.html`

- Validated and tested theoretical frameworks and hypotheses and commonly accepted methodologies and data are as yet largely missing, whereby practice-oriented adaptation research runs the risk of being driven by unproven assumptions and a lack of comprehensive and robust assessment.

- There is currently too much opportunity for unintentional convergence between scientists' and practitioners' knowledge and perspective, that reduces their ability to reflect and innovate.

- Interdisciplinary research shows that there are communicative and conceptual barriers related to disparate research backgrounds, concepts, and lines of inquiry which become more challenging to bridge into a common frame of knowledge and understanding when non-academic stakeholders with different motives, ideas, and goals are also involved.

- The current emphasis on practice-oriented, trans-disciplinary science for adaptation needs to be more open and collaborative. Currently it is often not very reflexive, nor attractive for the disciplinary sciences to be involved in.

- Interdisciplinary research suffers from a tendency to assume that the program objectives can be achieved by a one-size-fits-all approach in which stakeholder involvement is central.

These issues often polarize and seek to divide, rather than integrate knowledge for the benefit of all. To remedy this, and to ensure shared scientific and technological adaptation capacity, they envision a new generation of scholars (e.g., lateral "system thinkers", generalists or "integrators") who are educated to assist in addressing real-world problems and challenges. These scholars would require knowledge, abilities, skills and competencies to bridge current fundamental science domains as well as the theoretical to practical and natural science to social science divide (*Nature*, 2015b; Viseu, 2015)[28]. Such "integrators" would focus on breaking through heuristics and clarifying key concepts, engage in broader validation testing and fundamental explanatory inquiry linked with climate change and global adaptation issues, and seek to organize and integrate a new knowledge system that allows for a multiplicity of ontological perspectives, and a variety of validated methodologies that are better connected to addressing societal needs across spatial and temporal scales. They would seek to provide substantive novel insights and recommendations to support disciplinary, interdisciplinary, and trans-disciplinary research advancement hat could better engage others working within one or several of these knowledge domains. Moreover, Linkov et al. (2014) indicate that the current "resilience paradigm" needs to change, whereby resilience management needs to go beyond current risk management to address the broader complexities and uncertainties of large integrated systems. Integrators could be the agents of such a broad, integrated new paradigm of adaptation and resilience science whose mission it is to identify and implement

---

[28]Why integration matters:
http://www.nature.com/news/
integration-of-social-science-into-research-is-crucial-1.18355

real-world strategies of climate change (and other risk-related) adaptation (and mitigation) measures. This would complement and enhance current integrated risk assessment, modeling and management by exploring flexible responses, distributed decision-making, modularity, and redundancy, to reduce the loss of critical ecosystem structural integrity and functionality. Predicting the limit to resilience and the concomitant loss of ecosystem goods and services remains a broad, integrated and profound challenge. It requires one to address the degree of exploitation, stress and disturbance that different ecosystems can withstand without a critical loss of their resilience (Thrush et al., 2009).

1.3   Sustainability Frameworks, Metrics, and Indicators

A type of knowledge framework called a "sustainability framework" is a broad, structured conceptual model which represents a first level translation of sustainability principles and fundamental definitions (Figure 1.14). Similar to the ISO standards, such frameworks define high level objectives, intermediate and long-term targeted outcomes. Methodologies are required to translate frameworks into action in terms of metrics, indicators, and models. Decision-Support Software Tools (DSS) integrate knowledge within a given organization, industrial sector, or within a group of stakeholders to address a given utility need or application aim, combining metrics, indicators, and models together. One can distinguish between *offline* scientific research versus *online* operational DSS tools that are embedded within client-based/standalone or server-based/web-based portals as delivery modes.

A clear definition of the sustainable development goals and related policy commitments is urgently needed, in order to assess options for measuring and monitoring progress. At present, there is no agreement either on the definition of goals, targets, and indicators, or on assessment metrics. Nonetheless, progress on this front is being made by the Open Working Group on Sustainable Development Goals (SDGs). The Intergovernmental Science Policy Platform on Biodiversity and Ecosystem Services produces 182 different assessments, but many other organizations do not conduct such comprehensive assessments, nor regularly update them and any supporting databases. Institutional and country-specific sustainability frameworks have been developed and implemented with differing levels of requirements, data, and model-based assessment needs. Existing frameworks are based on a different set of assumptions, sustainability measures (i.e., criteria/sets of indicators), and accuracy of input datasets. A *metric* is defined as a standard of measurement, typically based on a quantitative estimation methodology, whereas a sustainability *indicator* measures the state of a metric that tracks progress toward sustainability, typically with a broader focus than metrics, and used to evaluate and motivate progress toward sustainability objectives (Ahi and Searcy, 2015, 2013). Indicator choice and weighting are inherently very subjective processes; depending on the value and importance judgments (Morse et al., 2001). There is often very little consistency between differ-

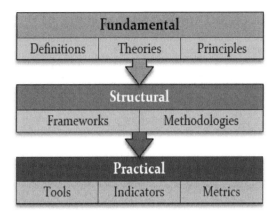

FIGURE 1.14: The hierarchy of sustainability elements: Concepts, defining structures, and practical measures (Tuazon et al., 2013).

ent frameworks and they are framed based more on what is available, than what is needed to reliably assess sustainability across all the industrial sectors. There are often inconsistencies in how impacts affect different resource sectors (Hitz and Smith, 2004). In this way, any toolbox for monitoring sustainable development progress requires optimization when integrating diverse knowledge from stakeholders, practitioners, and decision-makers.

Many organizations are involved in score card type assessments. Such approaches have been applied in a recent Review of Implementation of Agenda 21 and the Rio Principles as contained in the International Institute for Sustainable Development (IISD)'s Guidebook Food Security Indicator and Policy Analysis Tool (FIPAT) (IISD, 2014). There are several major steps (each containing key issues and challenges) that aid in building comprehensive, yet tractable and transparent, sustainability frameworks that are capable of integrating a desired set of target indicators, as shown in Figure 1.15 (Kouadio and Newlands, 2015). The intermediate steps of identification and prioritization of application areas, capacity building, and adoption issues, guidelines and technical constraints on data availability, comparability, cost, and quality are often not explored in-depth and given adequate quantitative analysis, before information is disseminated further. High quality data and predictive model output should be easily integrated as it becomes available, and requires a sustainability framework to possess sufficient flexibility and automation. There is a continual need for real-time, real-world capacity building to improve the availability and quality of data on sustainable development. High quality statistics are crucial both for setting targets and for monitoring progress via a robust accountability mechanism. Further investment in national statistical systems to integrate national data collections, and process and analyze data is needed, especially when such information and knowledge is diverse, disaggregated, and distributed. Better coordination is also needed between defining sustainable development goals and progress measurement (Schader et al., 2014; UNDESA, 2013). The development of the Adaptation Policy Framework (APF) is intended to help provide the rapidly

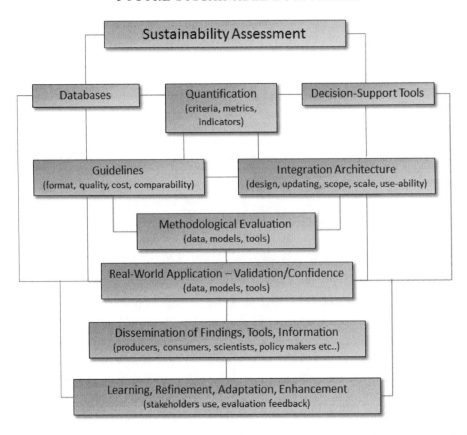

FIGURE 1.15: Major aspects in developing sustainable metrics (Kouadio and New-lands, 2015).

evolving process of adaptation policy-making with a much-needed road-map. Ultimately, the purpose of the APF is to support adaptation processes to protect—and enhance—human well-being in the face of climate change. The APF is built around four major principles that provide a basis from which individual and collective actions to adapt to climate change can be reconciled and integrated:

- Adaptation to short-term climate variability and extreme events serves as a starting point for reducing vulnerability to longer-term climate change
- Adaptation occurs at different levels in society, including the local level
- Adaptation policy and measures should be assessed in a development context
- Adaptation strategy and the stakeholder process by which it is implemented are equally important

The APF can be used by countries to both evaluate and complement existing planning processes to address climate change adaptation. As an assessment, planning and implementation framework, it lays out an approach to climate change adaptation that supports sustainable development, rather than the other way around. The APF is about practice rather than theory; it starts with the information that developing countries already possess concerning vulnerable systems such as agriculture, water resources, public health, and disaster management, and aims to exploit existing synergies and intersecting themes in order to enable better informed policy-making (Burton et al., 2004).

The System of Environmental-Economic Accounting (SEEA)[29] contains the internationally agreed standard concepts, definitions, classifications, accounting rules, and tables for producing internationally comparable statistics on the environment and its relationship with the economy. The SEEA's "Central Framework" is a multipurpose conceptual framework for understanding the interactions between the environment and the economy. By providing internationally agreed concepts and definitions on environmental-economic accounting, it is an invaluable tool for compiling integrated statistics, deriving coherent and comparable indicators and measuring progress toward SDGs. The UN Statistical Commission (UNSC) adopted the SEEA in 2012 as an international statistical standard at its 43rd meeting session. Its implementation in developing and developed countries is timely given the importance placed on integrated information for evidence-based policy making in the outcome document of the Rio+20 United Nations Conference on Sustainable Development. A new Integrated Reporting Framework and Integrated Reporting Database[30] have also been developed by the International Integrated Reporting Council (IIRC)[31] in 2013. This is a global coalition that includes regulators, investors, companies, accountants, and Non-Governmental Organizations (NGO) (IIRC, 2013).

A generalized, deterministic framework has been devised with an explicit set of metrics for agroecosystems or agricultural production systems (Tubiello and Rosenzweig, 2008) (refer back to Equation 1.2). This framework comprises biophysical indicators (soil, climate, water, biomass/yield), agricultural system characteristics (land, technology, irrigation, production), socio-economic indicators (rural welfare, poverty, nutrition, protection and trade, crop insurance) and climate policy (carbon sequestration potential, bioenergy, expansion, rotation/cropping system change, GHG target) as exposure, sensitivity, and adaptive capacity criteria, respectively, for assessing vulnerability. Metrics were also selected to span multiple spatial and temporal scales, i.e., from local to regional, national and global, and short-term and long-term information on impacts. While this framework includes important indicators across different types of agricultural systems, it applies more to food production than to bioenergy.

Frameworks have also been devised and proposed specifically to address bioenergy sustainability and provide recommendations for integrating specific environmental indicators with associated management pressures and environmental effects, and broader sets of sustainability development indicators (McBride et al., 2011; Pinter et al., 2005). Figure 1.16 provides an overview of a sustainability framework for agriculture involving spatial-based metrics (Kouadio and Newlands, 2015). A set of guiding principles for integrating spatial-based data and model-output into metrics for assessing sustainability have been proposed recently (Sayer et al., 2013). They prescribe ten principles: 1) continual learning and adaptive management, 2)

---

[29]SEEA, https://unstats.un.org/unsd/envaccounting/seea.asp
[30]IIRC examples, examples.theiirc.org/home
[31]IIRC, www.theiirc.org

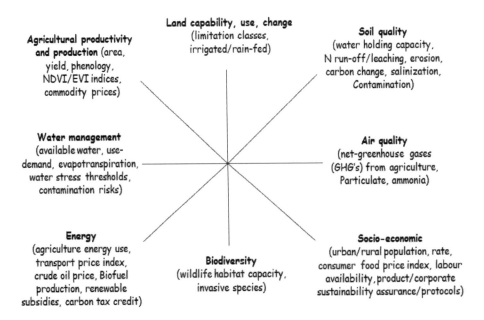

FIGURE 1.16: Overview of a set of ecosystem metrics and their leading indicators for assessing the sustainability of agricultural production systems (Kouadio and Newlands, 2015).

common concern entry point, 3) multiple scales, 4) multifunctionality, 5) multiple stakeholders, 6) negotiated and transparent change logic, 7) clarification of rights and responsibilities, 8) participatory and user-friendly monitoring, 9) resilience, and, 10) strengthened stakeholder capacity. Such principles reflect a realistic approach for developing spatial-based indicators, given that typically, indicators are developed and considered only when sufficient data and model-output becomes available. Sustainability requirements and needs for addressing a given problem or challenge differ widely between applications. To help in the convergence of diverse, interdisciplinary knowledge, participatory decision-making and monitoring approaches are needed that involve multiple stakeholders, in addition to core metric and sustainability decision analytics.

In tandem with the wide-ranging advancements made in identifying sustainability indicators, building sustainability frameworks and integrated reporting, large corporations, medium- and small-business enterprises are also seeking ways to build and enhance their business sustainability, by incorporating the objectives of sustainable development, namely social equity, economic efficiency, and environmental performance, into a company's operational practices. Manufacturing industries have shifted from end-of-pipe solutions to a focus on product life cycles, and integrated environmental strategies and management systems in an effort to achieve greater cost-saving supply-chain efficiencies and more sustainable production. Furthermore, efforts are increasingly made to create closed loop, cyclic production systems, and to adopt new business models (OECD, 2009). The Sustainability Framework of the World Bank's International Finance Corporation (IFC) (IFC, 2012) provides a platform for undertaking Sustainability Performance Reviews (SPR) and establishing a strong business case for corporate sustainability worldwide. However, while companies that compete globally are increasingly required to commit to and report on the overall sustainability performances of operational initiatives, indicator frameworks that are available to measure overall business sustainability may not effectively address all aspects of sustainability at the operational level. This is especially true in developing countries such as South Africa. Moreover, social criteria that are difficult to identify and quantify, often do not receive enough consideration (Labuschagne et al., 2005).

A recent review of the history of sustainable development and theoretical contributions and foundations is provided by Tuazon et al (2013). They detail a selected set of sustainability frameworks that have evolved over the past decade, namely: Cook's "The Natural Step" (NS) (Cook, 2004), Elkington's "Triple Bottom Line" (TBL) (Elkington, 1997) and Porritt's "Five Capitals" (FC) (Porritt, 2005) frameworks. While the TBL is framed around social, environmental, and economic equity goals, it builds back-casting (i.e., future change in the context of past change) into decision making and longer-term planning horizons that characterize the broader guidance and boundary conditions for achieving sustainability of the NS framework. The FC approach that evolved from TBL integrates accounting for technology and infrastructure as manufactured capital. "Capital" can be defined in a broad sense as any input into the production function of an individual, business, or organization. While capital was traditionally meant to include only the previously produced goods and non-financial assets that are used in the production of goods or services, a broader array of forms of capital are now widely recognized, separate from whether they are integrated into financial reports. The five kinds of capital are: human, social, built (manufactured or infrastructural), natural (environmental or ecological), and economic (financial capital)[32]. More details on these various forms of capital is also available from the Forum for the Future, an independent NGO working globally with business, government, and other organizations to solve complex sustainability challenges.

Singh et al. (2008) provide an excellent overview of sustainability assessment methodologies, discussing aspects, issues, and challenges linked with both individual and composite sustainability indicators. They provide a comprehensive listing and a summary of existing sustainability indices, and how they are formulated and applied in terms of statistical normalization, weighting, and aggregation. Indices they discuss include: the Living Planet Index, the Index of Environmental Friendliness, the Compass Index of Sustainability, the City Development Index, the Environmental Policy Performance Index and the Ecological Footprint. The indicators in sustainability frameworks can be highly coupled and involve dynamical interplay (Singh et al., 2008). The Compendium of Technical Support Team (TST) Issues UN Briefs is an inter-agency technical support team for the United Nations General Assembly Open Working Group on Sustainable Development Goals (SDGs) that provides further details and discussion pertaining to international goals and targets linked

---

[32]Five Capitals, `www.forumforthefuture.org/project/five-capitals/overview`

with poverty, food security and nutrition, sustainable agriculture, desertification and land degradation, water and sanitation, employment and social equity, education and culture, health and sustainable development, and human population dynamics (UNDESA, 2014).

A Federal Sustainable Development Strategy (FSDS) in Canada establishes national sustainability planning and reporting requirements. It has three key elements, namely: 1) an integrated, whole-of-government picture of actions and results to achieve environmental sustainability, 2) a link between sustainable development planning and reporting and the Government of Canada's core expenditure planning and reporting system; and, 3) effective measurement, monitoring and reporting in order to track and report on progress to Canadians (Environment Canada (EC), 2010)[33]. A set of environmental sustainability indicators in this strategy has been identified and integrated across four main themes: 1) addressing climate change and air quality, 2) maintaining water quality and availability, 3) protecting Nature, and 4) shrinking the environmental footprint. Indicators are cross-sectoral and include GHGs, a hazardous weather warning index, ambient levels of fine particulate matter, ozone, sulfur/nitrogen dioxide, volatile organic compounds (VOCs), the freshwater quantity and quality index, a protected land area, the ecological integrity of national parks (i.e., habitat loss and fragmentation, invasive species, pollutants and climate change, loss of large carnivores), and sustainable fisheries harvest level[34]. The Federal Sustainable Development Act requires a Sustainable Development Office within EC to develop and maintain systems and procedures to monitor progress on the FSDS's implementation. The Canadian Sustainability Indicators Network (CSIN)[35] also seeks to advance the best practices in measurement and sustainability indicator systems in Canada (and worldwide) in an effort to achieve progress toward global sustainable development. Through CSIN, sustainability indicator and reporting practitioners exchange ideas, data, and methods, and circulate announcements. Based on a "community of practice" approach, new and experienced practitioners share lessons learned and discuss relevant issues of theoretical, strategic, technical, and practical importance.

How to best use and integrate data and statistical metrics, and indicators is an important question that still remains unresolved and may have helped to bridge the divide between the perspective of Nordhaus (1973), who criticized the World Model as trying to generate meaningful insights on Earth dynamics without data. While there is clearly a need to integrate a sufficient amount of knowledge and update it so that it remains relevant and accurate to the current day, simply adding more indicators to a sustainability framework and associated methodology is, however, a misguided goal. A scientific basis to guide how to measure and reliably assess sustainability is required to deduce and optimize what knowledge is required. Here, statistics plays a central role in helping to better understand, communicate, and exchange findings between models of varying complexity. While many unique and

---

[33]EC, www.ec.gc.ca/dd-sd/default.asp?lang=En&n=9277C8B9-1

[34]EC indicators, www.ec.gc.ca/indicateurs-indicators/default.asp?lang=En

[35]CSIN, www.csin-rcid.ca/

competing sustainable development assessment frameworks, indicators, and metrics have been designed, very few have been extensively validated. Very sparse data and a broad set of model uncertainty estimates are often available. Moreover, many studies use subjective criteria in regard to what data to use, assimilate, and apply to address a given problem, without evaluating the levels of uncertainty associated with specific datasets and models, or their integration and combined output. Typically, additional bias and variance (i.e., uncertainty) introduced when data or model output is upscaling and downscaling to align with required spatial and temporal resolution and coverage for a given application, is also not considered or assessed (Newlands and Porcelli, 2015). Recently, Turner (2008) compared historical data for 1970–2000 (i.e., 30 years) with the "Limits to Growth" model scenarios, finding that data matches best with the "standard run" or Business-As-Usual Scenario (BAU). Global coupled socio-economic and environmental system collapses midway through the 21st Century and do not compare well with scenarios that involve comprehensive use of technology or stabilizing behavior and policies. This study also emphasizes the need to better understand and control pollution globally. Statistics is key to understanding non-equilibrium probability distributions, changing frequencies of occurrence, extreme anomalies in relation to long-term averages, and the spatial influence of atmospheric circulation patterns on the location and timing of extreme weather such as heat waves or droughts, that influence ecosystem sustainability decision-making (Hansen et al., 2012).

In recent years, there are many popular works on designing and implementing sustainability. In *Prosperity without Growth: Economics for a Finite Planet* Tim Jackson, a leading sustainability advisor to the British Government argues against continued economic growth in developing nations and provides a vision for human society that can flourish within the ecological limits of a finite planet (Jackson, 2009). In *The End of Growth* Jeff Rubin, who was the chief economist of CIBC World Markets for almost twenty years up to 2009, suggests that human ingenuity should not be underestimated, and that we have always found ways to adapt. He emphasizes that, "sustainability is not just an abstract notion, but is the governing idea behind the kinds of economies we need to foster" and that we must, "relinquish the past we've known. As the boundaries of a finite world continue to close in on us, our challenge is to learn that making do with less is better than always wanting more" (Rubin, 2012). In *Enough is Enough: Building a Sustainable Economy in a World of Finite Resources* Rob Dietz and Dan O'Neill, the executive director and chief economist of the Center for the Advancement of the Steady State Economy (CASSE) based in Arlington, Virginia, USA[36], describe how to establish a prosperous yet non-growing economy. This book was written to celebrate the legacy of Herman Daly, co-founder of the journal, *Ecological Economics*, and who was a senior economist with the World Bank, contributing many ground-breaking contributions in the study of economic development, population, resources and the environment. In *The Upcycle: Beyond Sustainability - Designing for Abundance*, Bill McDonough and Michael Braungarty provide inspiration for action on sustainability, with "Cradle to Cradle" integration—

---

[36]www.steadystate.org

a term that infers constant improvement of a product or system, and discuss how crucial creativity is alongside the need to think big, even if one has to act small, and about approaching problems with a bias for action (McDonough and Braungart, 2013). They also discuss achieving "zero-waste" by dividing materials into industrial, and ecological-based nutrients that need to be recycled and composted.

Having surveyed our adaptive capacity and existing approaches for assessing sustainability with new concepts, definitions and insights on the integrated perspective, we next take a look at how stressed our current ecosystems are, how we have perturbed their cycles, depleted finite resources, creating huge losses and wastes. This context reveals the drastic state and the broad extent of human impacts on ecosystems worldwide, and how they have a rapidly deteriorating ability to sustain their structure and function, and maintain resiliency. It is from this current context we then discuss future sustainability development pathways.

2

A Planet under Pressure

CONTENTS

> "Many aspects of climate change and associated impacts will continue for centuries, even if anthropogenic emissions of greenhouse gases cease. The risk of abrupt and irreversible change increases as the magnitude of the warming increases."

IPCC 5th Assessment Synthesis Report, Climate Change Beyond 2100, Irreversibility and Abrupt Changes, Section 2.4, IPCC (2014b).

2.1 The Rise of Anthromes

Large regions of the world possess distinct fauna and flora, and have adapted to climate and other conditions in unique ways. The five main biome types are ocean/aquatic, desert, temperate forest and tropical rainforest, arctic tundra and tundra, and grassland biomes. Each type has sub-types—for example, grassland biomes comprise Prairies, Steppe, and Savanna. Biomes represent the basic unit for describing patterns of ecosystem form, process, and biodiversity. Biomes contain many (often similar) ecosystems, even though system sizes and their ranges can vary. An "ecosystem" is distinguished from a biome as an ecological system or community of living organisms (e.g., plants, animals and microorganisms) and non-living components (e.g., air, water, sun, mineral soil) that interact together, relying on nutrient, water and energy storage and flows (or fluxes) driven by global biogeochemical cycles or pathways. These cycles recycle chemical substances as they move through the Earth system's biotic (biosphere) and abiotic (lithosphere, atmosphere, and hydrosphere) compartments involving interrelated biological, geological and chemical factors.

Biomes have been mapped according to regional variation in vegetation type and climate. Dominant vegetation types and climate both determine biomass productivity, subdominant plant forms, and fauna biodiversity. In terrestrial biomes, species diversity tends to increase with higher incident solar radiation (day length), temperature, humidity, soil moisture availability, and net primary productivity (NPP). Humans have re-engineered ecosystems to such an extent that we consume close to one-third of all available terrestrial net-NPP (Ellis and Ramankutty, 2008). Recognizing this, Ellis and Ramankutty (2008) have generated the first characterization of terrestrial biomes based on global patterns of sustained, direct human interaction with ecosystems, identifying 18 different anthropogenic biomes or "anthromes" based on empirical analysis of global population, land-use, and land-cover[1]. Anthromes involve human and ecosystem interaction and are therefore mosaics comprising ecosystem processes that are a function of regional population density, land-use, biota, climate, terrain, and geology. Anthromes include dense settlements (urban and mixed settlements), villages, croplands, rangelands, forests and wildlands. Anthromes cover more than 75% of the Earth's ice-free land and comprise almost 90% of terrestrial NPP and 80% of global tree cover, covering more of the Earth's surface than wild or pristine ecosystems. According to their biome re-analysis, in 2008 (i.e., world population of 6.4 billion people), 40% lived in dense settlements (82% urban), 40% lived in village biomes (38% urban), 15% lived in cropland biomes (7% urban), and 5% lived in rangeland biomes with 5% urban and forested biomes containing 0.6% of the global population. Our terrestrial biosphere has been fundamentally re-engineered, re-structured and re-distributed by human activity and interactions with ecosystems (Alessa and Chapin III, 2008). Global patterns of species composition and abundance, primary productivity, land-surface hydrology, and the biogeochemical cycles of carbon, nitrogen, and phosphorus, have all been substantially altered (see Ellis and Ramankutty (2008) and referenced therein).

Peel et al. (2007) have generated a new global map of climate, based on the Köppen-Geiger climate classification using updated, long-term monthly precipitation and temperature time series from 4279 stations (i.e., 3650 with precipitation and 944 with temperature quality-controlled records) distributed around the world. Each climate zone is identified according to different criteria selected from the mean annual precipitation and temperature, the hottest/coldest monthly temperature, the total number of months temperature is above $10°C$, and the driest and wettest months in summer and winter. They utilized all the historically available station data, quality-controlled and interpolated it into two-dimensional (i.e., latitude and longitude) thin-plate spline surfaces (with tension) and then gridded at the 0.1° by 0.1° resolution across each continent[2]. Globally the dominant climate class by land area is arid B

---

[1]ecotope.org/anthromes/maps/

[2]ARCGrid Rasters, www.hydrol-earth-syst-sci.net/

(30.2%) followed by cold D (24.6%), tropical A (19.0%), temperate C (13.4%) and polar E (12.8%) (see Peel et al. (2007) for precise zonal criteria and definition). Rubel and Kottek (2010) have also recently generated global maps of climate zones under the Köppen–Geiger climate classification spanning 1901–2100, consisting of a historical period (1901–2002) based on observational data and future period (2003–2100) based on ensemble projections of a global climate model (GCM) scenario output (Rubel and Kottek, 2010)[3]. They report that major shifts in these climate zones are very likely to occur, with the largest shifts between the main classes of equatorial climate (A), arid climate (B), warm temperate climate (C), snow climate (D) and polar climate (E) of 2.6–3.4 % (E to D), 2.2–4.7 % (D to C), 1.3–2.0 (C to B) and 2.1–3.2% (C to A).

In addition to climate, changes in disturbance, soil moisture regimes, higher temperature, and a rising atmospheric carbon dioxide (CO_2) concentration will drive changes in terrestrial NPP in the future, further disturbing and shifting biomes. A recent analysis by Weinzettel et al. (2013) reveals that affluence is driving today's displacement of land use worldwide (Weinzettel et al., 2013). While higher-income countries require high productive land per capita, they displace a large fraction of their land use through international trade (close to 6% of global land demand), even though they have more land available than low-income countries on a per-capita basis. As income doubles, equivalent land and ocean area footprints were found to increase by one third (on a per capita basis). This is driven mainly by international trade and rising national imports, which were found to increase proportionally to income. Nonetheless, country-specific human footprints (land per capita) as the amount of biologically productive land required to satisfy the consumption per average inhabitant vary widely with respect to the 2007 global average of 2.7 gha/p (i.e., 18 billion gha) and the global *biocapacity* (average per capita) of 1.8 gha/p (i.e., 11.9 billion gha)[4]. Ecological footprints are now regionally at 0.4 gha/person (gha/p) for Bangladesh and Pakistan, 5.8 gha/p for Finland, and 6.7 gha/p for Norway. The European Union's (EU's) land footprint was 2.5 gha/p (16% of global total), followed by the United States (13% of the global total or 3.5 gha/p), China (12%; 0.77 gha/p), and India (8%; 0.55 gha/p). Canada has the world's 8th highest ecological footprint per capita, according to the 2012 Living Planet Report (WWF, 2012).

Biodiversity loss can cause ecosystems to become stressed or degraded, and to collapse. An index of biodiversity called the Living Planet Index (LPI) measures the health of ecosystems by tracking population trends (WWF, 2014, 2012, 2010). According to the most recent Living Planet (2014) Report by the World Wildlife Fund (WWF), this index of biodiversity has declined globally by 52% between 1970 and 2010, based on trends in 10,380 populations of 3,038 mammal, bird, reptile, amphibian, and fish species. The change in this index is, however, very different between tropical and temperate zones, attributed to different rates and timing of

---

[3]176 maps for 1901–2100 period are available at `koeppen-geiger.vu-wien.ac.at/shifts.htm`
[4]The global hectare (gha) quantifies both the Ecological Footprint of people or activities and the biocapacity of the Earth or its regions. It represents the average productivity of all biologically productive areas on Earth in a given year.

land-use change and the lagging of associated habitat destruction and degradation i.e. 36% in the temperate zone and 56% in the tropical zone overall. Within terrestrial ecosystems, the terrestrial LPI has declined by 36% since 1970. For marine ecosystems, the LPI has declined by 39%. Within freshwater ecosystems, the LPI has declined by 76%, far larger than the marine or terrestrial LPIs. The rate of species extinction has increased by 100–1000 times the background rates that were typical over Earth's history, reaching a global average extinction rate of 100 E/MSY (i.e., extinctions per million species-years) with currently 25% of species in major taxonomic groups threatened with extinction. Extinction rates vary from the highest to lowest for fungi and protists, plants, vertebrates and invertebrates[5]. This program currently manages data on over 73,000 species, but this number is set to increase substantially in the next few years, in meeting a 160,000 species target. The latest statistics are available[6]. The average global extinction rate is projected to increase another 10-fold within the current century (Mace et al., 2014; Cardinale et al., 2012; SCBD, 2010). In 2014, the UN Conservation of Migratory Species of Wild Animals (CMS), a group comprising 900 experts from 120 countries, has added 21 bird, fish, and mammal migratory species to their protection list, including the polar bear. The direct and indirect consequences of the future-projected large shifts in biome areas, ecosystem biodiversity, and species extinctions that are intertwined with climate change and human activity remain very difficult to unravel and quantify.

A brief synopsis and synthesis of latest knowledge, statistics, evidence, and international consensus regarding the current state of our ecosystems and the availability of non-renewable resources follows next. Current outlooks and projections for 2020, 2050 and beyond are highlighted based on anticipated, plausible consequences of our current state, activities, and understanding of accelerating climate change, population and other global change drivers and pressures.

2.2   Stressed Ecosystems

Ecosystems provide humans with food, forage, bioenergy, and pharmaceuticals and are essential to human well-being (Power, 2010). Modern agriculture is a dominant activity whereby humans have impacted ecosystems. Abiotic (i.e., non-living) negative stresses originate from the environment, such as extremes in temperature, precipitation, wind, and irradiance, whereas biotic stresses are due to living organisms (e.g., bacteria, fungi, viruses, parasites, insects, weeds). Hoffman and Hercus (2000) discuss stress as an evolutionary force, while also recognizing the limits to adaptation. They discuss how biotic and abiotic (i.e., climate) stresses can speed up favorable adaptation, but are dependent on reproductive output, genotypic and phe-

---

[5]IUCN Red List of Threatened Species, www.iucnredlist.org
[6]IUCN Summary Statistics,
   www.iucnredlist.org/about/summary-statistics#TrendsInBiodiversityStatus

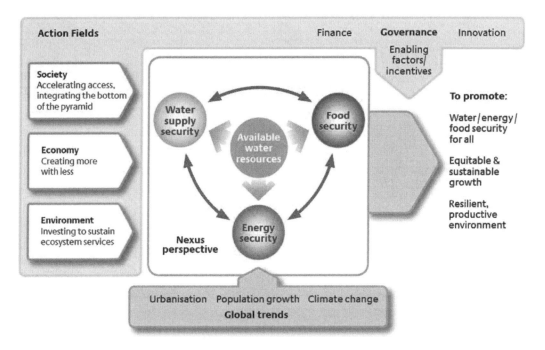

FIGURE 2.1: The water, energy, and food security nexus (Hoff, 2011).

notypic trade-offs, genetic diversity/variability and mutations, leading to complex shifts in species morphological and physiological traits. Modern agriculture, broadly defined, now includes crop and livestock production, forestry and agroforestry, fisheries and aquaculture, greenhouse horticulture, including renewable energy farming (e.g., algal and biomass bioenergy, solar and wind farms).

The *nexus* approach is a platform that brings together related disciplines and sectors based on the recognition of the importance of the interconnections of resources and their sustainable use (UNU-FLORES, 2014)[7]. The nexus approach partitions an international or national sustainability approach or framework, and can help focus sustainability action within a more regional or more local context. This constitutes a multi-scaling approach to complement levels of human organization and interaction with our environment. When it is difficult to find sustainable solutions, the regional actions and activities being considered in a nexus grouping can be assessed from all three sustainability dimensions (UNU-FLORES, 2014; Hoff, 2011). An overview of the food-water-energy-climate change nexus summarizing major linkages and interactions is shown in Figure 2.1. Supply-chains for food, energy, and water are becoming ever more integrated and interdependent. The close link between the lack of food (supply and availability) with water-related constraints is related to the large volumes of water that exists as vapor flow (i.e., evapotranspiration, ET) required in plant growth, ranging on average between 1,000 and 3,000 m^3 t^{-1} dry matter grain yield for the world's dominating cereal crops (Rockström et al., 2007; Rockström,

[7](UN University (UNU), Institute for Integrated Management of Material Fluxes and of Resources (FLORES))

2003). We have changed the world's biomes and ecosystems so rapidly and extensively, that the current state of our ecosystems is very likely detrimental to both the health of our environment and our own health (Lathers, 2013; Kinney, 2008; Shea et al., 2008; St. Louis and Hess, 2008; Patz et al., 2007; Schiedek et al., 2007; Patz et al., 2005). Children and the elderly, including the poor, are most vulnerable to a range of climate-related health effects whether in developed or developing nations. In 2014, the United States through the EPA introduced President Obama's Clean Power Plan to cut pollution from large power plant emitters of greenhouse gases and other pollutants and to provide future climate and health benefits estimated at US $55–$93 billion by 2030. This action is anticipated to avoid 2,700 to 6,600 premature deaths and 140,000 to 150,000 asthma attacks in children[8]. Such actions help to reverse current alarming climate change and human health trends. Nonetheless, many changes to our atmosphere and ecosystems may now be irreversible, as they are already set in motion and have momentum. If irreversible, adverse conditions to human health and survival may continue to persist. If reversible, adverse health conditions may also persist, especially if pollution reductions are episodic and not applied globally (Kintisch, 2014).

If sufficient mitigation action does not occurring globally, we lose the potential to reverse or minimize detrimental changes that are occuring and will occur, including our ability to adapt in time, according to the Millennium Ecosystem Assessment (MEA) (IPCC, 2014b; MEA, 2005). One of the most visible impacts is evident from changes in land-use and the conversion of land for agriculture, housing, urban, and industrial use. Broader-based climate change impacts, and interrelated species invasions, increases in nitrogen deposition and soil erosion rates, are less visible in built-up, urban environments (Bardgett, 2011). Since 1950, the number of people living in urban areas (i.e., cities) and migrating out of rural areas has increased enormously around the world[9]. In 2014, 54% of the world's population lived in urban areas, a proportion that is expected to increase to 66% (i.e., 2.5 billion more people) by 2050 UNDESA. In North America, 82% of the population lives in urban areas, comparing to 73% in Europe, while 40% in Africa and 48% in Asia live in rural areas. Urban areas are projected to continue to rise in Africa and Asia to 56% and 64% by 2050, respectively. As urbanization continues, the global (emerged ice-free) land area of 13.2 billion hectares (ha) continues to change in its distribution and use. Human impacts and indirect influences, are at the same time, changing the suitability of land for different uses (i.e., linked with ecosystems goods and services), now and in the future, beyond just a consideration of regional climate and soil fertility and productivity conditions. As of 2002, forested land was 28% (3.7 billion ha) of global land area, with grasslands and woodlands (includes rangelands, shrubland, pasture land, and cropland sown with pasture trees and fodder crops) comprising 35% (4.6 billion ha) and 25% sparsely vegetated and barren, settlement and infrastructural and inland water bodies (FAO, 2011b).

Pongratz et al. (2008) have conducted a global-scale reconstruction of global

---

[8]EPA, `www2.epa.gov/sites/production/files/2014-05/documents/20140602fs-overview.pdf`
[9]World Resources Institute, `www.wri.org`

land-cover and agricultural land (i.e., cropland and pasture) over the last millennium, since AD 800. Their reconstructed land-cover is in general agreement with the recorded history of agriculture, human expansion and migration, showing that up to AD 1700, temperate and tropical broadleaf deciduous forests were most severely affected by crop cultivation, while large areas of natural grassland were used as pasture. Their assessment provides findings that are important for calibrating current land cover with pre-industrial land cover change. An assessment of global permanent cropland area over the past 300 years using data from 1850 and onward, by Ramakutty and Foley (1998) provides an estimate of cropland area as 12–13% of global land area (1.6 billion ha or 16 Gha). This assessment combines national and regional agricultural inventory data and Discover satellite-derived land-cover data aggregated to the 10 km scale. It considered cropland that is harvested, land temporally idle to be fallow or due to crop failure, land for pasture and permanent crops and agroforestry (Ramakutty and Foley, 1998; Loveland and Belward, 1997). Within the past half-century, land converted to agriculture for cultivation globally has increased by 159 million ha (Mha) since 1961, with 12% of global land area being used for the cultivation of agricultural crops. Recall that this estimate is approaching the 15% of the global ice-free land surface converted to cropland that is considered a planetary boundary threshold (Rockström et al., 2009a,b). The recent cropland estimate was obtained from applying the International Framework for Land Evaluation called the Agro-Ecological Zones (AEZ), a comprehensive methodology developed by the Food and Agriculture Organization of the United Nations (FAO) that identifies land management options integrating land resource inventories and the evaluation of biophysical limitations and potentials (FAO, 2011b; Fischer et al., 2002).

Land available for cultivation varies across the world and its regions, having differing levels of suitability for cropland as well as other uses. In addition, land use changes face additional ecosystem constraints. For instance, despite more than 1.6 billion ha of 4.4 billion ha (or 36%) (excluding protected areas) of currently uncultivated land that could, theoretically, be cultivated and brought into crop production, real-world ecosystem constraints impose greater restrictions linked with ecological fragility, low soil fertility, toxicity, disease reducing potential yields and land productivity, including protected areas, biodiversity value and potential biodiversity loss, and carbon sequestration constraints (e.g., forests), including socio-economic factors such as a lack of storage, transport and other infrastructure, and high opportunity costs. As a consequence, more than 75% of our global land surface remains unsuitable to be used as rainfed cropland, because of severe productivity constraints and regional climate conditions being either too cold (13%), too dry (27%), too steep (12%), or with poor soils (40%) (FAO, 2011b; Fischer et al., 2002). Global land cover assessments show that while cropland did historically increase rapidly, it is now no longer increasing, and has essentially stagnated at around 1.5–1.6 G ha globally. As Thenkabail et al. (2010) highlight, the so-called "Green Revolution" during 1955–2005 led to increases in productivity per unit of land by crop breeding of high-grain or high-oil yielding, fast-growing and pest-resistant varieties, alongside crop intensification and improved cropland management (e.g., controlled drainage), and large increases in nutrient, water and crop protection inputs through irrigation, nitrogen and phosphorus fertilizers and herbicides, pesticides and controlled drainage (Thenkabail et al., 2010; Thenkabail, 2010). However, because these inputs are costly yet necessary to circumvent reductions in land productivity and soil fertility within agricultural areas, in many parts of the world, productivity has not increased and even declined, such that many African nations that were previously net food exporters are now food importers (UNDESA, 2014). Globally, 5.2 billion pounds of pesticides were used in 2006–07, with herbicides accounting for the largest portion of total use, followed by other pesticides, insecticides, and fungicides[10]. Cropland productivity gains have come at a cost, and have introduced new problems, such as salinization, the over-use of fertilizer causing eutrophication and large dead zones in coastal areas and lakes, and over-use of herbicides-pesticides-insecticides causing leaching, soil erosion, surface and ground water pollution and pollination loss. As agriculture has intensified, land has become degraded. *Degradation* is defined as, "The reduction or loss in the biological and economic productivity and complexity of terrestrial ecosystems, as well as in the ecological, biochemical and hydrological processes that operate in them" (UNCCD, 1996).

Despite the increase in food production during the Green Revolution, malnutrition still poses a major public health and development challenge and is estimated at 868 million people worldwide who suffer from undernourishment. While this is 20% less than decades earlier, it still represents 15% of the developed world (FAO, 2011a; MEA, 2005). While more than half of food calories are supplied by only three crops—wheat, corn/maize and rice (UNDESA, 2014), the future potential and need to adapt new cultivars of these crops, and the introduction of others, will require adequate diversity in plant genetics.

2.3   Perturbed Cycles

Human activity has impacted global biogeochemical cycles—storage, flows, and their strength of interaction (Peñuelas et al., 2012). Understanding how biogeochemical cycles have been and can be perturbed, alongside how they interact, forms a basis on which to understand how ecosystems maintain resilience. Perturbations in these cycles across spatial and temporal scales influence how fast and integrated ecosystems behave, respond and adapt.

2.3.1   Water

Water security is defined as "the capacity of a population to safeguard sustainable access to adequate quantities of acceptable quality water for sustaining livelihoods, human well-being," and, "Human well-being has multiple constituents, including basic material for a good life, freedom of choice and action, health, good social

---

[10]EPA, www.epa.gov/opp00001/pestsales/07pestsales/usage2007.htm

relations, and security" (UN, 2013). Currently, 70–80% of all water used by humans goes to producing our food, of which roughly 65% is "green water" use for growing food in 1.13 Gha on cropland that is rainfed and supplies soil moisture within the unsaturated soil zone, while 35% goes to "blue water" use for growing food in 400 Mha of irrigated croplands using water from reservoirs, barrages, lakes, rivers, and groundwater stored in deep aquifers. "Grey water" is wastewater or "sullage" from sewage flow that originates from clothes washers, bathtubs, showers and sinks, but it does not include wastewater from kitchen sinks, dishwashers or toilets with gross fecal coliform contamination that generally has high concentrations of organic matter termed black water. While grey water represents about 60% of the total globally-averaged household generated wastewater, it is far less contaminated with fecal coliforms and some chemical pollutants from bathing and laundry, microbial and chemical contamination. Grey water composition is highly variable, and requires proper treatment to remove potentially harmful substances, but can be reused for watering gardens and sports fields, for ornamental fountain use, waterfalls, landscaping, lawn sprinkling systems for irrigation, car washing and toilet flushing. Nonetheless, the collection, treatment, and recycling of grey water can still pose potential health risks (World Health Organization (WHO)) (WHO, 2006). Wastewater is commonly discharged into rivers, lakes, or seas with little (if any) treatment, but can be used as an energy source (biogas and biochar) for making wastewater treatment plants energy self-sufficient (UNESCO, 2014b).

Regional climate change variability and stochasticity in rainfall frequency and amount, coupled influences of air temperature, inter-annual reductions of snow-melt (spring run-off) introduce considerable uncertainty in agriculture's water use. The rise in irrigated land has been to further increase crop yields under more intensive agriculture and land management practices, converting predominately rainfed land into irrigation land, despite the widely varying land suitability constraints of rainfed land linked with aridity and soil water availability. Blue water use for irrigating croplands has increased by 50% (i.e., from 139 Mha to 301 Mha). At the same time, urbanization and recreation, and industrialization and biofuel water use continue to increase, comprising nearly 11% and 19% of water available globally, respectively.

Worldwide focus is now shifting from the Green Revolution to the Blue Revolution on growing more food per drop of water or "crop per drop" through Precision Agriculture (PA). PA seeks, among many goals, to increase water and nutrient-use efficiency across agricultural landscapes. In addition, global food and water security issues are being addressed with the benefits provided by satellite, remote-sensing information in mapping, monitoring and assessing land use, land use change, and environmental footprints[11]. A new remote-sensing operational method for identifying disturbed land areas, their degradation risk level and potential impact on the water (i.e., hydrological) cycle, including feedback effects on climate for large areas has been recently developed and validated by Garcia et al. (2008) that offers new opportunities for attaining higher crop-per-drop (García et al., 2008). Vorosmarty et al., (2010) discuss current global threats to water security. While large investments

---

[11]Croplands, `www.mdpi.com/journal/remotesensing/special_issues/croplands/`

in irrigation and water technology continues to help increase food production, freshwater biodiversity loss within habitats associated with 65% of continental discharge are now moderately to highly threatened (Haddeland et al., 2014; Vorosmarty et al., 2010). In 2013, the FAO released Version 5 of their digital global map of areas based on sub-national irrigation statistics (pre-1990 and 1990–2010). This map identifies areas around the world that are being irrigated using surface or groundwater, including areas equipped for irrigation. The accuracy of this geospatial information depends highly on the available area statistics, and it is deemed highest within Asia, Northern America and Southern and Western Europe, and lowest for most countries in Sub-Saharan Africa, in Eastern Europe and the Russian Federation (Siebert et al., 2013, 2010). Zwart et al. (2010a,b) have devised an integrated approach that combines satellite remote-sensing data and model-based spatial prediction (using the WATer PROductivity or WATPRO model) to generate a global map of water productivity associated with growing wheat (rainfed and irrigated). This integrated approach evaluates the regional and subregional use of water resources, making use of global datasets of precipitation and reference evapotranspiration to identify regional water use efficiencies.

The total volume of water on Earth is about 1.4 billion km^3 or $1.4 \cdot 10^{18}$ m^3. The volume of freshwater resources is around 35 million km^3 (2.5% of the total volume). The total usable freshwater supply for ecosystems and humans is about 200,000 km^3 and is thus less than 1% of all freshwater resources (UN, 2013). The global water cycle consists of the oceans, water in the atmosphere, and water in the landscape with water in all three phases, i.e., water vapor, liquid water, and ice. Large amounts of latent heat are released during the phase change to liquid water; therefore the water cycle is closely linked to the energy cycle as well as the carbon, nitrogen, and phosphorus biogeochemical cycles. The cycle is closed by the fluxes between these reservoirs. Although the amounts of water in the atmosphere and river channels are relatively small, fluxes are high (Oki et al., 2004) (refer to Figure 2.2). According to the Comprehensive Assessment of Water Management in Agriculture today's food production requires a consumptive water use of about 6800 km^3 yr^{-1} (Hanjra and Qureshi, 2010; Molden, 2007). Currently, green water use in rainfed agriculture is increasing at the rate of $\sim$5,000 km^3 yr^{-1} (i.e., cubic km per year) and this rate may have to jump to $\sim$7,500 and again to 12,400 km^3 yr^{-1} by 2030 and 2050 (i.e., by 50% and 148%), respectively, to ensure food security.

Currently, $\sim$4,000 km^3 yr^{-1} of blue water resources are withdrawn, whereas blue water consumptive use is $\sim$2,600 km^3 yr^{-1} (Rockström et al., 2009a, 2007). 60% of the blue water that we withdraw, flows back through rivers and groundwater, with 40% consumed by plants as they transpire and evaporative losses. Rost et al. (2008) compare published estimates of green and blue water consumption differentiated between natural ecosystems, cropland and grazing land, lakes and reservoirs, and human consumption (Rost et al., 2008). Human consumptive use of blue water for irrigation is anticipated to increase at a rate of 25–50% or 400–800 km^3 yr^{-1} by 2050. The upper limit of accessible blue water resources is estimated at $\sim$ 12,500–15,000 km^3 yr^{-1} (with water scarcity occurring when withdrawals exceed 5,000-6,000) (Rockström et al., 2009a, 2007).

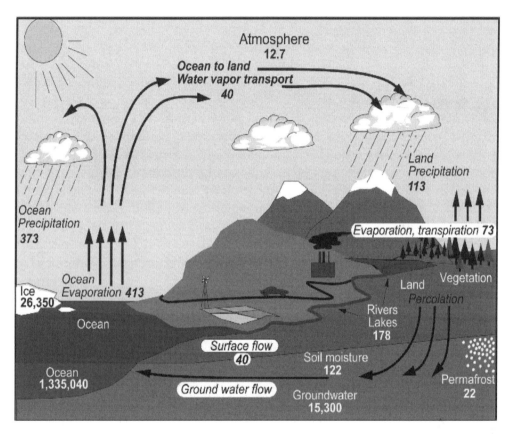

FIGURE 2.2: The global water (hydrological) cycle involving the oceans, the atmosphere, and terrestrial biosphere. Water exists in all three phases, i.e., water vapor, liquid water, and ice. Although the amounts of water in the atmosphere and river channels are relatively small, fluxes are high. Estimates of the main water reservoirs, given in 10^3 km^3, and the flow of moisture through the system, given in 10^3 km^3 yr^{-1}, equivalent to Eg (10^{18}g yr^{-1}) (Trenberth et al., 2006).

Most assessments of global water resources have focused on surface water, but groundwater is the primary source of drinking water worldwide. Globally, the rate of groundwater use is increasing by 1–2% per year (UNESCO, 2014b). The depletion of groundwater in relation to the rate of natural renewal required to support ecosystems is a major question and concern (Gleeson et al., 2012; Morrison et al., 2009). Aquifers have an important strategic value as accessible over-year stores of water in a relatively stable condition without any evaporative loss. Percolating water in aquifers is naturally decontaminated along diffuse recharge and circulation pathways (FAO, 2011a). Gleeson et al. (2012) define our groundwater footprint (GF) as the area required to sustain groundwater use and groundwater-dependent ecosystem services. They estimate this footprint globally, based on average rainfed (i.e., natural) and irrigation (i.e., artificial) recharge rates between 1958–2000. They find that large aquifers are critical to agriculture (especially in Asia and North America) and that humans are over-exploiting them, in part, because of changing precipitation patterns, rising surface (e.g., evaporative) water loss, pollution and depletion levels. This compares with the use of available groundwater that requires disinfection treatment for microbial contaminants and cleaning to remove chemical contamination than surface water. The size of the global groundwater footprint is currently $(131.9 \pm 24.9) \cdot 10^6$ km^2, far larger than the actual area of hydrologically active aquifers and is dominated by only a few countries (i.e., including the United States, China, Pakistan, Iran, India, Mexico, and Saudi Arabia)[12]. Almost 60% of the people living in regions where groundwater availability or groundwater-dependent surface water is unsustainable, are in India and China.

Many countries are relieving the pressure on their available water resources and future water scarcity by, not importing real water, but through international trade and net-imports of virtual water (Hoekstra and Hung, 2002). The water that is used in industrial (e.g., agricultural) production and contained in an output product is called "virtual water." Virtual water explains how physical water scarcity in countries in arid regions is relaxed by importing water-intensive commodities. In this way, the "real-world" water cycle is quite different from the natural water cycle, even on the global scale (Oki et al., 2004). The global volume of international virtual water flows in relation to trade in agricultural and industrial products averaged 2320 billion m^3 yr^{-1} across 1996–2005 (Mekonnen and Hoekstra, 2011)[13]. Oki et al. (2004) estimate global virtual and real water flows in 2000 associated with major cereal (wheat, rice, maize, and barley) trade, estimating the total virtual water trade (imported virtual water) for such commodities was 1,140 km^3 yr^{-1} (2000), corresponding to 680 km^3 yr^{-1} of real water, and represents a water saving of 460 km^3 yr^{-1}.

Projections are "what-if" scenarios, whereby not all assumptions hold, but instead, boundary conditions are tracked and expected to hold continuously. In this way, projections describe the future in terms of a bounded pathway leading to it.

---

[12]UN Educational, Scientific and Cultural Organization (UNESCO) Groundwater,
  `www.whymap.org/whymap/EN/Products/products_node_en.html`
[13]Water footprint, `www.waterfootprint.org`

This compares to "future predictions" or "forecasts," where initial conditions are specified, and all assumptions are expected to hold and approximate reality. Ensemble forecasts involve perturbing initial conditions. When a projection is branded "most likely" it becomes a forecast or prediction. A forecast is often obtained by using both probabilistic (i.e., stochastic) and deterministic models, possibly a set of these, outputs of which can enable some level of confidence to be attached to the projections. In contrast a projection or projection scenario is a coherent, internally consistent and plausible description of a possible future state of the world. It is not a forecast; rather, each scenario is one alternative image of how the future might unfold. A projection may serve as the raw material for a scenario, but scenarios often require additional information (e.g., about historical baseline or benchmark conditions). A set of scenarios is often adopted to reflect, as well as possible, the range of uncertainty in projections. Other terms that have been used as synonyms for scenario are "characterization," "storyline," and "construction."[14] While projections are associated with statistical likelihoods based on historical observations, reported and recorded evidence, there are, nonetheless, still significant uncertainties associated with both climate and weather-based forecasts and projections in terms of how both humans and our Earth system will interact, respond, and adapt (Slingo and Palmer, 2011; Von Storch and Zwiers, 2002). The IPCC Representative Concentration Pathway (RCP) represent four different evolution patterns for atmospheric greenhouse gas emissions and concentrations, land-use changes and the emission of air pollutants (e.g., ozone and aerosols), but do not consider future natural forcings of the Earth system, such as volcanic eruptions, changes in some natural sources (e.g., methane CH_4 and nitrous oxide N_2O emission), total solar irradiance, or adaptive responses by the Earth system to potential inter-related cascading impacts. They were developed using Integrated Assessment Model (IAM)s and a wide range of climate model simulations to project future consequences of the climate system that consists of five main interacting components: the atmosphere, the hydrosphere, the cryosphere, the land surface, and the biosphere. These latest IPCC (AR5) scenarios compare to Special Report on Emissions Scenarios (SRES) used in previous IPCC assessments (e.g., AR4), except that RCPs, include a consideration of climate policy (IPCC, 2014b; van Vuuren et al., 2011; Moss et al., 2010). RCP8.5 is comparable to SRES A2/A1F1 as a very high GHG emissions scenario. The two stabilization scenarios are RCP6.0 and RCP4.5, are comparable to SRES B2 and B1, respectively. There is no equivalent scenario for the low GHG forcing scenario RCP2.6 in SRES.

Gerten et al. (2011) have projected future water requirements and availability (2070–2099) in a spatially-explicit way, developing a global water scarcity indicator (Rost et al., 2008). They assume per capita water requirements to produce a balanced diet representing a benchmark for hunger alleviation of 3000 kilocalories per capita per day[15] comprising 80% vegetal food and 20% animal derived-products. They also assume that a country was considered water-scarce if its water availability falls below the water requirement for the specified diet. They find that per capita

---

[14]IPCC, `www.ipcc-data.org/guidelines/pages/definitions.html`
[15]1 kilocalorie is 4184 joules

water availability will diminish in many regions of the world along with calorie-specific water requirements. These findings are based on future projections using a complex vegetation-hydrology model (Lund-Potsdam-Jena managed Land model, LPJmL), and 17 Global Climate Model (GCM) projections under the IPCC SRES emission scenarios (A2 and B1). This study finds that the global number of people living in water-scarce countries will rise to 3.5 billion (B1 scenario) and 6 billion (A2 scenario), respectively, meaning that the fraction of the global population living in water-scarce countries will rise from 28% at present to 43%–50% (A2) by the 2080s. There is a greater than 90% probability that per capita water availability will decline by over 10% in countries in Europe and northern Asia. Africa, Middle and Near-Eastern countries will remain in a water scarce state, and there is a high risk that North America and Australia will see their per capita water availability decline by 10% (Gerten et al., 2011). Similarly, global projections out to 2050, based on climate change scenarios developed from Hadley Centre climate simulations (HadCM2 and HadCM3), indicate that by 2025, 5 out of 8 billion people (or 62.5%) will be living in countries experiencing water stress. By 2025, these model projections estimated that around 5 billion people, out of a total population of around 8 billion, will be living in countries experiencing water stress (i.e., using 20% of their available water resources) (Arnell, 1999).

In 2013, the International Institute for Applied Systems Analysis (IIASA) highlighted that there was a crucial need to look at the future of water resources as a cross-cutting system issue. Previously, no comprehensive, global assessment has integrated the different water uses for food, energy, industry, domestic and environmental services to provide insights about how to prioritize competing needs and how we could generate synergies in the future (Kabat and Contestabile, 2013). Long-range global water scenarios as a contribution to the 5th (2014) World Water Development Report (WWDR) (WWDR4) from World Water Assessment Programme (WWAP) of the UNESCO (UNESCO, 2014b,a, 2012) have explored the major interactions between energy and water. While we consume vast amounts of water to generate energy, we also consume vast amounts of energy to extract, process, and deliver clean water (Morrison et al., 2009). Water is required in the production of conventional (i.e., fossil-fuel-based), unconventional, and renewable energy (e.g, solar, geothermal, biogas, biochar, hydropower, energy crops).[16] While the world moves to more renewable sources of energy and away from using fossil-fuel sources to reduce atmospheric GHGs, the interaction between water and energy, will become stronger, with a higher water risk. Biodiesel, bioethanol, and lignocellulosic energy crop production requires far higher water withdrawal and consumption than conventional gas, coal, and shale gas (UNESCO, 2014b). Hydraulic fracking is a example of new emerging risks associated with unconventional gas exploration and development. Optimization of hydraulic fracturing requires a system-wide approach that considers the transport, installation, and powering of industrial operations, in addition to the integrated analysis of its potential environmental impacts. In 2012, the EPA launched new research and a strategic multi-agency collaboration to

---

[16]Measured in toe or tonne of oil equivalent units (1 toe = 11.63 MWh = 41.9 GJ)

better understand the potential impacts of hydraulic fracturing on drinking water resources, with detailed investigation of water risks associated with each stage of the fracturing water cycle (EPA, 2012)[17]. Hydraulic fracturing and horizontal drilling for the extraction of natural gas requires typical water injection volumes of 246,000 liters (65,000 US gallons) for coal-bed methane (CBM) production, but far higher amounts of 50 million liters per well (13 million gallons) for shale gas (EPA, 2012). Fracturing fluids that are injected contain about 90% water, 8% sand used as a proppant, and 2% chemicals (acids, surfactants, biocides, and scaling inhibitors). Fracking produces large volumes of wastewater with high salinity, and increases the potential risk of water contamination from naturally occurring radioactive materials in drinking water resources, as well as in ecosystems released from wells (UNESCO, 2014b; EPA, 2012).

Huntington (2006) has reviewed the current state of science regarding historical trends in hydrological variables, including precipitation, runoff, tropospheric water vapor, soil moisture, glacier mass balance, evaporation, evapotranspiration, and growing season length. Climate change and global warming will result in increases in evaporation and precipitation, leading to the hypothesis that one of the major consequences will be an intensification (or acceleration) of the water cycle. Such findings are consistent with the latest integrated findings of the IPCC. According to the latest RCP scenarios, global mean surface temperature increases projected in the future—with baseline anthropogenic scenarios, being those without additional mitigation, are from ~3.7 to 4.8°C above the average for 1850–1900 associated with a median transient climate response, and from 2.5 to 7.8°C when uncertainty in climate is included (IPCC, 2014b). This rise in temperature is associated with intensification in the water cycle and of rainfall in time and space. Rainfall is anticipated to become far more non-uniform causing extreme floods and droughts, and to increase in higher latitudes and the equatorial Pacific and mid-latitude wet regions, but to decrease in the dry subtropical regions. Global glacier volume (not including Antarctica) is projected to decrease by 15–35% to 35–85%, depending on RCP scenario. Sea-level rise during 1901–2010 was larger than the previous two millennia, rising by 0.17–0.21 m, and is projected to increase between 0.26–0.82 m from 2081–2100. The uptake of CO_2 in our atmosphere is being absorbed by our oceans. This is resulting in ocean acidification that has subsequently decreased the pH of ocean surface water by 0.1 since the industrial era. Increases in pH are projected to lie within 0.06 to 0.32 by 2081–2100, depending on the RCP scenario. Oceans continue to absorb more than 90% of all energy accumulated between 1971–2010 of the energy stored in our climate system (IPCC, 2014b). Marine and terrestrial ecosystems differ in their connectivity and environmental gradients. The ocean environment is characterized by strong vertical and horizontal gradients in several abiotic factors, such as light, turbulence, concentrations of dissolved elements, oxygen, hydrostatic pressure and temperature, some of which show diurnal and seasonal fluctuations, notably in light levels and temperature (Reusch, 2013). Such stratification is anticipated to become more severe as the oceans warm, such that large areas of open and coastal and marine

---

[17]EPA, www2.epa.gov/hfstudy

ecosystems will become hypoxic (oxygen-poor) (Reusch, 2013; Diaz and Rosenberg, 2008).

Extreme climatic and/or weather events that combine to produce heat-waves, droughts, floods, hurricanes/cyclones, and wildfires are anticipated to become more frequent and intense (IPCC, 2014b). The assessment, prediction and forecasting of such risks, possibilities and their cumulative impacts, will need to be integrated into our future decision-making. Here, it is important to distinguish events linked to weather, which reflects short-term conditions of the atmosphere, to those linked with climate, as the average daily weather for an extended period of time at a certain location (Travis, 2014). Hitz and Smith (2004) have identified, in a review of the global impacts of climate change, that the majority of impact assessments assume change in average climate conditions, not in its variance, and assume too simplistic assumptions regarding how ecosystems will adapt. They indicate the need for greater consistency and sophistication in ecosystem impact assessment and forecasting to be more representative of real-world impact uncertainty. They recommend improving statistical metrics so that they are more meaningful across industrial sectors for policy decision-makers. They also recommend that spatial interpolation (i.e., aggregation) and scaling of data inputs and model-output, needs to be better validated, with spatial uncertainty better assessed. Littell et al. (2011) also have examined uncertainty in climate-driven ecological models and how they can best be used to reliably inform adaptation to climate change.

Terrestrial Net Primary Productivity (NPP), as atmospheric carbon dioxide continues to rise, may saturate, and become less of a sink for CO_2 in coming decades. In C_3 plants the activity of rubisco, a carbon-fixing enzyme, levels off between 800–1000 ppmv CO_2. To reach rubisco, CO_2 molecules must diffuse through two consecutive segments of a continuous pathway in the leaves of C_3 plant species—one that involves stomatal diffusion through gases, and mesophyll diffusion through liquids and lipids, such as cell walls, plasmalemma, cytosol, chloroplast envelope membranes, and stroma. The diffusion "path length" of mesophyll diffusion is generally shorter than that of stomatal diffusion. Recently, Sun et al. (2014) find that plants may absorb more CO_2 than previously thought. They have determined responses that better represent and estimate mesophyll diffusion can account for a 16% correction that is large enough to explain the persistent over-estimation of crop growth rates of historical atmospheric CO_2 by many Earth system models that do not explicitly represent internal drawdown. Without such correction, models overestimate the amount of CO_2 that is available for carboxylation and underestimate the photosynthetic responsiveness to atmospheric CO_2. To improve the accuracy and reliability of simulated exchanges of CO_2, water, and energy between the Earth's land surface and the atmosphere, Still et al. (2003) have integrated multiple data sources to generate a global distribution map of C_3 and C_4 plants. Data on crop fractions, national harvest crop area data and satellite remote-sensing data were utilized in their analysis. Considerable spatial heterogeneity exists in their distribution, which predicts the global coverage of C_4 vegetation and C_3 vegetation to be 18.8 and 87.4 million km^2, respectively.

In addition to C_3 plants, the future productivity of C_4 vegetation (i.e., grasses) is also anticipated to change as CO_2 levels rise. Albertine et al. (2014) report that pollen and allergen production in grass can increase by 50% per flower under elevated levels of CO_2, regardless of ozone (O_3) levels. Scaling this up based on estimates of increased pollen production and the number of flowering plants per treatment, they estimate that airborne grass pollen concentrations could increase in the future by up to 200%. This provides evidence for significant coupled impacts of climate change involving vegetative growth and human airborne allergies worldwide. Such impacts are anticipated to become more prevalent, whereby 10–30% of the global population may be affected by allergic rhinitis and more than 300 million may be affected by asthma. Pollen from grass species, which are highly allergenic and occur worldwide, elicits allergic responses in 20% of the general population and 40% of atopic individuals (Albertine et al., 2014). Alongside anticipated increases in pollen production for C_4 species, decreases in the survival of pollinators of C_3 species are anticipated (a see-saw effect). Klein et al. (2007) have studied the importance of pollinators such as bees, wasps, thrips, birds, flies, and beetles in changing landscapes for agricultural crops worldwide. For animal-mediated pollination for crops that are directly consumed by humans, they find that pollinators are essential for 13 crops, with highly pollinator dependent production for 30, moderately for 27, slightly for 21, unimportant for 7, and of unknown significance for the remaining 9 crops. The crops studied include: canola, apples, grapefruit, coffee, sunflower, tomato, and macademia nuts. However, the chronic exposure of bees to two pesticides, neonicotinoid and pyrethroid, at field-level concentrations has been found to impair the natural foraging behavior, and increase worker mortality, leading to significant reductions in brood development and colony success (Gill et al., 2012).

Bardgett (2011) provides new evidence is mounting that the transfer of carbon through roots of plants to the soil is changing, as an example of adaptive, plant-soil feedback to climate change. Such change may play a primary role in regulating ecosystem responses to climate change and its mitigation, having large consequences for ecosystem carbon cycling (i.e., given that soil is the third largest global store of carbon and, together with plants, contains around 2.7 times more carbon than the atmosphere). This therefore suggests that there is an enormous potential to exploit plant root traits and soil microbial processes that favor soil carbon sequestration. There is contentious debate on the real-world potential to increase the capacity of soils to sequester carbon from the atmosphere and hence mitigate climate change (Falkowski et al., 2000). Bardgett (2011) highlights how both the loss and gain of carbon in soil are strongly regulated by plant-microbial-soil interactions. The sensitivity of the Earth system to microbial evolution is also emphasized by recent findings of Rothman et al. (2014), who have studied the most severe biotic crisis and species extinction associated with a severe disruption of the Earth's carbon cycle in the fossil record attributed to massive Siberian volcanism (i.e., end-Permian extinction within 20,000 years beginning 252.28 million years ago, Ma). They have proposed that the global carbon cycle was disrupted by the emergence of a new microbial metabolic pathway that enabled efficient conversion of marine organic carbon to methane. Such methanogenic expansion was catalyzed by nickel from the volcanic event based on carbon isotope analysis, and the phylogenetic analysis of methanogenic archaea and nickel concentrations in South China sediments.

2.3.2 Nutrient (carbon, nitrogen, phosphorus)

An overview of global ecosystem storage and flows of carbon (C) is shown in Figure 2.3. On fast time-scales, plants and trees absorb carbon out of the atmosphere by photosynthesis and release it back through respiration. On slow time-scales, soils absorb carbon from the decomposition of biomass/plant material, where it may reside for years, decades, or centuries, before being decomposed by soil microbes and released back into the atmosphere.

The Global Carbon Project (GCP), a community research effort, has integrated data into *The Global Carbon Atlas* which now serves as a platform to explore and visualize the most up-to-date data on carbon fluxes resulting from human activities and natural processes, alongside outreach, emissions and research components[18]. A special issue of the open-access Biogeoscience journal features the latest estimates from regional carbon assessment and latest findings from carbon cycling research[19]. As highlighted by RECCAP[20], the characterization of estimates of regional carbon budgets and processes is inherently a statistical task, whereby almost all quantities used or produced are realizations or instances of probability distributions (Enting et al., 2012). Enting et al. (2012) detail statistical issues involved in carbon cycling modeling and CO_2 emissions assessment. For example, the covariance structure of inventory-based emission estimates needs to be treated statistically as spatially and temporally ranging datasets and data products contain sampling (i.e., measurement) variability, different degrees of time stationarity, different scaling mean and variance effects requiring factor corrections, adjustments for variable-selection with integration of auxiliary indices, and geostatistical inversion error. Inversion (often matrix inversion) methods use indirect inference where the direction of inference is the opposite of real world causality, such as when estimating net fluxes from the integration of observed surface emissions or concentration data and modeled transport (e.g., Bayesian probabilistic or reactive-diffusion-advection deterministic-type models). There are big data challenges in obtaining reliable emission covariance estimates, given the complexity and size of geospatial environmental data. For example, for 1° by 1° resolution grid, with a record length of 20 years and 52 weeks and 5 processes at each grid node, there are $3 \cdot 10^8$ flux values that need to be estimated (Enting et al., 2012). Also, reference zones and boundaries are often misaligned, creating problems for spatial partitioning and spatial decomposition, and introducing statistical bias and ambiguity. This is, in part, due to natural fluxes from biomes and ocean areas underlying anthropogenic (i.e., human) emissions measured within political census divisions. In addition to measurement uncertainties, scaling factors and other statistical corrections, there are also uncertainties related

---

[18] Global Carbon Project, www.globalcarbonproject.org/
[19] Global carbon cycling, www.biogeosciences.net/special_issue107.html
[20] Carbon Cycle Uncertainty in Regional Carbon Cycle
    Assessment and Processes (RECCAP)

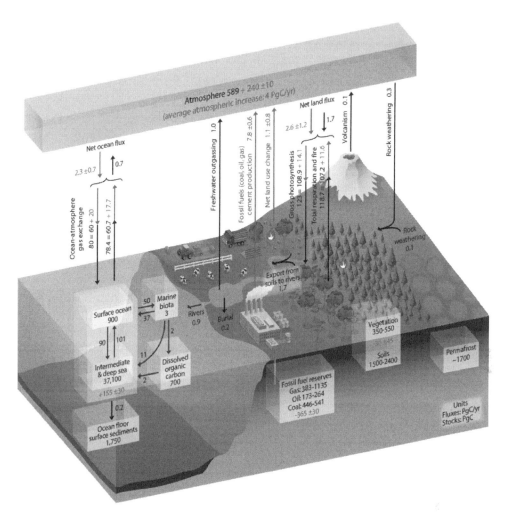

FIGURE 2.3: The global carbon (C) cycle showing stocks (units of petagrams or PgC, where 1 PgC = 1015 gC and is equal to 1 gigaton or 1 GtC ∼ 3.7 Gt CO₂) and annual carbon exchange fluxes (PgC per year). Black numbers and arrows indicate reservoir mass and exchange fluxes estimated for the time prior to the Industrial Era, circa 1750. Grey arrows are: Fossil fuel and cement emissions of CO₂, Net land use change, and the Average atmospheric increase of CO₂ in the atmosphere, also called "CO₂ growth rate." The uptake of anthropogenic CO₂ by the ocean and by terrestrial ecosystems, often called "carbon sinks" are the red arrows part of Net land flux and Net ocean flux. Grey numbers in the reservoirs denote cumulative changes of anthropogenic carbon over the Industrial Period 1750–2011 (see Figure 6.1, Page 471 IPCC (2013) and references therein).

to our knowledge of the carbon cycling process linked with ocean CO_2 flux, land-use change, and interaction effects due to the fact that all biotic sinks for CO_2 require other nutrients in addition to carbon (Falkowski et al., 2000). Canadell (2010) recommends that carbon science be more strongly integrated within other domains in support of environmental sustainability and a portfolio for broad, integrated research that specifically advances a major data science and statistical goal—to create a capability to detect and attribute changes in carbon fluxes operationally, especially from the largest and most vulnerable carbon pools in the Earth system, such as permafrost carbon, peatlands, methane hydrate deposits, tropical forests, and the Southern Ocean sink (Canadell, 2010; Canadell et al., 2010). Schuur et al. (2015) provide an overview of new insights on the anticipated impact of climate change and the warming of high-latitude (within Arctic and sub-Arctic) regions of the Earth on the release of carbon and methane from permafrost carbon. Permafrost carbon is carbon accumulated deep in permafrost soils that can be mineralized by soil microbes and converted to CO_2 and CH_4 within time-scales of years to decades. Future increases in carbon emissions from permafrost land regions are anticipated to be more gradual and sustained than abrupt and massive. The conversion of forested land for agriculture within the tropics is also of particular concern, as tropical vegetation currently stores $\sim$340 billion Mt C, which is 40 times more than annual global fossil fuel emissions[21] (i.e., approximately 8738 million Mt C) (Boden et al., 2010). This vast tropical carbon reservoir is at risk because currently 10.5% of the tropics is cropland, and future cropland expansion is projected to be greatest in such biomes. Moreover, for each unit of land cleared, the tropics lose nearly two times as much carbon ($\sim$120 versus $\sim$63 t ha^{-1} yr^{-1}) and produce less than 50% of the annual crop yield compared with temperate regions (1.71 versus 3.84 t ha^{-1}yr^{-1}) (West et al., 2010).

The global nitrogen (N) cycle (see Figure 2.4) has been perturbed by humans, creating reactive N at a rate two times larger than natural terrestrial N. From a real-world perspective, it comprises an integration of natural, intended, unintended, and substantially perturbed fluxes (UK Centre for Ecology and Hydrology (CEH)) (Sutton et al., 2013). Reactive N is dominated by the production of ammonia for fertilizing crops, in animal feed, and in food additives, as well as the use in other industries. In agroecosystems, a portion of N transfers into waterways through soil erosion, runoff, and leaching, a portion is deposited over land that can increase forest CO_2 sink potential, and is released to the atmosphere through the combustion of fossil fuels in the form of nitrous oxide (N_2O), NH_3 and NO_x that increase radiative forcing as well as influence tropospheric ozone O_3 and aerosol chemistry in the atmosphere (IPCC, 2013). Nitrogen fixation and primary productivity (i.e., growth of zooplankton and phytoplankton) within the ocean can be limited by wind-borne iron fluxes (i.e., aeolian iron) that are coupled to land-use and water cycling (Falkowski et al., 2000; Fung et al., 2000).

Agriculture is dependent on phosphorus derived from phosphate rock, which

---

[21]Carbon Dioxide Information Analysis Center (CDIAC),
   `cdiac.ornl.gov/trends/emis/glo.html`

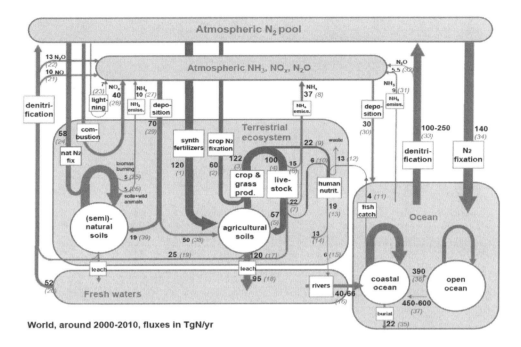

World, around 2000-2010, fluxes in TgN/yr

FIGURE 2.4: The global nitrogen (N) cycle, 2000–2010, comprising natural, intended and unintended or substantially perturbed fluxes (Tg N per year) (Sutton et al., 2013).

is a non-renewable resource, and global reserves may be depleted in 50–100 years time (Batjes, 2011; Cordell et al., 2009). Bennett et al. (2001) estimate the production of phosphate rock globally to be 19.8 Tg yr^{-1} (1995–96) with a total amount of P of roughly 18.5 Tg yr^{-1} that is mined being added into surface soils annually.[22] More recent estimates on phosphorus fertilizers sourced from mined phosphate rock are lower at 15 Tg yr^{-1} or MT P yr^{-1} (see Figure 2.5) (Cordell et al., 2009). Only 4% of P that is mined is not added into soils and is instead used to produce products such as flame retardants, paper, glass, plastics, rubber, pharmaceuticals, petroleum products, pesticides, and toothpaste. Agriculture fertilizer additions to achieve the same productivity are rising as soil nutrients and their fertility is depleted (i.e., a situation of diminishing returns). The rate of fertilizer P necessary to overcome P deficiency increases nonlinearly with increasing P sorption capacity of the soil. In this way, P builds up in upland soils, leading to water quality issues. The leaching of P into freshwater lakes leads to excess P production or eutrophication leading to the excessive growth of algae and other aquatic plants, as well as phytoplankton blooms that are harmful to humans and other aquatic organisms in relation to their foul odors and the oxygen depletion creating hypoxic zones. This can harm aquatic life, nearshore coral reefs, create shellfish poisoning of humans, lead to a loss of biodiversity, and extirpate native plants (Bennett et al., 2001).

A global assessment of soil phosphorus retention potential (or alternatively soil

[22] 1 teragram = 1 million metric tons

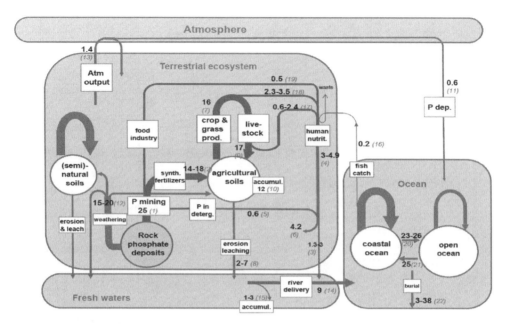

World, around 2000-2010, fluxes in TgP/yr

FIGURE 2.5: The global phosphorus (P) cycle, around 2000–2010, comprising natural, intended and unintended or substantially perturbed fluxes (Tg P per year) (Sutton et al., 2013).

sorption potential) has recently been conducted (Batjes, 2011). Despite its global coverage, this assessment unravels the complexity in soil P sorption/mobilization processes, but relies on broad assumptions by sorting the world soils into four classes of soil P retention potential, differentiated by soil pH, soil mineralogy, and clay content. Worldwide interest in the use of biological agents (e.g., inoculants), such as phosphorus (P)-solubilizing microorganisms is increasing, because they offer significant improvements for crop growth and development, while providing new opportunities for simultaneous biocontrol and reduction in detrimental environmental impacts linked with overuse of phosphate fertilizers (Vassilev et al., 2006). Crop uptake of P can be increased by increasing P soil concentration and/or the activity of root-zone soil microbes, and there are many pathways leading to higher availability of P in the rhizospheres. The primary pathways involve root growth promotion and/or rhizosphere acidification through hydrolytic enzymes, organic acids, and siderophores (iron-metabolizing compounds secreted by microorganisms) (Richardson and Simpson, 2011). Very few inoculants have been tested in agricultural field trials to determine their effectiveness in increasing crop yield. A recent statistical analysis of multi-year, multi-site field trial experiments, conducted across major corn (maize) producing agricultural cropland in the United States (2005–11) provides new insights on increasing yield in response to inoculation with the phosphorus-solubilizing fungus, *Penicillium bilaiae* Chalabuda (Leggett et al., 2014). Despite significant inter-annual variability, Leggett et al. (2014) determine that inoculation is most effective at increasing corn yield in fields with low or very low soil P status. At higher levels of soil P, the yield increases more with inoculation. No significant correlation between the yield response and soil acidity (i.e., pH) was detected, indicating also that pH reduction (i.e., through organic acid or proton efflux) was unlikely to be the primary pathway for better P availability. Variance in the yield increases under inoculation occur due to different P fertilization histories for the different field locations, as well as transient (e.g., rainfall) and topographic effects.

It is crucial to understand how our perturbed carbon (C), nitrogen (N) and phosphorus (P) cycles dynamically behave, scale in time and space, along with the complexity of their interactions. Biogeochemical cycling of C, N, P and S (sulphur) is strongly mediated by biotic factors such as species composition because of their role in producing, mobilizing, and metabolizing substrates. Producers (decomposers) and consumers (herbivores) have different requirements for nutrients, so a species with a higher N:P requirement tends to recycle N at a relatively lower concentration than a species with a lower N:P requirement, so that the ratio of nutrient release from a given substrate is determined by the nutrient requirements of different species. Different biogeochemical mechanisms and reactions are involved, so that these elements are not cycled in simple stoichiometric ratios. For example, organic matter in soils is represented by plant debris (i.e., litter) in various stages of decomposition through to humus, and there is a strong interaction between litter C:N ratios, and rates of carbon decomposition and net N mineralization. Soil organic matter is decomposed by soil microorganisms for energy (i.e., C) and the mineralization of N is a byproduct, as N is bonded to the C skeleton of organic compounds (Eviner and Chapin, 2003; Likens et al., 1981) (Scientific Committee on Problems of the Environment (SCOPE)). Emissions of GHGs are also strongly dependent on the relative rates of mineralization-mobilization and nitrification-denitrification within soils, as discussed in the next section. Phosphorus (P) is not bound to the C skeleton of organic matter, but linked with C via ester bonds and P mineralization via enzyme activity that separates P from organic matter. Sulfur (S) is also released from organic matter (intermediate between N and P), as it is present in organic matter in C and ester chemical bonds, and is released by microbes metabolizing C. Plant community composition can also influence the rates of biogeochemical cycling by altering microclimate conditions such as temperature, acidity (pH), soil moisture, and oxygen (O_2) concentrations (i.e., aerobic and anerobic conditions). Increased levels of O_2 concentration in the soil enhances the decomposition of organic matter, increases nitrification rates, decreases methane (CH_4) formation, and lowers the consumption of SO_4 decreasing mobilization of Iron (Fe) and P mineralization (Eviner and Chapin, 2003). Most phosphorus in soils is present in a form that is unavailable to plants or microbes, and the P cycle is strongly regulated by non-biological processes, so it is decoupled from C and N cycling in terrestrial ecosystems. In marine ecosystems, sulphate (SO_4^{2-}) is very abundant in sea-water and is a major oxidizing agent for organic matter in anaerobic marine sediments, so is responsible for regenerating oxygen in the form of carbon dioxide. In terrestrial ecosystems, sulfur is relatively scarce (Bolin et al., 1983).

2.3.3   Other biogeochemical cycles (methane, sulfur, mercury)

The global budget of atmospheric methane (CH_4) is $\sim$500–600 Tg CH_4 yr^{-1}, but a comprehensive statistical quantification of the flux of methane through trophic food chains is not yet available (Conrad, 2009). About 70% of the total global methane (CH_4) source (600 Tg) is biological production (i.e., biogenic) by methanogens that are anaerobic members of the Archaea (single-celled and functionally similar to bacteria living in freshwater wetlands, man-made lakes, and rice fields), including rumin fermentation in livestock (cattle, sheep and other ruminants), termite digestive systems, and solid waste treatment facilities. The anthropogenic contribution to methane emissions now accounts for 50–65% of total emissions, of which the fossil fuel component integrating anthropogenic emissions related to leaks in the fossil fuel industry and natural geological leaks is 30% (IPCC, 2014b, 2013). The latest IPCC AR5 report identifies that, "after almost one decade of stable CH_4 concentrations since the early 1990s, atmospheric measurements have shown renewed CH_4 concentrations growth since 2007." The drivers of such renewed growth have not been statistically determined, on top of current methane budget historical estimates (2000–2009). This is, in part, because methane emissions are highly variable in space and time, much like nitrous oxide. The methane budget for the decade (i.e., 2000–2009, bottom-up estimates) is 177–284 Tg(CH_4) yr^{-1} for natural wetlands emissions, 187–224 Tg(CH_4) yr^{-1} for agriculture and waste (rice, animals, and waste), 85–105 Tg(CH_4) yr^{-1} for fossil fuel related emissions, 61–200 Tg(CH_4) yr^{-1} for other natural emissions including, among other fluxes, geological, termites, and fresh water emissions, and 32–39 Tg(CH_4) yr^{-1} for biomass and biofuel burning. Microbial methane oxidation under anaerobic (anoxic) and aerobic (oxic) conditions and other environmental factors driving microbial community composition and activity also contributes to global methane emissions.

The interaction of the global sulfur and methane cycles is very strong. It is estimated that industrial mining and combustion have deposited sulfate mercury across the globe, substantially reducing the natural freshwater wetland methane source by 5 Tg (terragrams) below pre-industrial levels (or the total global emission from wetlands estimated at 30 Tg by 50% to 15 Tg) (Schimel, 2004). Sulfate-reducing bacteria also can out-compete methanogens for substrates. Future global emissions will be strongly dependent on global temperature rise and changes in the geographical distribution of wetland areas (O'Conner et al., 2010) producing feedback effects. The combination of methane emission from wetlands, permafrost thawing, and the destabilization of marine hydrates indicates that methane emissions will increase in the future. O'Conner et al. (2010) discuss major knowledge gaps on future methane emission and the potential for catastrophic emissions taking place in the context of integrated Earth system modeling. In particular, they explain that large amounts of methane are stored in marine hydrates stored within deep sediments, as well as in shallow waters (e.g., Arctic Ocean). There are large statistical uncertainties in methane hydrate inventories and the prediction of the time-scales associated with anticipated increases in heat penetration in the ocean and sediments, as well as the fate of methane once it is released within seawater. While CH_4 is removed from soils through oxidation of sufficiently aerated soils, in the atmosphere CH_4 is primarily removed by photo-chemical reactions involving hydroxyl (OH) radicals. This later process has a residence time or turnover time of less than 10 years in the troposphere, a much shorter duration than N_2O, estimated to be 131–141 years (Prather et al., 2012). This further highlights the strong role of bioaccumulation, as it largely resides not in the atmosphere, but within ecosystems. In the stratosphere as well as the marine boundary layer (MBL), chlorine and oxygen radicals are involved in chemical reactions that remove methane. The MBL over our oceans is about 2–3 km thick and is a layer of our atmosphere (within the troposphere) that is directly influenced by the presence of the ocean's surface. It has a critical role in regulating surface energy and moisture fluxes and in controlling the convective transfer of energy and moisture to the free atmosphere, and can be used to identify large-scale trends in atmospheric GHGs[23].

Humans have perturbed the biogeochemical cycle of mercury (Hg) by coal burning, mining, and industrial activities, which bring mercury from long-term sedimentary storage into the atmosphere and have increased the amounts of mercury by a rate factor of 3–5. Mercury pollution poses global human health and environmental risks and is transported (i.e., mobilized, circulated, deposited) globally (Selin, 2009). It circulates in the atmosphere and when it oxidizes it is deposited into marine and terrestrial ecosystems, where it bioaccumulates up the trophic food chains. It eventually, after centuries and millennia, returns to deep-ocean sediments via methylation ocean processes. Mercury poses high risks because when it converts to methylmercury in aquatic and terrestrial ecosystems—it becomes a neurotoxin. Bioaccumulation of methylmercury (MeHg) in predatory fish can elevate concentrations by a factor of a 10^6 relative to water. High levels have been measured also in biota, bioaccumulating within Arctic ecosystems. A recent study by Stern et al. (2012) that focuses on how climate change influences arctic mercury, has elucidated how the extensive loss of sea-ice in the Arctic Ocean and the concurrent shift of plants from perennial to annual plant communities are promoting primary productivity changes, shifting food webs via changes in animal diets and movement behavior in response to changing sea-ice regimes, altering mercury methylation and demethylation rates, and changing mercury distribution and transport across the ocean-sea-ice-atmosphere interface. Historically, natural mercury emissions originate from volcanic and enriched land areas called global mercuriferous belts, situated along plate tectonic boundaries as primary geological sources that comprise the total estimated source amount of ~ 500 Mg yr^{-1}, with global geothermal emissions of 60 Mg yr^{-1}. However, human activity and anticipated climate change is already significantly perturbing the mercury cycle in regions that are most susceptible to rising global temperatures. Current anthropogenic emissions of mercury to the atmosphere range from 2200-4000 Mg yr^{-1} (Selin, 2009). Despite the environmental and health risks, many of the land-atmosphere and ocean-atmosphere cycling and oxidation processes are not well understood, with uncertainty in time and space

---

[23]NOAA's Earth System Research Laboratory (ESRL),www.esrl.noaa.gov/gmd/ccgg/mbl/

poorly quantified. Instead, current atmospheric inventories simply assume a constant conversion factor of 2.12 PgC/ppm (IPCC, 2013).

There are 80,000–100,000 chemicals on the global market (UNEP, 2012)[24]. From 2000 to 2010, chemical production in China and India (BRICS group of countries) grew at an average annual rate of 24% and 14%, respectively, whereas the growth rate in OECD countries like the United States, Japan, and Germany was between 5 and 8%. OECD member countries as a group still account for the bulk of world chemical production, but developing countries and countries with economies in transition are increasingly significant, such that OECD projections indicate that by 2020, developing countries will be home to 31% and 33% of global chemical production and consumption, respectively (UNEP, 2012). Productivity in most biomes is limited by N, and in some by available P, as they limit plant/crop growth and soil carbon storage. Wang et al. (2010) has used a global model of these cycles to derive the global distribution and uncertainty of N or P limitation on the productivity of terrestrial ecosystems at steady state under present conditions. This model comprises different storage components and identified fluxes used to derive a system of nonlinear differential equations alongside empirically-based equations quantifying various system losses, uptakes, and biochemical mineralization factors. Within the terrestrial biosphere, global model estimates indicate that there is a total amount of C of 2767 Gt C (i.e., 19% plant, 4% litter and 77% soil organic matter, SOM). The total amount of N is 135 Gt N (i.e., 94% stored in the soil, 5% live biomass, 1% litter). The total amount of P is 17 Gt (i.e., 33–67% stored in SOM, depending on P mineralization assumptions). Plot-, field- and regional-scale measurements monitoring the ecosystem response to elevated CO_2, increased rate of soil respiration and N deposition enable models to be better validated and more reliable.

2.4 Harmful Emissions

2.4.1 Greenhouse gases

"The Greenhouse Effect" occurs when heat generated from sunlight at the Earth's surface is trapped by certain gases and is prevented from escaping through the atmosphere. Water vapor is the largest contributor to the natural greenhouse effect, and has a leading role in driving the Earth's climate (roughly 2–3 times greater than CO_2) being strongly controlled by air temperature in driving condensation and precipitation, with a mean residence time of 10 days. Water vapour is not strongly influenced by emissions and vice versa (within the stratosphere, above about 10 km in altitude). Human activities that produce water vapor contribute less than natural ones, but do occur as a result of increased evaporation from irrigating crops and power-plant cooling. Increases in water vapor are not considered to contribute to radiative forcing within the tropospheric (i.e., below 10 km in altitude) (IPCC,

---

[24]UNEP's Global Chemicals Outlook

2013). Romps et al. (2014) have recently proposed that water vapor influences precipitation rate via convective available potential energy (CAPE), explaining an increase in the lightning flash rate, and 77% of the variance in time-series of total cloud-to-ground lightning over the US. When this data is integrated into an ensemble of climate models, lightning strikes are predicted to increase $12 \pm 5\%$ per $^\circ$C of global temperature increase, and about 50% over this century.

Human CO_2 emissions originate from the combustion of fossil-fuels (termed FFCO2) in the solid (e.g., coal) $\sim$39%, liquid (e.g., petroleum)$\sim$37%, and gaseous (e.g., natural gas) forms $\sim$19%. Natural gas flaring and cement manufacture also contribute significantly (roughly 1% and 5%, respectively) to global emissions. Cement industrial manufacturing converts calcium carbonate to lime with CO_2 as a byproduct that is emitted to the atmosphere. It is one of the largest non-combustion related, industrial sources. While cement production has supporting data and statistics, emissions and their uncertainties from acid and steel production are not generally available. This is because of incomplete data inventories. Emission data inventories for assessments generally integrate available governmental national or sector-based inventories or energy statistics alongside private-corporate based surveys and reporting. With incomplete data, there is, in turn, a lack of sufficient reliability in production and emission assessment statistics (Andres et al., 2012). Furthermore, global emissions estimates rely on energy data and globally-averaged baseline "emission factors" (so-called Tier 1 emission factors). Global emission statistics are computed from energy consumption at the country level, converted to emissions by fuel type, such that increases and decreases in emissions closely follow changes in GDP corrected for improvements in energy efficiency (Friedlingstein et al., 2010).

Global trends and variability may not significantly capture major regional-scale trends and variability, as countries develop and apply national estimates separately, and can correct them for regional-scale variability in different ways (i.e., Tier 2 data-corrected, and Tier 3 data-corrected and model-corrected). Andres et al. (2012) highlight issues and challenges related to comparing estimates across multiple emission datasets due to a lack of consistency of assumptions and boundary conditions. Just as with our perturbed water cycle, global-scale trade influences the transfer of carbon between regions both physically or virtually as embodied in production processes ("virtual carbon"), and represents a substantial fraction of global carbon emissions—whether emissions that are related to the production of goods and services in one country are consumed in another, alongside the highly distributed supply-demand in fossil-fuel, petroleum-derived and harvested wood, crops, and livestock products (Peters et al., 2012). Production- and consumption-based inventories, emission estimates, and derived-statistics all may need to more reliably account for trade imports and exports, especially as such trade is global and will continue to expand. The ultra-high-resolution GEOS-5 computer model GEOS-5 coupled with new observational data from the Orbiting Carbon Observatory-2 (OCO-2) satellite by the Global Modeling and Assimilation Office at NASA's Goddard Space Flight

Center shows how CO_2 in the atmosphere is transported all across the globe without adhering to political boundaries[25].

The *concentration* of carbon dioxide (i.e., CO_2) in our atmosphere has increased by 40% from 278 (circa 1750), to 390.5 ppm[26] based on 2011 levels. During the same time duration, concentrations of atmospheric methane (CH_4) and nitrous oxide (N_2O) have also increased by 150% (i.e., 722 ppb to 1803 ppb[27]) and 20% (271 ppb to 324.2 ppb), respectively. These current concentrations of atmospheric GHG exceed any level measured for at least the past 800,000 years, the period covered by ice cores, and net CO_2 emissions[28] on average at a rate that exceeds any observed rate of change over the previous 20,000 years (IPCC, 2013). GWP is defined as the ratio of the time-integrated radiative forcing from the instantaneous release of 1 kg of a trace substance relative to that of 1 kg of a reference gas (i.e., set to be CO_2) denoted GWP100 in $CO_{2,eq}$ units, whereby 1 kg of CH_4 and 1 kg of N_2O, each have the same GWP as 21 kg and 310 kg of CO_2, respectively. Net-emission estimates of the IPCC integrate forcing due to all GHGs (i.e., both halogenated gases and tropospheric ozone), aerosol loading and land-surface albedo change. Aerosol loading from particulates, ozone, and oxides of sulphur and nitrogen influence the energy radiative balance, by scattering incoming radiation back to space (i.e., an atmospheric *cooling* effect), while also acting as a heat source or elevated heat pump causing surface cooling. Aerosol loading also influences the water cycle by affecting cloud precipitation and shifting precipitation patterns (IPCC, 2013). These mediating influences, and their role in partially offsetting the warming contribution of black carbon absorption of solar radiation, makes aerosols the largest component of uncertainty in total radiative-forcing estimates (IPCC, 2014b). Land-use changes have increased the land surface albedo, which leads to a radiative forcing of between –0.25 to –0.05 Wm^{-2}, however, due to different assumptions on the albedo of natural and managed land surfaces (e.g., croplands, pastures) as well as changes in surface temperature, albedo estimates have a considerable range of statistical uncertainty.

Anthropogenic CO_2 *emissions* to the atmosphere were 545 ± 85 PgC (1750-2011), increasing at an average rate of 2.0 ± 0.1 ppm yr^{-1} between 2002 and 2011. Fossil-fuel combustion and cement production together contributed 365 ± 30 PgC, with land-use change (i.e., deforestation, afforestation and reforestation) contributing 180 ± 80 PgC (IPCC, 2013). Since 1750, almost 50% of emissions have remained in the atmosphere (240 ± 10 PgC), with the rest being removed from our atmosphere by ocean and terrestrial sinks and stored in the natural carbon cycle reservoirs— ocean reservoirs have stored 155 ± 30 PgC, while terrestrial vegetation biomass and soils that were not directly affected by land-use change, have stored 150 ± 90 PgC (IPCC, 2013). Agricultural GHG emissions also include methane (CH_4) and nitrous oxide (N_2O) and have a higher global warming potential and more potent

---

[25]United States National Aeronautics and Space Administration (NASA),
`www.nasa.gov/content/goddard/a-closer-look-at-carbon-dioxide/`
[26]ppm - parts per million
[27]ppb - parts per billion
[28]units: CO_2 equivalent, denoted $CO_{2,eq}$ that integrate these emissions accounting for their relative "global warming potential (GWP)

impact in changing our climate than CO_2 alone. Methane is produced from the digestive processes of herbivores and management of livestock manure, while nitrous oxide is released from agricultural soils due to soil cultivation. Agroecosystems can be both a source and sink of GHGs because they can sequester carbon in the soil or in agroforestry biomass across longer time-scales.

The UN Framework Convention on Climate Change (UNFCCC)[29] was adopted in 1992 and has since been ratified by 195 parties. The Kyoto Protocol was adopted in 1997 to implement the UNFCCC and entered into force in 2005. It laid down targets for the reduction or limitation of GHG only in developed countries and transition economies. In 2007, the Parties initiated work aimed at drawing up a post–2012 climate agreement, applicable to all emitters of greenhouse gases. The Copenhagen political accord of 2009, the Conferences of Cancun (2010), Durban (2011) and Doha (2012) have laid the foundations of a new international regime, supplementing the existing instruments in the framework of the UNFCCC and the Kyoto Protocol. In 2014, the European Union (EU) set a goal to cut greenhouse gases by 40% by 2030 below the level in 1990, the US set a new target of cutting our net greenhouse gas emissions 26% to 28% below 2005 levels by 2025, and President Xi of China announced new plans to increase the share of renewable energy and nuclear power to roughly 20% by 2030.

In 2015, Canada committed to further reduce GHG emissions by 30% below 2005 levels by 2030, targeting emissions reductions to a level of 515 Mt $CO_{2,eq}$, going beyond the earlier commitment of 17% below 2005 levels (i.e., 736 Mt $CO_{2,eq}$), by 2020 and the previous reduction target of 611 Mt $CO_{2,eq}$ level. Such reductions are anticipated to be achieved through new national-wide regulations to reduce methane that leaks from industrial processes and pipelines, and by eliminating emissions from the chemical and fertilizer industry and natural-gas fired electricity power plants. Nonetheless, increasing Alberta oilsands production (i.e., contributing 8% to the national total GHG emissions in 2011 and close to 40% of the increase in GHG emissions between 1990 and 2011), are projected to increase by 102 Mt from 2005 to 2030. Oilsands emissions are expected not only to increase alongside increases in production, but also to out-pace improvements in emissions intensity within the period up to 2030. At a meeting of the Group of 7 leaders (G7) in June 2015, Canada pledged, along with other member countries, to the ambitious goal of fully ending fossil-fuel use by 2100 rather than a firm commitment in 2050 to a nonrenewable-to-renewable transition-type commitment to form a low-carbon economy with relatively light use of fossil fuels 50 years earlier (i.e., by 2050). Also in 2015, a coordinated multi-province strategy was furthered that offers new opportunity and potential for Canada to better align its current climate and energy policies, to transition away from its carbon-intensive fossil fuel use and production into a green/clean technology-based economy based on clean technology, renewable energy sources and industrial bioproducts (i.e., biofuels, bioenergy, biomaterials and biochemicals).

On December 12, 2015, Canada and 194 other countries reached the "Paris Agreement" at the 21st session of the Conference of the Parties (COP) and the 11th session

---

[29]UNFCCC, `https://unfccc.int/2860.php`

of the Conference of the Parties to the United Nations Framework Convention on Climate Change (UNFCCC). This agreement is an ambitious and balanced agreement to fight climate change and move the world on a pathway to a low carbon future through mitigating and reducing GHGs, developing and implementing national adaptation action plans. It specifically aims to limit global average temperature rise to well below 2°C and pursue efforts to limit the increase to 1.5°C [30]. As of January 29, 2016, Canada GHG emissions are projected to be 768 $CO_{2,eq}$ in 2020 and 815 $CO_{2,eq}$ in 2030. The commitment to reduce GHG emissions by 30% below 2005 levels by 2030, now serves as a floor for possibly, a higher, more aggressive reduction target. A pan-Canadian framework is now being built involving strong collaboration from the federal, provincial, and territorial governments, including regional stakeholder working groups. With a multi-scale implementation and action plan, a national climate change strategy setting national targets is considered to be far more effective and viable over the longer-term. Independent provincial carbon-pricing and cap-and-trade policies aim to better manage, account and reduce Canada's carbon footprint, but significant hurdles exist in ensuring that they are environmentally effective. While a carbon price or cap-and-trade policy applied Canada-wide would provide full consistency in carbon pricing, individual provincial-based strategies may, nonetheless, provide sufficient market competitiveness and incentives, and a needed balance between flexibility and consistency for different sectors of the economy to reduce GHG emissions (Chipanshi et al., 2015; Ecofiscal, 2015; Flanagan, 2015; Newlands and Townley-Smith, 2012; Newlands et al., 2012; Newlands, 2008, 2007; Smith et al., 2007).

Updated global emissions data statistical analysis, nonetheless, reveals large "emission gaps," despite significant pledges from major industrial nations. The 2014 UNEP Gap Report of the UNEP and the World Resources Institute (WRI) indicates that by 2020 global emissions must decline, and be 15% by 2030 and 50% by 2050, with respect to 2010 levels. It also shows that global carbon neutrality must be achieved by mid-to-late century (2055–2070) to mitigate the most severe anticipated impacts of climate change and to ensure global mean temperature does not rise higher than 2°C (UNEP, 2014)[31]. The WRI has devised "The Greenhouse Gas Protocol" comprising a "Mitigation Goal Standard" and a "Policy and Action Standard" to help governments design better policies and emission reductions goals, and measure progress against them. This protocol provides an accounting and reporting standard methodology for national and subnational greenhouse gas reduction goals that has been developed through an international stakeholder process and piloted in several jurisdictions[32]. Here statistical methods for sensitivity analysis, validation of applied risk assessment methodologies against available data, and other aspects of statistical support for corporate goals are critical for extending and enhancing the site- or regionally-specific application of such standards. Kander et al. (2015) have proposed an important improvement in Consumption-Based Carbon Account-

---

[30]UNFCCC, Paris Agreement, http://unfccc.int/2860.php

[31]UNEP Emissions Report 2014, www.unep.org/emissionsgapreport2014/

[32]GHG Protocol and Mitigation, Policy and Action Standards, www.ghgprotocol.org

ing (CBA) to better account for different technologies within national trade export sectors, thereby reflecting more accurately how national GHG policy changes and international trade affects total emissions. They have shown how trade considerations change the global map of carbon footprints significantly, with India, Indonesia and Brazil remaining low emitters (< 2 t CO_2/cap), many EU countries changing from high to medium emitters, China, Mexico, and Russia changing from low to medium emitters, and the US, Canada, and Australia remaining high emitters (>16 t CO_2/cap).

Typically, agriculture is a sink for CO_2; however, it acts as a source when CO_2 is released to the air through the decomposition of plant matter and soil organic matter (SOM). The source of methane is enteric fermentation in ruminant animals and anaerobic decomposition of stored manure. Nitrous oxide is released from organic and inorganic fertilizers, crop residue decomposition, and during the cultivation of organic soils and from manure storage. Denitrification and nitrification by microorganisms are the principal processes of nitrous oxide production in soils, with physical factors (e.g., soil water content via water-filled pore space (WFPS) in relation to precipitation, interflow, drainage and evapotranspiration), and biological factors (e.g., microbial activity changing oxygen consumption and availability) that lead to large temporal and spatial variation in emission rates (Chipanshi et al., 2015). Synder et al. (2009) have reviewed the effect of N source, rate, timing, and placement, in combination with other cropping and tillage practices on GHG emissions. They conclude that intensive crop management systems do not necessarily increase GHG emissions per unit of crop or food production, and they may spare natural areas from conversion to cropland as well as allow conversion of selected lands to forests for GHG mitigation. This study also highlights that with GHG emission data being obtained primarily across several weeks to a few months, the ability to accurately determine agroecosystem level management effects, and to calibrate and validate model-estimates, is greatly limited. Also, site and weather were identified as explaining significant differences among fertilizer nitrogen (N) sources of N_2O agricultural emissions, supporting the idea that more stochastic, statistical modeling approaches may be necessary to better track N_2O emissions and explore the complex interactions between N species with environmental variables (D'Odorico et al., 2003; Porporato et al., 2003).

Nitrifying bacteria oxidize ammonium (NH_4) to nitrite (NO_2) and nitrate (NO_3) under aerobic conditions, and denitrifying bacteria reduce nitrate (NO_3) to N_2 under anaerobic conditions, usually requiring organic C (refer to Figure 2.6). In this way, N_2O is controlled by the interplay of two main loops: the mineralization-immobilization and the nitrification-denitrification loops. Nitrification and denitrification rates may be highly variable with nitrite, nitrate and N_2O being produced and consumed across a complex pathway, whereby emission occurs if production-consumption rates do not balance each other. Microbial community dynamics in response to soil temperature, oxygen, and water, the forms of C substrate utilized by microbial groups, and the linkage of microbial populations to decomposition products from microsite to bulk soil conditions, all require further research, despite the N cycle being one of the most studied biogeochemical cycles (Müller and Clough,

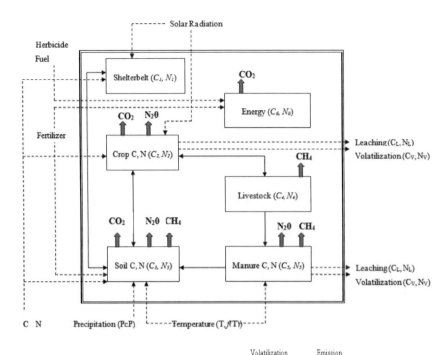

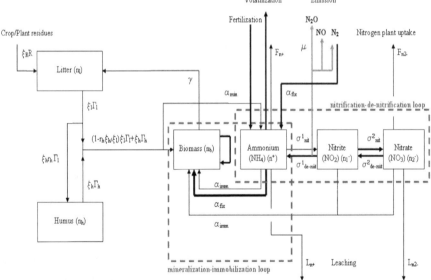

FIGURE 2.6: Top: Main ecosystem components of carbon and nitrogen cycling within agroecosystems, showing major type of GHG emission released from each component. Bottom: Complexity in the storage, flow and their dynamic interaction in generating nitrous oxide N_2O emissions from agricultural landscapes. Major storages for carbon (litter, humus and C in biomass) and nitrogen (N in biomass, ammonium, nitrite, and nitrate) are included. The nitrification-denitrification versus mineralization-immobilization loops are highlighted.

2013). Many environmental factors, such as pH, soil texture, moisture, temperature, and oxygen availability have a role in regulating the overall transformation of soil N and rate of N_2O emission; however, these same factors may control the partitioning of N and the availability of carbon/biomass reductants, and reactive ammonium, nitrite, nitrate. In some cases denitrifiers reduce NO_3 to N_2 without producing any N_2O, but in other cases large amounts are produced. In addition to direct N_2O emission from soil, they also occur indirectly by volatilization (i.e., ammonia, NH_3) or the leaching of nitrate (NO_3), with plumes that can be formed and transported long distances between livestock and cropping production systems. Ammonia (NH_3), a reactive gas with implication on health and greenhouse gas enhancement originates from manure (resulting in CH_4 under anaerobic decomposition and N_2O under aerobic conditions) and inorganic chemical fertilizers. Beneficial Best Management Practices (BMP)s reduce ammonia emission through nutrient testing, optimizing the timing, application of liquid manure and fertilizer, and increasing the ability of farms to store and apply manure effectively. Congruently, changing livestock diets, treating manure, and improving the method, timing, and rates of manure application, significantly reduce N_2O emissions, directly and indirectly.

To more effectively mitigate GHG emission from livestock and manure, recent findings highlight the need to strike the right balance in terms of which mitigation options are selected, where they are applied, at what scale (i.e., farm, regional, national and global scales), and to ensure that animal agriculture will be able to satisfy the growing global demand for food alongside minimizing harmful environmental impacts (McAllister et al., 2011). This involves an approach that integrates multiple datasets and agroecosystem-level models for the ecosystem-level assessment of GHGs to guide mitigation and adaptation policy making. Similarly, Thornton (2010) suggest that, "Trade-offs are likely to become increasingly important, between livestock breeding for increased efficiency of resource-use, knock-on impacts on fertility and other traits and environmental impacts such as methane production. Whole-system and life-cycle analyses (cradle-to-grave analyses that assess the full range of relevant costs and benefits) will become increasingly important in disentangling these complexities." GHGs associated with livestock supply chains add up to 7.1 Gt of carbon dioxide equivalent ($CO_{2,eq}$) per year, or 14.5% of all human-caused GHG releases, with the main sources of emissions being feed production and processing (45%), emission from digestion by cows (39%), and manure decomposition (10%), with the remainder attributable to the processing and transportation of animal products (FAO, 2013)[33]. Typically, GHGs are determined at the plot/field/farm or individual enterprise scale, so there are major questions on how to upscale historical estimates and future predictions, while integrating local and regional effects. GHG emissions could be potentially reduced further, specifically, by quantifying emissions with currently large associated uncertainty: 1) direct/indirect impacts of volatile organic carbons (VOCs), dissolved organic/inorganic carbon fluxes, more extensive measurement of CH_4 and N_2O emissions incorporating atmospheric advective-reactive-diffusion transport mechanisms, and more comprehensive coverage of the

---

[33]FAO report, `www.fao.org/3/a-i3437e/index.html`

dominant crops and different cropping systems (Osborne et al., 2010). There is also currently a great need to improve the spatial resolution involving intensive sampling, and the application of new technology for semi-continuous emissions measurement.

2.4.2   Stratospheric ozone

Ozone in our stratosphere protects living organisms at the Earth's surface against the harmful solar ultraviolet radiation (UVB and UVC) from the sun. It has a globally averaged tropospheric lifetime of approximately 23 days, and plays a major role in the energy budget as it absorbs both solar UV and terrestrial IR radiation acting as a strong greenhouse gas (Staehelin et al., 2001). At the Earth's surface, ozone is an air pollutant adversely affecting human health, vegetation, crop yield and quality (Cooper et al., 2014). Chlorofluorocarbons (CFCs), chlorofluoromethanes such as CF_2CL_2 and $CFCL_3$, non-methane volatile organic compounds (VOCs), nitrous oxides (NO_x), chlorine monoxide, chlorine- and bromine-containing volatile-gases and carbon monoxide all breakdown by photochemical reactions at sufficiently long wavelengths, producing chlorine atoms that deplete ozone O_3. Concentrations of these components have considerably been reduced since ozone depletion was first identified in the late 1970s and first reported by Farman et al. (1985). Ozone depletion in the Antarctic was further confirmed to be declining on broad, regional-scales by NASA's Nimbus 7 satellite measurements (Stolarski et al., 1986). Nonetheless, ozone depletion has reportedly substantially slowed, since a major decline in spring-time atmospheric ozone concentration over Antarctica was first observed. This is a positive outcome of global action that was taken through the Montreal Protocol on Substances that Deplete the Ozone Layer that was legally-binding first in 1989, but has since undergone further annual adjustments and amendments to strengthen and better enable more effective implementation[34]. A recent special issue on Ozone is available[35].

Increased atmospheric CO_2 concentrations are expected to cause cooling of the lower stratosphere and this could enhance the formation of polar stratospheric clouds, which convert potential ozone-depleting species to their active forms. Model-based findings show that doubling CO_2 concentration within the winter stratosphere of the Northern hemisphere leads to the formation of an Arctic ozone hole of a size that is comparable to that observed over Antarctica, with nearly 100% local depletion of lower-stratospheric ozone (Austin et al., 1992). Cooper et al. (2014) have integrated both historical and more recent observational datasets (i.e., in-situ and satellite remote-sensing data) to map ozone trends and distributions globally. They report the heterogeneity of sources, sinks, and lifetimes produce a global tropospheric ozone distribution that is highly variable by season, location, and altitude. Considerable uncertainty remains regarding ozone mixing ratios and trends in remote areas such as the oceans or across sparsely sampled continental regions like Africa, the Middle East, South America, and India. Observations indicate that *tropospheric* ozone has increased globally, with ozone doubling in Europe between 1950

---

[34]Ozone, `ozone.unep.org/new_site/en/montreal_protocol.php`
[35]Ozone Hole, `www.nature.com/nature/focus/ozonehole/`

and 2000, and significant seasonal variability in the upper tropospheric ozone (most often peaking in spring) within the northeastern USA, the North Atlantic Ocean, Europe, the Middle East, northern India, southern China and Japan (Cooper et al., 2014). Such variability may be linked to the control the transport of ozone by the El Niño Southern Oscillation (ENSO), the Pacific North American (PNA) and the North Atlantic Oscillation (NAO) climate teleconnections. The Tropospheric Ozone Assessment Report (TOAR) aims to provide new and enhanced global metrics for climate change, human health, and crop/ecosystem research as part of the third-phase focusing on "atmospheric chemistry research toward a sustainable world" of the International Global Atmospheric Chemistry Program (IGAC), first formed in 1990[36]. This program addresses the international concern over rapid changes observed in Earth's atmosphere, and integrates global surface and tropospheric ozone distribution and reports on ozone trends and variability. TOAR reports will include a broad set of ozone statistics, impact assessment metrics and the associated uncertainty to provide a consistent and comprehensive quantification and to support future research and observations.

2.5 Depletion of Finite Resources

The global demand for natural resources has doubled since 1966, that we are now using the equivalent of 1.5 planets to support our activities, or 50% more natural resources than the Earth can sustain (WWF, 2010) (World Wildlife Fund (WWF)). Natural resources are derived from the environment. Some of them are essential for our survival, while most are used for satisfying our needs. They may be biotic (living, organic) or abiotic (non-living and non-organic), and they may exist as separate entities or comprise many entities that are aggregated and inseparable. Some are ubiquitous (e.g., sunlight, air), existing everywhere around us, while others are partitioned (e.g., land, water). The majority of Earth's natural resources are nonrenewable and finite, which means that if we use them continuously, they are exhaustible, and will eventually run out. This is because non-renewable resources either form slowly or do not naturally form in the environment—such as fossil-fuel coal, petroleum and natural gas. Extracting and using such resources is becoming more and more costly, despite the reserve amounts, which could be used in the future, and stocks, which are considered accessible only with new technological breakthroughs. Renewable resources are those that are naturally replenished as long as their rate of replenishment/recovery exceeds their rate of consumption to ensure they are not depleted. Because of their role in contributing to rising GHG (i.e., carbon dioxide, methane, and nitrous oxide) and human-induced climate change, it is urgent that we transition to using renewable resources derived from land/soil, water, forests, plants, and animals. We must also primarily use energy from the sun, wind, wave, biomass, and geothermal sources.

---

[36]IGAC, `www.igacproject.org/TOAR`

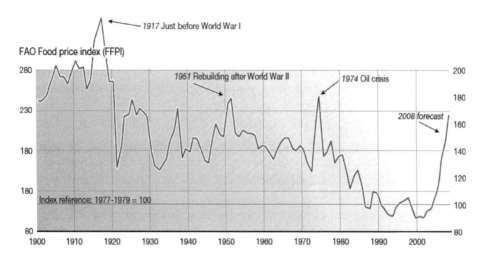

FIGURE 2.7: FAO Food price index (FFPI). Changes in the prices of major commodities from 1900 to 2008 reveal a general decline in food prices, but with several peaks in the past century, the last and most recent one the most extreme. The recent world food crisis is the result of the combined effects of competition for cropland from the growth in biofuels, low cereal stocks, high oil prices, speculation in food markets and extreme weather events. The crisis has resulted in a several-fold increase in several central commodity prices, driven 110 million people into poverty and added 44 million more to the already undernourished. Information on the role and constraints of the environment in increasing future food production is urgently needed. While food prices are again declining, they still widely remain above 2004 levels. (Cartographer/Designer: Hugo Ahlenius, GRID-Arendal, UNEP (2009)).

Water, nutrients, and energy are used in our food production systems to help provide us with food, energy, fiber and other ecosystem services. Economic trends in global agricultural markets are defined by changes in supply and demand for the production of agricultural commodities, yet there are a myriad of interrelated environmental, social, technological, and resource-based considerations that underlie such economic trends (De Vries, 2013).

Changes in the prices of major commodities from 1900–2008 show a historical decline, but with several large rises within the last century, with marked increases in recent years (Figure 2.7), despite increasing supply in cereal and meat production and the use of associated fertilizer, irrigation and pesticides. According to the United States Department of Agriculture (USDA), numerous factors have contributed to the recent rise in food commodity prices, such as: slower growth in production and more rapid growth in demand, and a reduction of world supplies of grains and oilseeds over the last decade (Figure 2.8), increasing global demand for biofuels feedstocks, extreme weather in 2006–07 within major grain- and oilseed-producing areas, rising energy prices, changes in inflation, and growing foreign exchange holdings by major food-importing countries, as well as policies adopted recently by some exporting and importing countries to mitigate their own food price inflation (Trostle, 2008). De-

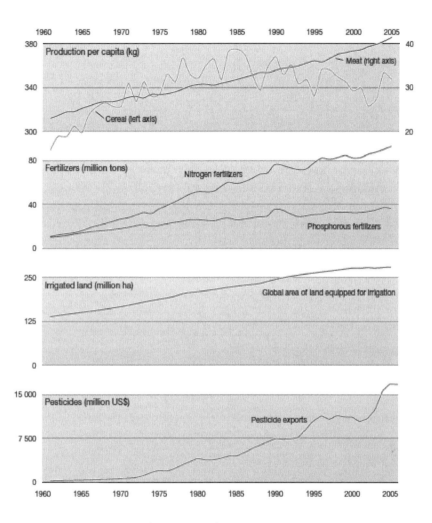

FIGURE 2.8: Global trends (1960–2005) in cereal and meat production, use of fertilizer, irrigation, and pesticides. (Cartographer/designer: Hugo Ahlenius, GRID-Arendal) (See UNEP (2009); Tilman et al. (2002) and references therein).

spite the competition between food and energy for land, water, nutrients and other resources, integrated economic and environmental studies still predict that there could be a 20–60% rise in food prices by 2050, depending on the type of food, that is predominately due to declining yields brought upon us by climate change (Nelson et al., 2013; FAO, 2010; Nelson et al., 2010). As agricultural production systems integrate agronomic (e.g., climate, soils, crops, and livestock) and economic elements (e.g., material, labor, energy inputs, food and services outputs), they are affected by socio-economic and cultural processes at local, regional, national, and international scales, including markets and trade, policies, trends in rural/urban population, and technological development (The International Forum on Assessing Sustainability in Agriculture (INFASA)) (Tubiello and Rosenzweig, 2008; INFASA, 2006). Climate change and extreme events are perturbing not only our water, and nutrient cycles, but also the response of our socio-economic system regulating supply-demand. Understanding the interconnections, risks, and beneficial trade-offs is key to ensuring greater efficiency and equability of agricultural resource supply, alongside the long-term sustainability and resilience of our agricultural systems. There is also increasing pressure for multi-criteria based decision-making due to more stringent market-driven sustainability requirements and greater public awareness of environmental risks (Schwarz et al., 2002).

2.5.1 Fossil-fuel energy reserves

Fossil-fuel resources can be partitioned into potential, actual available for depletion/exploitation, a reserve for future use, and the stock that is accessible with technological advancement. Moreover, some additional proportions of a resource may exist, but not be known, not having been surveyed. One may also need to consider a long-term reserve amount of an ecological resource. Natural resources can be further partitioned according to many additional characteristics related to their human use—such as their location/origin, scale/boundaries, quantity/rights/ownership, time-scale for "exhaustability," environmental risk level associated with extraction and other uncertain trade-offs, including sustainability development performance. Despite various resource partitioning, there exists a fundamental overarching constraint: renewable resources are naturally replenished, if their rate of replenishment/recovery exceeds that of the rate of consumption to ensure they are not depleted. Non-renewable resources either form slowly or do not naturally form in the environment. A resource is termed "non-renewable" when its rate of consumption exceeds its rate of replenishment and/or recovery.

The International Energy Outlooks (IEO) projects that world energy consumption will grow by 56% between 2010 and 2040 (EIA, 2013). The statistical analysis of data from a variety of sources indicates that the global production of *conventional* oil has reached a plateau. Non-Organization of Petroleum Exporting Countries (OPEC) oil plateaued up to 2005, with the price of oil continuing to rise by 15% annually (Kerr, 2012). The recent global economic recession during 2008 marked a period of rapid decline in oil prices (Ratti and Vespignani, 2015). 2015 projections from the International Energy Agency (IEA) for 2030 indicated that global

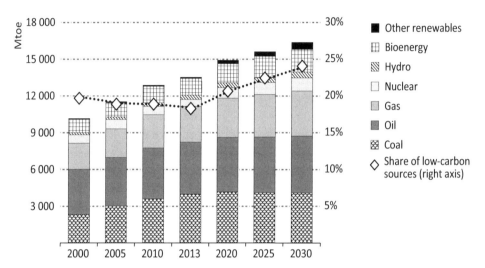

FIGURE 2.9: a) Global primary energy demand by type (INDC Scenario based on INDCs from countries currently accounting for 34% of global energy-related emissions and policy-related intentions of others) (Mtoe = million tonnes of oil equivalent; Gt = gigatonnes). Petroleum liquids include crude oil, lease condensate, natural gas plant liquids, bitumen, extra-heavy oil and refinery gains. Other liquids include gas-to-liquids, coal-to-liquids, kerogen, and biofuel. In the INDC Scenario, annual global energy- and process-related GHG emissions will increase from 37.5 Gt $CO_{2,eq}$ in 2013, to 40.6 Gt $CO_{2,eq}$ by 2030, and the global carbon budget, consistent with a 50% chance of keeping a temperature increase of below 2°C, is exhausted around 2040 (OECD/IEA, 2015) (©2015 OECD/IEA. World Energy Outlook Special Report 2015: Energy and Climate Change, IEA Publishing. Licence: www.iea.org/t&c/termsandconditions), b) Global CO_2 emissions by sector source, OECD Environmental Outlook Baseline, 1980–2050. Energy transformation includes emissions from oil refineries, coal, and gas liquefaction (Marchal et al., 2011).

oil production was to increase to about 99 Mb/d and that if this target is met, world oil production may exceed 2000 Gb (billion barrels) in just a span of 135 years (OECD/IEA, 2015; Hughes and Rudolph, 2011).

With OPEC countries increasing global production in 2016, the West Texas Intermediate (WTI) and Brent oil price fell by 50% in just a few months, to below $US 30/bbl for the first time since 2003. Yet, the IEA's Medium-Term Oil Market Report for 2016 now projects global oil supply will increase by 4.1 mb/d (2015-2021), a projected rate that is far less than 11 mb/d over 2009-2015[37]. Murray and King (2012) argue that the global oil market has tipped or phase-shifted into a new dynamic state whereby production is now inelastic and unable to respond to real rising consumption demands, causing large swings, and uncertainty in global oil markets and oil price due to changes in the energy supply-chains. In addition, the integrated production of conventional with natural gas liquids (NGLs), and non-conventional oil continues to increase (Hughes and Rudolph, 2011). The adoption of natural gas (NGL) and its consumption growth rate of 1.7% annually, is driven by increasing supplies of "tight gas" from fracking, shale gas, oil sands, and coalbed methane.

Renewable energy and nuclear power continue to grow by 2.5% per year as the fastest-growing energy sources globally (increasing by around 30% by 2030). Fossil-fuels (i.e., non-conventional and "redefined" to include mixture with renewable bio-fuel) are projected to very likely continue to supply almost 75% of our energy consumptive use to 2030, whereby all low-carbon options collectively will account for approximately 25% of the primary energy demand in 2030 (Figure 2.9). Some analyses suggest that there could be slower than expected production of global oil supply in 2030 scenarios of 75 Mb/d (i.e., Mb - million barrels per day) (OECD/IEA, 2013; Aleklett et al., 2010). As highlighted by Murray and King (2012), globally we get $55 \cdot 10^{18}$ J of useful energy from $475 \cdot 10^{18}$ J of primary energy from fossil-fuels, biomass, and nuclear power plants combined, so that even if fossil-fuels are still in our energy mix in the next decade, reducing major energy losses and inefficiencies in the converting, transmitting, and use of energy should be a major global goal. New supplies are situated increasing distances from consumer markets, and exploration increasingly relies on technological advancement (Longwell, 2002).

2.5.2 Rare-Earth elements (REEs)

World energy consumption is on the order of $5 \cdot 10^{20}$ J per yr (White, 2012). According to White (2012), while silicon is abundant at 46%, the rare-earth elements (REEs) (i.e., scandium, yttrium and the 15 lanthanides) are critical to computers and rare earth magnets for electrical generators in wind turbines. REEs have unique properties and there are little to no other alternatives for their application and use in modern technologies such as electronics and renewable and rechargeable energy generation and storage, energy efficient lights (lamp phosphors), electric cars, auto catalysts, permanent magnets in wind turbines, and other specific military and

---

[37]IEA, http://www.iea.org/bookshop/718-Medium-Term_Oil_Market_Report_2016

aerospace applications (Lehmann, 2014). The world's supply of indium, used in display devices as indium tin oxide (ITO) is declining at such a rapid rate that it will likely be depleted in less than 20 years (White, 2012). In this way, REEs are becoming increasingly important in the transition to a green economy. Davies (2011) highlights that Western governments are awakening slowly to the threat of losing access to key elements with 14 identified raw materials and metal groups, including all the REEs and platinum group metals, and antimony, beryllium, cobalt, fluorspar, gallium, germanium, graphite, indium, magnesium, niobium, tantalum, and tungsten.

The global REE market is dominated by China, having a global share in mine production of about 80%, down considerably from roughly 95% in 2011, when China imposed an export quota on REEs (Lehmann, 2014). Two large open-pit mines of an estimated 8% in REE associated with carbonatite deposits are coming into production in Mountain Pass (USA) and Mount Weld (Australia). Lehmann (2014) emphasizes that REEs are not rarer than other industrial base metals such as lead, zinc, copper, or tin, but scarcity is better attributed to decades of low REE prices and consequent lack of exploration and extraction. Nonetheless, he indicates that due to an increasing competitive REE market, very few of the more than 200 current exploration projects around the world will likely be viable to generate a longer-term supply.

Global demand will continue to increase, given that REEs are crucial for many industrial products and technologies, within both non-renewable and renewable pathways (e.g., wind turbines). These elemental resources are, however, finite. In addition, their use has potentially harmful environmental impacts over their life-cycle of extraction, use and waste. In the future, they will very likely either be too costly to access and extract or they will simply run out. There need to be greater policy actions taken now to increase REE recycling and its efficiency is required, as current use of Earth's elements in our industrial processes and products does not sufficiently take into account accessibility linked with their current and future demand (Golev et al., 2014; Davies, 2011).

2.6   Crop Yield Gaps, Food Losses, and Wastes

2.6.1   Crop yield gaps

Reducing crop yield gaps (i.e., between the current average and potential/maximum attainable yield) is critical to circumvent potential plateaus, diminishing or declines in yields in the future arising from an accumulation of factors, including limiting land, water, nutrient resources, and rising fertilizer and other energy costs, and the uncertain effects of climate change and climate variability. As Long (2000) explains, model projections, that indicate the fertilization effect of rising CO_2 will offset crop yield losses arising from the impacts of climate change and variability (e.g., increased temperature and decreased soil moisture), are derived from outdated enclosure experiments. Recent findings from experimentation using free-air concentration enrichment (FACE) technology and large-scale trials of major grain crops, produce yields that are 50% less than enclosure estimates. This casts doubt on the reliability of the 2050 projections, for example, that rising CO_2 levels will offset crop yield losses, but instead will further increase crop yield gaps. Moreover, episodic extreme events such as floods and heat waves create hot-spots where crop productivity and crop yield/production are drastically reduced, because even short episodes of extreme temperatures and precipitation can cause crops too much stress during their reproductive stage (Pionteka et al., 2014).

Teixeira et al. (2013) conducted the first global assessment of heat stress risk for four key agricultural crops (wheat, maize, rice, soybean) using the FAO/IIASA Global AEZ and IPCC AR4 GCM A1B SRES emission scenario out to 2071–2100. Their global assessment finds that there is a substantially high risk of yield loss within continental regions at high latitudes within the Northern Hemisphere (40–60 °N latitude). Temperate and sub-tropical agricultural areas are also very likely to experience substantial crop yield losses due to extreme temperature episodes. Parry et al. (2004) compare crop yield losses for a broader set of the HadCM2/HadCM3 climate model and SRES emission forcing scenarios (i.e., A1FI, A2, B1, B2) for 2020, 2050 and 2080. The A1FI scenario (i.e., 810 ppm CO_2), as expected with its large increase in global temperatures, exhibits the greatest decreases in yield, regionally decreasing by -3 to -7%, and globally by -5% (2080). Yield losses within developed and developing countries are largest under the A2 scenario, and smallest under the B1 and B2 scenarios.

Currently, average yields in rainfed and irrigated (i.e., wheat, rice, maize/corn) agroecosystems are $<50\%$, and 80% of predicted yield potential, respectively (Lobell et al., 2009). Ray et al. (2013) have integrated 2.5 million agricultural statistics across 13,500 political units across the world in tracking the yield of maize/corn, rice, wheat, and soybean (i.e., which currently produce two-thirds of global agricultural calories) and has determined that these crop yields are increasing at annual rates of 1.6%, 1.0%, 0.9%, and 1.3%, but that such rates are far less than the 2.4% yr^{-1} rate that is required to double global production by 2050. Current yield trends are thus insufficient to double global crop production by 2050. An integrated assessment of many studies covering a wide range of regions and crops shows that the negative impacts of climate change on crop yields have been more common than the positive impacts with high confidence (IPCC, 2014b). The latest IPCC AR5 projected impacts of climate change on crop yield vary across crops, regions and adaptation scenarios. A total of 10% of projections for the 2030–2049 period predict crop yield gains of more than 10%. About 10% of the projections show yield losses of more than 25%, compared with the late 20th century. Global temperature increases of 4°C or more, combined with increasing global food demand, are anticipated to pose large risks to food security, regionally and globally.

Further statistical analysis and modeling to quantify and reduce uncertainties linked with biogeochemical cycling and temperature stress effects is urgently needed to better understand the effects of climate change on agricultural production and to devise targeted adaptation strategies. This is despite strong agreement of strong negative effects on agricultural production under anticipated climate change, across seven Gridded Global Climate Model (GGCM)s, five GCMs, and four RCPs, in reported findings of the Agricultural Model Intercomparison and Improvement Project (AGMIP) and Inter-Sectoral Impacts Model Intercomparison Project (ISI-MIP) (Rosenzweig et al., 2014). The performance of statistical methods for estimating crop yield gaps has been reviewed by van Ittersum et al. (2013). The difference between theoretical yield levels and actual yields define the yield gaps, and precise spatially explicit knowledge is needed to better guide sustainable intensification of agriculture. Data gaps can be addressed using an integrated approach that combines available data with predictive models. Specifically, there is a need for accurate field-scale yield data alongside agricultural management or agronomic information, as well as theoretical yield predictions using crop yield and/or agroecosystem models that are well-calibrated and validated to enable field-scale estimates to be statistically upscaled to the regional, national, and global scales. A statistical meta-data analysis of 362 published organic and conventional comparative crop yields indicate that organic yields are on average 80% of conventional field-scale/plot yields, but variation is substantial (standard deviation 21%) (de Ponti et al., 2012). They suggest that the average yield gap between conventional and organic systems may be larger than 20% when upscaled to the regional-scale when crop rotation and nutrient availability variances are taken into account.

Soil salinity is a major environmental constraint to crop yield production. It is estimated that 89% of the world's water is salinated and 50% of the world's agricultural land is threatened by encroaching seawater and trace amounts of sodium chloride (NaCl). An estimated 45 million ha of irrigated agricultural land has soils that are significantly saline, with 250 million people living on soil degraded by salt due to floods and sea-water intrusion (Roy et al., 2014)[38]. Current findings indicate that, in general, crops respond to salt toxicity in the soil, primarily during two phenological or development phases—an *osmotic* phase (i.e., stomatal closure, with concomitant increases in leaf temperature and the inhibition of shoot elongation) and an *ionic* phase involving the bioaccumulation of salt in the older leaves leading to causing their premature senescence. Three main salt stress tolerance mechanisms are available to plants: tissue tolerance, osmotic tolerance and ion exclusion within roots, but many aspects of osmotic sensing and signaling mechanisms and how such mechanisms are used by different crop species is currently not well understood (Deinlein et al., 2014; Roy et al., 2014). Statistical-based field survey sampling to evaluate crops developed by genetic marker-assisted selection (MAS) or transgenic approaches are critically required to enable a reliable determination of alternative strategies for improving the salinity stress tolerance of specific crop species and the relative effects of modifying tolerance traits on crop yield and yield loss.

---

[38]Soil salinity, `www.theguardian.com/science/2014/oct/18/`
`humble-potato-poised-to-launch-food-revolution`

2.6.2   Trophic instability, food losses, and wastes

A third of all food produced (1.3 billion tonnes per year) is lost or wasted globally (Gustavsson et al., 2011). Projected global waste generation (millions of tonnes per day) out to 2100, relative to past and current levels is shown in Figure 2.10 (Gustavsson et al., 2011). Globally, waste production has risen ten-fold as societies have become more urban and affluent, whereby we are now generating waste faster than other environmental pollutants, including GHGs, and by 2025 our waste may double (Hoornweg et al., 2013). Meat, poultry, and fish, typically require the largest amounts of energy to produce, while dairy and vegetables have the highest embedded energy (Cuéllar and Webber, 2010). According to Cuéller and Webber (2010), "research is necessary to obtain more recent and accurate accounts of the energy used in fisheries, aquaculture, food packaging, disposal, and commercial food preparation. An updated and comprehensive study of food waste in the US food system and globally that accounts for waste in the fishing industry, on the farm, and during food processing is also necessary." Hoornweg et al. (2013) explain how waste production must peak this century otherwise population growth and urbanization will out-pace waste reduction with disastrous human health and environmental impacts and consequences. Different countries waste different amounts and types of food during its harvesting, production, distribution, and use/consumption (refer to Figure 2.11). As a country becomes richer, the composition of its solid waste changes. More affluence generates more packaging, imports, electronic waste, and broken toys and appliances. For this reason, solid waste amount can thus be used as a proxy for the environmental impact of urbanization, as less than 5% stems from waste management, which includes emissions from collection trucks, landfills, and incinerators. How many mobile phones are discarded is also now a measure of affluence.

Global fisheries bycatch may amount to 40% of the world's catch, totaling 63 billion pounds per year, with 17–22% of the catch in the US discarded every year (Keledjian et al., 2014). Global fisheries catches (i.e., landings) have been declining since the late 1980s by ~0.7 million tonnes per year based on the statistical analysis of FAO global catch statistics (1950–1999), corrected for over-reporting and the estimation of illegal, unreported or unregulated (IUU) catches (Pauly et al., 2002; Watson and Pauly, 2001). As Pauly et al. (2002) detail, just as in agriculture, crop yield models and gaps are based on actual observed yields compared to theoretically attainable potential yields; in fisheries, single-species population assessment models have assumed fishing effort at some optimum level is the maximum sustainable yield, but such optimal effort levels have been rarely implemented. An observed trend of human exploiting marine ecosystems and "fishing down marine food webs" has been linked with a global decline of 0.05–0.10 trophic levels (TL) per decade based on the analysis of global fisheries landings. Such findings indicate that the removal of large long-lived fishes from ocean ecosystems causes tropic cascades and population outbursts of certain species that were previously suppressed by predation. Examples of such species are invertebrates (i.e., squid and jellyfish). Mean trophic level (TL) (also known as the Marine Trophic Index, MTI) comprises a sustainability metric for exploited marine ecosystems, as a measure of the overall health and stability of

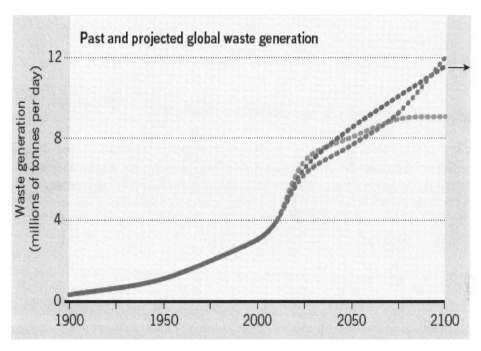

FIGURE 2.10: Past and projected global waste generation. Three scenario projections to 2100 for waste generation spell different futures. In the first Shared Socioeconomic Pathway scenario (SSP1) the 7 billion population is 90% urbanized, development goals are achieved, fossil-fuel consumption is reduced and populations are more environmentally conscious. SSP2 (upper scenario curve) is the "business-as-usual" forecast, with an estimated population of 9.5 million and 80% urbanization. In the SSP3 (bottom scenario curve), 70% of the world's 13.5 billion people live in cities and there are pockets of extreme poverty and moderate wealth, and many countries with rapidly growing populations. The SSP2, SSP3 scenarios both project waste to peak sometime after 2100. Only a scenario with a smaller, wealthier world population and more environmentally-friendly consumption behaviors, enables peak garbage to occur this century (Hoornweg et al., 2013).

a marine ecosystem or area. It is a proxy measure for overfishing and an indication of how abundant and rich the large, high trophic level fish are (Pauly and Watson, 2005). In ecosystem terms, it is essentially a measure of the overall structural robustness of a food pyramid or food web comprising predator-prey diet compositions. TL specified for individual organisms in a food web provides a scale that allows for relative comparisons between species, which in general could be members of different groups, such as producers, herbivores, carnivores, omnivores, and decomposers. Recently, the human trophic level (HTL) and the position of humans within the food web was statistically calculated by integrating ecological theory, demography, and socio-economics (Bonhommeau et al., 2013). Findings from this study indicate that humans have an ecosystem trophic level similar to anchovies or pigs (i.e., global median HTL in 2009 was 2.21±0.13, increasing 3% since 1961). Humans have a dominant influence and role in perturbing ecosystems through changes in land use, biogeochemical cycling, biodiversity, and climate, yet these findings indicate that we have a mid-range TL. In addition, the findings show that humans are not at the same level as apex predators (highest TL of 5.5) that survive almost entirely by eating meat/as carnivores. Nonetheless, the portion of meat and fat (i.e., poultry, pork) in our diets is rising, particularly in countries like China, India, and Brazil, and is contributing to increasing our TL. Such increase in TL is juxtaposed against the rapid, global increase in human food loss and waste. Interestingly, although countries have very diverse diets, countries can still be statistically distinguished by dietary trends into just five major groups.

The portion of marine food in human diets has increased consistently, alongside the global decline in the mean TL of marine fisheries catches. Bonhommeau et al. (2013) also report significant association between HTL and World Development Indicators[39] on the interrelationship of our socio-economic, environmental, and health conditions, and changing dietary patterns. In ecosystems, trophic structural changes may lead to alternate, multiple steady states (MSS), making the management and restoration of marine ecosystems and ecosystem-based management of populations (i.e., rather than single-species based management) much more complex and risky. Pauly et al. (2002) also identify that the concept of sustainability applied to marine ecosystems and fisheries management may be fundamentally flawed, because these systems are now so perturbed, and populations of fish and other marine organisms so reduced, by industrial-scale fishing. Instead, these ecosystems need to be rebuilt, applying the principles and practices of restoration ecology. Integrated, statistical assessment, modeling and forecasting are likely essential for establishing and maintaining marine protected/conservation areas (MPAs), ensuring reductions in fishing mortality, and conserving and regulating highly-migratory fish species populations and top predators that are now under severe population decline (and possibly extinction). Stochastic models will be necessary to integrate ultrasonic tracking, satellite tagging, aerial survey, catch data and vessel movements, and other satellite remote-sensing information on marine populations and ecosystems. Such

---

[39]World Development Indicators,
   data.worldbank.org/data-catalog/world-development-indicators

data enable building multi-scale ecosystem assessment models—from individuals, to schools/aggregations, populations to the ecosystem-, basin-wide and global scale. Statistical, cross-scale insights on individual fish behavior and adaptation, spatial metapopulation dynamics and mixing of shoaling/schooling populations under pressure to collapse their population range, geospatial tracking of changes in life-history migration routes, and changing thermal regimes/preferred habitat in response to cchanging ocean circulation patterns may transform our understanding of marine population dynamics.

According to the FAO report (2011) "Global Food Losses and Food Waste," the causes of food losses and waste in low-income countries are attributable to financial, managerial and technical limitations in harvesting techniques, storage and cooling facilities in difficult climatic conditions, infrastructure, and packaging and marketing systems. In contrast, the causes of food losses and waste in medium- and high-income countries are a result primarily of consumer behavior and integrated supply-chain co-ordination problems. For example, farmer-buyer sales agreements may contribute to quantities of farm crops being wasted, quality standards reject food items not perfect in shape or appearance, alongside insufficient purchase planning and expiring "best-before-dates" also cause large amounts of waste. It is estimated that the per capita food waste by consumers in Europe and North-America is 95–115 kg/year, while for Sub-Saharan Africa and South/Southeast Asia it is only 6–11 kg/year (Gustavsson et al., 2011). While close to one-third of the global food supply each year is lost or wasted across all stages of the agri-food supply-chain system due to pests/disease, weather events/climate conditions, infrastructure and distribution, food retail and services and by consumer households, in Canada alone, losses and wastes of food amounted to 6 billion kg, or 29.4% of the national food supply in 2010 (where retail and household food loss and waste accounts for 9.1% and 20.3% of this total, respectively).[40,41] Hubacek et al. (2014) describe the concept of "teleconnections" to describe the remote (spatial) linkages between local consumption embedded in its local context, and remote (i.e., regional, national and global-scale) environmental impacts. They detail how spatially-explicit footprint models can be built by applying "big data" statistical methods and approaches for massive geospatial data collection, integration, and analysis for estimating consumption patterns at fine spatial scales in the context of geographical conditions, and socio-demographic (i.e., lifestyle), infrastructural factors. Reducing losses in the food chain and increasing agricultural efficiency are also likely to contribute significantly to averting a future phosphate crisis (Cordell et al., 2009). Moreover, less than 1% of REEs in end-of-life consumer products are being recycled, yet that is essential to ensure the future supply of the most critical rare earths (Binnemans et al., 2013).

---

[40] An Overview of the Canadian Agriculture and Agri-Food System, 2015 (Agriculture and Agri-Food Canada).

[41] With 6 billion kg of food lost/wasted in 2015, per-capita food-waste loss is 168 kg/year, based on an estimated Canadian population of 35,702,707 people (Statistics Canada, Jan 1, 2015).

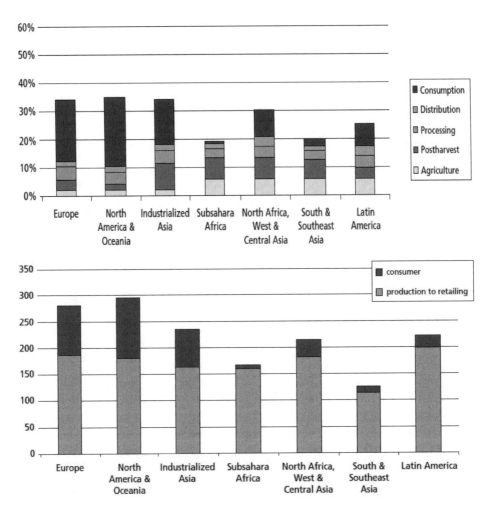

FIGURE 2.11: Food lost after harvest and food wasted along the distribution and consumption chain, or food wastage, has a dual negative environmental impact: undue pressure on natural resources and ecosystem services and pollution through food discards. Top: Initial global production lost or wasted by type. Bottom: Per capita food losses and waste (kg/yr) at consumption and pre-consumption stages. Within the global context of increasingly scarce natural resources, more than one-third of the food produced today is not eaten, which is about 1.3 billion tonnes per year (Gustavsson et al., 2011). Food and Agriculture Organization of the United Nations, 2011. Reproduced with permission.

2.7   Future Sustainable Development Pathways

A major challenge facing society is our ability to produce food, energy, fiber and other material efficiently and sustainably, to meet growing resource demands without harming the environment (e.g., soil, water and air quality). An "ecosystem" can be defined as a spatially and functionally coherent unit of agricultural activity that includes both living and nonliving components and their interactions. Natural and managed ecosystems provide humans with food, forage, bioenergy and pharmaceuticals and are essential to human well being. These systems rely on ecosystem services provided by natural ecosystems, including pollination, biological pest control, maintenance of soil structure and fertility, nutrient cycling, and hydrological services (Power, 2010). Preliminary assessments indicate that the value of these ecosystem services to agriculture is enormous and often under-appreciated. Ecosystems also produce a variety of ecosystem services, such as regulation of soil and water quality, carbon sequestration, support for biodiversity and cultural services. Depending on management practices, agriculture can also be the source of numerous disservices, including loss of wildlife habitat, nutrient runoff, sedimentation of waterways, GHG, and pesticide poisoning of humans and non-target species. The trade-offs that may occur between provisioning services and other ecosystem services and disservices should be evaluated in terms of spatial scale, temporal scale, and reversibility, such that as more effective assessment, modeling and forecasting methods become available, the potential for win-win scenario outcomes increases (Gasparatos et al., 2011; Power, 2010).

Understanding the relative environmental and financial performance of ecosystem goods and services (i.e.,, including emerging bio-based renewable energy and products), and their performance relative to reference "conventional" systems, is critical to prioritize industry and government research and development, and the implementation of reliable and robust sustainable development pathways and public policies. Rapid rates of development have a wide array of both immediate and long-term consequences for environmental and human health and welfare. In many cases, solutions and actions to address one immediate threat have only raised another. For instance, the misinformed, over-use of toxic substances with insecticides, herbicides and pesticides or built-in resistance via genetic modifications of agricultural crops have typically targeted one invasive species (Hellmann et al., 2008), changing animal migration patterns and the spread of infectious diseases like malaria (Patz et al., 2005), without preventing or stopping concurrent and recurrent invasions driven by warming temperatures and climate change. More aggressive pathogens of crop plants, such as in the case of the fungal disease, wheat yellow/stripe rust (*Puccinia striiformis* f.sp.*tritici*) can overcome the effect of resistance genes in hosts rapidly, and new more aggressive and virulent strains can produce higher amounts of spores, germinate at higher temperatures, and disperse around the globe following wind trajectories, causing major wheat crop epidemics (Hovmöller et al., 2008; Brown and Hovmöller, 2002). In particular, the increased risk of crop epidemics by many pathogens is favored when large regions of the same host crop/s are grown, as is the case in modern agricultural crop production strategies that increasingly rely on genetic/bioengineering for crop improvement and protection from insects and disease. Such host homogeneity increases the susceptibility to pathogen geno-types, compared to more heterogeneous, diversified crop production systems that offer only smaller, isolated niches for pathogens (Elad and Pertot, 2014; Juroszek and von Tiedemann, 2013; Shaw, 2009; Krupinsky et al., 2002). The health and quality of our environment decreases with the increased soil, water, and air contam-ination, land degradation and erosion, nutrient leaching and run-off into waterways. Such trade-offs and seesaw effects highlight the need to recognize and understand the complex interrelationships that exist within ecosystems.

As we shift our goals and activities towards a more sustainable developmen-tal path, we must avoid potentially damaging, "band-aid", temporary solutions. There has been poor international governance over the globalization and intensi-fication of water use, and future water security will require a clear shift in emphasis toward solutions-oriented approaches (Pahl-Wostl et al., 2013; Vörösmarty et al., 2013). Detrimental environmental impacts lead to cumulative impacts on human health (St. Louis and Hess, 2008). Health impacts arise and spread through both direct and indirect consequences of climate change, such as: the use of genetically-modified foods (Kuiper et al., 2001; Uzogara, 2000), the prevalence and rise of allergic disease (Shea et al., 2008; Hadley, 2006), air particulates, the use of toxic chemicals and contaminants (Noyes et al., 2009; Kinney, 2008; Schiedek et al., 2007), the re-duced biodiversity, genetic variation and increased species extinction risk (Thomas et al., 2004), microbes (Schimel, 2004), drug-resistant bacteria from the use of an-tibiotics for livestock (Thornton, 2010), pharmaceuticals and endocrine disrupting chemicals (EDCs) (Lathers, 2013), and ocean pollution by microplastics causing ad-verse effects in marine organisms (Law and Thomson, 2014). We must, therefore, shift to a sustainable development pathway in an integrated way, both in solving immediate pressures, and ensuring human and ecosystem health over the long-term.

Geoengineering and bioengineering aim to modify large-scale biological-chemical-physical environment systems or control our climate so as to slow or undo human-induced climate change. But such large-scale human engineering approaches have large-scale risks and unknowns, including major practical, financial, and ethical questions and dilemmas. It is not known whether such modifications would cause a catastrophic shift in ecosystems and the atmosphere. Such approaches are at odds with the precautionary approach toward managing and solving environmental prob-lems (Schneider, 2008). Moreover, stabilizing climate change is largely an energy problem, because conventional fossil-fuel energy sources emit carbon dioxide, a ma-jor GHG, to the atmosphere. Advanced technology such as nuclear fission, fusion, hydrogen production, storage and transport, superconducting electric grids may play a role in our energy future—but a broad-based range of intensive research and de-velopment is urgently needed to produce technological options that can allow both climate stabilization and economic development (Hoffert et al., 2002). Pressures and demands for renewable resources are expected, in the future, to continue to rise. So too are the risks. Resource constraints and interrelationships must be understood and recognized. Relying on just a single renewable energy resource and development pathway is risky, and instead, a path of integrated development is needed. Because of strong inter-dependence between water, food, and energy cycles, as well as between ecosystem health and soil, water and air health, it is imperative that renewable resources be used equitably and wisely—monitoring, protecting and conserving them. Along with ensuring ecological balance, we must also fundamentally recognize that we share a moral and ethical responsibility to use resources equitably and sustain them for future generations, as well as acknowledging the indigenous rights of past generations. An integrated understanding of our past, current and future environmental, economic and social and institutional considerations must link our short-term and longer-term sustainability goals.

Fuglestvedt et al. (2014) have shown, for the first time, that the potential disruptive impacts of future volcanic eruptions on our climate could be counteracted or minimized/damped through deliberate emissions of short-lived greenhouse gases. Unlike geoengineering exploration to mitigate warming from long-lived greenhouse gases, designed emissions to counter temporary cooling would not have the disadvantage of needing to be sustained over long periods. Nevertheless, implementation would still face significant challenges. Another geoengineering technology being explored is "artificial photosynthesis" (AP technology) that uses sunlight to split water into oxygen and hydrogen fuel, and is an attempt to take the biggest source of energy we have—the sun and store it in a densest way possible to generate energy. Recent progress has been made using titanium dioxide protective coatings on high-performing photoabsorbers (e.g., silicon), leading to big gains in the stability required for any possible commercial viability, and with reduced costs in its fabrication, this artificial system could become cost-effective (Marshall, 2014; Nature, 2013). However, AP may require large amounts of REEs, unless energy gains and cost savings can be made using the more abundant iron oxide, and it may not be feasible over the long term, when one accounts for the low abundance of such resources. AP is already considered more efficient than biofuels, and it produces hydrogen as a zero-emission fuel with the only byproduct as water. Many see future sustainable fuel production involving algal-derived or AP-derived hydrogen fuels, but this would require major changes in energy infrastructure.

Natural, autotrophic microalgae are also capable of achieving very high efficiencies in converting solar energy into biomass as a feedstock (i.e., greater than terrestrial oilseed energy crops like Camelina) for a future algal-based biofuel industry. Furthermore, autotropic microalgae utilize CO_2 as their carbon source, offering a possible new CO_2 sequestration (i.e., carbon capture) pathway. Using open pond and closed tubular photo-bioreactor (PBR) systems, an upgraded "green diesel" blend stock can be made involving a conversion process called hydrotreating. Better microalgal strains capable of sustaining high growth rates at high lipid content need to be found. There is, in addition, a wide lack of consensus on the near-term economic viability of algal biofuels, due to uncertainties (Davis et al., 2011). Because algae are commonly genetically engineered, major issues remain regarding human exposure to algae-derived toxins, allergens, and carcinogens from both existing and Genetically Modified Organisms (GMO) (Menetrez, 2012). Recently, the US National Renewable Energy Laboratory (NREL) within the US Department of

Energy (DOE) has explored the role of iron-rich proteins or iron-sulfur-containing ferredoxins (FDXs) in algae metabolism (Peden et al., 2014). They identify two FDXs within algal chloroplast that promote electron transfer to and from hydrogenases. Given that there is competition for photosynthetic reductant among different pathways, by controlling electron transfer/distribution to specific FDXs (and depending on other cellular conditions and requirements) they show that algae (such as *Chlamydomonas reinhardtii*) can produce hydrogen (Peden et al., 2014).

Globally, renewable energy (RE) account for an estimated 13% or 64 EJ yr^{-1} (2008) of the total 492 Exajoules (EJ) of primary energy supply and it continues to rise. More than 30 EJ yr^{-1} of this contribution involves the use of biomass. A recent report that provides an overview of the electricity used by the global Information and Communication Technology (ICT) ecosystem or "Global Digitial Ecosystem" comprising the Internet, Big Data and the Cloud, handheld devices, ICT embedded in machines, data centers, networks and factories, finds that it now uses $\sim 1{,}500$ TWh of electricity annually, equal to the electricity power generation of Japan and Germany combined, 50% more than energy associated with global aviation, and approaching 10% of global electricity power generation. Future energy projections based on 164 global scenarios from 16 different large-scale integrated models indicate that by 2030, RE will supply 17–43%, and by 2050, will reach 27–77% or 173–400 EJ yr^{-1}. Scenario estimates vary highly between different models and their assumptions. They also have different uncertainty associated with them, as the scenarios do not fully represent a random sample suitable for rigorous statistical analysis nor a full RE portfolio of options, whereby ocean energy is only considered in a few scenarios (IPCC, 2012). Median values for all existing electricity power generation technologies using RE resources (i.e., using biopower, photovoltaics, solar, geothermal, ocean wave, wind) range from 4 to 46 g $CO_{2,eq}$/kWh, while those for fossil-fuels range from 469–1,001 g $CO_{2,eq}$/kWh (excluding land use change emissions)(See IPCC (2012), Figure SPM.8, Page 19). A synthesis of life-cycle estimates of GHGs produced from bioenergy alternatives indicates that RE renewable transportation technologies produce the highest GHG emission ($40 - 80$ kg $CO_{2,eq}$/GJ), and heating applications produce the lowest (< 10 kg $CO_{2,eq}$/GJ), with electrical power generation or "green power" being a mid-range alternative ($5 - 40$ kg $CO_{2,eq}$/GJ) (REAP, 2008). An overview of the variety of currently commercialized and developing bioenergy pathways from biomass feedstocks involving thermochemical, chemical, biochemical and biological conversion routes to heat, power, combined heat and power (CHP) and liquid or gaseous fuels is shown in Figure 2.12. The use of available marginal and degraded lands for bioenergy production could help mitigate GHG emissions. For some options (e.g., perennial grasses, woody plants, etc.), net gains of soil and aboveground carbon can be obtained. Global biofuel production amounted to 46 million tons of oil-equivalent (Mtoe) that is, slightly more than 2% of the total fuel used in road transport, mainly in the form of bioethanol and biodiesel in major producing countries such as the United States, Brazil, and the EU. Reliable baseline assessments and integrated energy forecasts will be required to identify how long world oil production will continue to grow and when it will eventually peak, plateau and then decline as the world transitions in the future to fully renewable energy production and consumption.

Biochar is a solid material obtained from the carbonization of biomass by pyrolysis or gasification (i.e., heating woody biomass to high temperatures in the absence of oxygen). Biochar may be added to soils with the intention of improving soil functions and to reduce emissions from biomass that would otherwise naturally degrade to greenhouse gases. Biochar also has appreciable carbon sequestration value, a higher calorific value than its original feedstock, and can be used to improve the structure and fertility of soils (IPCC (2012) and references therein, Chapter 2). Biochar produced in traditional kilns, burning without smoke and the release of GHG emissions, has a low bulk density, reducing transport costs. It is commonly commercialized as a fuel for domestic cooking and heating or is used in the metallurgical industry. The global production of biochar is more than 44 million tons (2005)[42] (Laird et al., 2009). However, a synthesis of the latest research and knowledge on biofuels indicates that there is a need for studies on how biomass resources may be distributed globally over various demand sectors, as there are insufficient spatially-explicit allocation forecasts and projections on the consequence and impacts of different biomass pathways. The net avoidance costs per tonne of CO_2 for biomass usage depend on various factors, including the biomass resource and supply (logistics), and conversion costs that critically depend on improved/advanced technologies and the cost of fossil fuel (IPCC, 2012). A life-cycle (i.e., broad and integrated) perspective, considering feedstock, conversion and the end use of the energy/products is required to provide a framework for favorable decision making, while minimizing unintended negative consequences. More comprehensive, client-focused sustainability assessments that integrate available data and models on inter-linked ecosystem goods and services are needed to better inform decision-makers and stakeholders.

Bioenergy markets will provide major business opportunities, environmental benefits, and rural development on a global scale, with economy-of-scale efficiencies and cost benefits. Recent assessments indicate that if worldwide production is to reach 300 EJ this century, at US\$ 5–10 per GJ this would represent a market of \$1.5–3 trillion yr^{-1} (IEA-Bioenergy, 2007). Issues with fossil-fuel and renewable energy subsidies and the costs of transitioning to new sustainable development pathways and markets however, remain largely unresolved beyond just devising general global or national road-maps (OECD, 010b; GSI-IISD, 2007). The re-allocation of subsidies is likely required to support green growth and address rising global mitigation and adaptation costs. A 2014 report by the Overseas Development Institute (ODI) and Oil Change International,[43] provides a synthesis of data and statistics on fossil-fuel subsides. It reports that the Group of Twenty (G20) made up of governments and central bank governors from the top 20 major world economies is giving US\$ 88 billion to support fossil-fuel exploration, despite pledging to phase out these subsidies back in 2009 (Bast et al., 2014). Such subsidies are in the form of investment by state-owned enterprises (SOEs) (US\$ 49 billion yr^{-1}) national subsidies delivered

---

[42]Biochar, `www.biochar-international.org/`
[43]Overseas Development Institute (ODI), `www.odi.org`

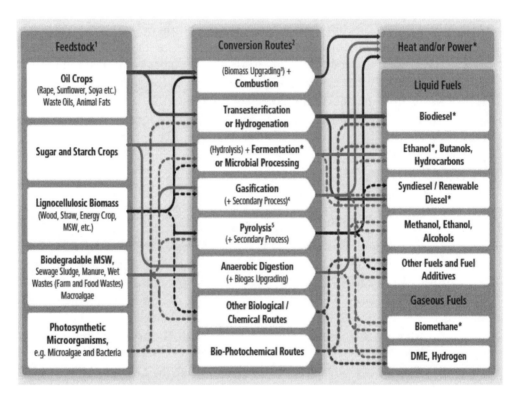

FIGURE 2.12: Schematic view of the variety of commercial (solid lines) and developing bioenergy routes (dotted lines) from agricultural and forestry biomass feedstocks through thermochemical, chemical, biochemical and biological conversion routes to heat, power, combined heat and power (CHP) and liquid or gaseous fuels. Commercial products are marked with an asterisk (Chum et al., 2011) (Figure TS 2.3, Page 49).

through direct spending and tax breaks (US\$ 23 billion yr^{-1}), and public financing from multilateral development banks (MDBs) and financial institutions (US\$ 16 billion yr^{-1}). A working paper report from the International Monetary Fund (IMF) released in May, 2015, provides a new, updated estimate of regional and global fossil-fuel subsidies at US\$ 5.3 trillion yr^{-1} that better accounts for higher *post-tax* energy subsidies. While "pre-tax" consumer subsidies arise when the price paid by consumers is below the cost of supplying energy, "post-tax" consumer subsidies arise when the price paid by consumers is *below* the supply cost of energy. The later subsidies account for additional corrective tax associated with environmental damage that arises from energy consumption and an additional consumption tax applied to all consumption goods. Renewable energy subsidies on the other hand, reached US\$ 101 billion in 2012, higher than global fossil-fuel subsidies, with 60% of this within the EU countries. Global subsidies to renewables are projected to increase to over US\$ 220 billion by 2035 (OECD/IEA, 2013). Global capital investment in digital economy infrastructure is now over US\$ 5 trillion, and is anticipated to increase by another US\$ 3 trillion within the next decade (Mills, 2013). Current subsidies to support transition to renewables are, however, far less than the estimated support needed, as the required cumulative investment is estimated to be US\$ 6.5 trillion to transition to using renewable energy technologies between 2013 to 2035 (only 5% of which is for biofuels). The use of renewables in producing electricity will likely account for 62% of such investment, with new transmission and distribution lines of US\$ 260 billion needed for their integration (OECD/IEA, 2013). An estimated US\$ 90 trillion will be invested in infrastructure over the next 15 years, with a portion now migrating toward green investments, and since 2006, well over US\$ 1 trillion has been invested in new renewable energies. At the United Nations Secretary General's Climate Summit in 2014, pledges and commitments announced indicate that commercial banks will provide US\$ 30 billion in new climate finance by the end of 2015 by issuing green bonds and other innovative financing initiatives. Also, the global insurance industry intends to double its green investments to US\$ 82 billion by the end of 2015 and announced it would increase the amount placed in climate smart investments to 10 times the current amount by 2020. As well, 24 leading global producers of palm oil and other food commodities traders aim to contribute to the goal of zero net deforestation by 2020 and to work with governments, private sector partners and indigenous peoples to ensure a sustainable supply chain[44].

Industrial oilseed production is very likely the next major emerging market linked with industry demand for bioproducts, biofuel and feed. Industrial oilseeds are particularly favorable because they can be introduced in crop rotation on land unsuitable for food-based crop production, and leverage the existing value chain infrastructure to scale rapidly and maximize the economics of production. Camelina, for example, provides benefits over traditional petroleum-based fuel, because it reduces GHG by up to 80%, reduces sulfur dioxide, and is not competitive with food pro-

---

[44]Christiana Figueres (UNFCCC), Bloomberg BNA, November 04, 2014,
http://www.bloomberg.com/news/2014-11-04/
-remarkable-opportunity-for-global-economy-in-upcoming-climate-talks-figueres.
html

duction because it can be grown in rotation with wheat and on marginal land. Also, alternative or industrial oilseeds are rotational crops grown along with wheat on land that would otherwise be left fallow, thereby enabling farmers to diversify their crop base and reduce mono-cropping that reduces yields and degrades soil quality. These crops also offer higher drought, heat, and frost tolerance, including higher pest-resistance. Industrial oilseed production could provide significant additional revenue from acreage with a low-cost input crop with two end user markets—the oil for renewable jetfuel and "meal" for livestock and dairy industries, or as bio-oil for industrial lubricants, biopesticides, or bioplastics. The global market for jet fuel alternatives alone is estimated to be a $212 billion market opportunity. Biojet is the only "drop-in" alternative (i.e., termed drop-in hydro-treated renewable jet (HRJ) fuel) available for the aviation industry. Aviation has no other alternative to liquid fuel and must be compatible to the Jet A-1 aviation kerosene fuel standard, which further highlights the crucial role, importance, and significance that agriculture has in supporting the future renewable and sustainable aviation industry.

Global emissions of CO_2 attributable to fossil-fuel products and their use have surpassed 10.9 Gt (2007), with roughly 60% (6.6 Gt) of global emissions from the transport sector. Sustainable, renewable fuel sources are essential to enable continued growth of the international aviation industry, to reduce the potential impact of rising volatility in petroleum prices on operating costs, to increasing societal concerns on the environmental impacts of aviation, and to meet current GHG reduction targets. Climate change policy will likely continue to change the economics of using jet kerosene alongside CO_2 carbon cost allowances in favor of using biofuel derived from non-food sources. The replacement of fossil energy with bioenergy and biofuel must involve detailed statistics on water requirements, along with other environmental metrics. Gerbens-Leenes et al. (2009) have analyzed the water footprint (WF) of bioenergy and find that the WF of biomass-generated electricity or "bioelectricity" is smaller than that of biofuels because it is more efficient to use total biomass for electricity or heat than a fraction of a crop (its sugar, starch, or oil content) for biofuel. Bioethanol water use footprint is smaller than biodiesel's, with the range of WF (units of litres, L) varying between 1400–20,000 L of water per L of biofuel, depending on what feedstock is used.

Human "climate migrants" or "climate refugees" may require nations to enact new immigration and refugee policy and reforms as people are forced to move away from risky "hotspots" or to leave areas following a disaster to move to less risky areas offering more opportunity, prosperity, and safety, as available resources decline and adverse changes occur. Entire coastal urban areas (e.g., cities) may have to move as sea-level rises and as cyclone activity (i.e., globally-averaged at 1 degree of latitude per decade) migrates poleward (Kossin et al., 2014). Whether as a result of social-economic or environmental drivers, societies will increasingly become more urban. The ecology of increasingly complex urban and industrial systems will need to be understood, re-designed, and re-engineered adaptively to ensure that materials are conserved, urban metabolism in terms of material and energy storages and flows are reliably assessed and monitored, rising expectations are managed, and environmental quality and human-health targets are achieved (Hoornweg et al., 2013; Kennedy and Hoornweg, 2012).

Animals are also expected to migrate to where resources are more plentiful and conditions less adverse, with increased ocean and terrestrial species migration taking place, but constrained by geographical limits to species-range shifts (Burrows et al., 2014). Reusch (2013) has recently reviewed global climate change in the oceans and its impacts on species' physiology, distribution, community dynamics, latitudinal habitat and dispersal, and life-history migration range shifts, causing phenological mismatches and changing trophic interactions and marine ecosystem functioning. He discusses plastic versus adaptive responses to global change in our oceans. Marine species should possess large standing genetic diversity and hence display a high evolutionary potential (phenotypic plasticity), defined as the ability of a genotype to produce different phenotypes in different environments. There are, however, few real-world examples where individual plasticity (i.e., the capacity of an individual to adjust its phenology to environmental variables) and an evolutionary gene flow response have been observed that prevents or slows down local adaptation to (for example) thermal-based habitat regimes of marine animals, or an earlier timing of breeding and migration in avian populations (Charmantier and Gienapp, 2013; Reusch, 2013).

Climate change is anticipated to create new risks, limiting our future adaptive choices, alongside amplifying existing risks. It will present "a new normal" that is more dynamic, involving the intensification of water, energy, and nutrient cycles, with more "severe, pervasive and irreversible impacts" (IPCC, 2014b). This will require reliable, consistent, clear, transparent and comprehensive mitigation and adaptation response strategies. Sound governance that builds and maintains equity, justice, fairness, and peace worldwide will also be of paramount importance. The World Bank has identified some guiding principles on establishing green growth strategies that include placing a cost on carbon (refer to Figure 2.13, Table 7.3, page 161 (WBG, 2012)). Designing, transitioning, and fully implementing such strategies globally with minimal disruption or detrimental unintended consequences remains highly complex and uncertain. Adaptation and mitigation can also create other risks as well as benefits, resulting in both incremental changes to more fundamental, transformative ones in real-world ecosystems (IPCC, 2014b). The latest reports of the IPCC provide an internationally relevant, comprehensive, state-of-the-art, integrated scientific assessment of historical and future changes in our Earth system. It provides key recommendations for policy decision makers on potential mitigation and adaptation options and strategies to minimize risks, alleviate current and future impacts of anticipated climate change predominately driven by human activity and atmospheric GHGs. It provides directives for governments, societies, and citizens to walk a new path toward a long-term sustainable future for the Earth system, within which we can survive and develop, along with the environment on which we fundamentally depend (IPCC, 2014b).

Our sustainable future will require Sustainable Development Tools (SDT)s that enable global to regional to local integrated assessment, modeling, and forecasting. Statistical threshold (e.g., quantile and extreme-value), change-point and spatio-

		Local and immediate benefits	
		LOWER (Trade-offs exist between short-and long-term or local and global benefits)	**HIGHER** (Policies provide local and immediate benefits)
Inertia and/or risk of lock-in and irreversibility	**LOWER** (action is less urgent)	• Lower-carbon, higher-cost energy supply • Carbon pricing • Stricter wastewater regulation	• Drinking water and sanitation, solid waste management • Lower-carbon, lower-cost energy supply • Loss reduction in electricity supply • Energy demand management • Small-scale multipurpose water reservoirs
	HIGHER (action is urgent)	• Reduced deforestation • Coastal zone and natural area protection • Fisheries catch management	• Land use planning • Public urban transport • Family planning • Sustainable intensification in agriculture • Large-scale multipurpose water reservoirs

FIGURE 2.13: Some guiding principles for establishing green growth strategies (WBG, 2012) (©2012, World Bank. Creative Commons Attribution license (CC BY 3.0 IGO), `http://hdl.handle.net/123123123123/123License`).

temporal trend detection methods, variable- and model- selection methods linked with sensitivity, validation and optimization, geospatial scaling, interpolation and extrapolation methods, stochastic methods that integrate new auxiliary indices, causal network models, and many other statistical uncertainty frameworks, hybrid methodologies and automated algorithms—will all be needed to address sustainable development issues, applications, and challenges. This will pose challenges for statistics, requiring sufficient problem-specific context and interdisciplinary knowledge, to synthesize, process and manipulate large, complex datasets or "big data", to apply methods under non-stationary, non-Gaussian distribution assumptions, Advances in causal-based statistics (as opposed to associative regression-based approaches) and complex network models will also be needed to explore higher-order, complex interactions and spatio-temporal dependencies. Future road-maps must assess the alternative pathways available for assessing agroecosystem health and the management will need to involve comprehensive integrated assessment approaches that involve new metrics, and integrated models—both involving the substantial application of statistics (Zhu and Ruth, 2013; Zhu et al., 2012). Future agriculture will see further crop diversification and adaptation and competition and complex, ethical debates and dilemmas between land-use for food and energy (Rosillo-Calle, 2012; Thompson, 2012; GSI-IISD, 2007). For this reason, statistics will be key to improved risk communication with stakeholders and the public.

Adapting and advancing statistical methods and embedding them within sustainable development analytical research and operational decision-support tools will likely require human and machine learning intelligence to process, sort, and utilize big data (i.e., large complex datasets). Maybe, in the far future, a "global brain" or core of Artificial Intelligence (AI) as a central data repository and global operating system will exist for guiding the sustainable development of our Earth system, based on a fundamental set of CAS rules. Here "artificial intelligence" is "The theory and development of computer systems able to perform tasks normally requiring human intelligence, such as visual perception, speech recognition, decision-making, and translation between languages (*Oxford Dictionary*), or similarly, "The simulation of intelligent behavior in computers and the capability of a machine to imitate intelligent human behavior" (*Merriam-Webster Dictionary*). Until then, we will have to continue to operate in a largely decentralized, dis-aggregated and distributed way (see Gasser and Huhns (1989) on distributed AI in science and society). They will also require fundamental rules linked with our best understanding of CAS. They will need to be deployed and interpreted consistently and developed cooperatively and collaboratively. SDTs will track, monitor and predict integrated risks across societies, economic markets, and industrial processes to guide multi-scale sustainable development mitigation and adaptation response decision-making. Potential opportunities and risks associated with future advances in machine-learning and artificial intelligence linked with sustainability are further detailed and discussed in the next chapters.

3

Modeling Complex, Adaptive Systems

CONTENTS

"Mother Earth is in trouble, at least for humanity. ... Should we turn all our energies to worrying about elevated temperatures, the rise of sea level, and the potential loss of coastal communities? Or do more immediate problems, such as malnutrition, pollution and the spread of disease take precedence? ...To have any hope of dealing with such a complex combination of threats to our survival, we must study the Earth as an integrated physical and biological system. By understanding what makes that system work, we will understand how it can fail, thereby finding a way to prioritize actions and maintain the Earth's ability to continue to nurture and sustain us."

Simon A. Levin, Princeton University © 2000 Simon A. Levin, Fragile Dominion: Complexity and the Commons, Reprinted by permission, Perseus Publishing, a member of Perseus Books Group.

3.1 Systems Theory and Consilience

3.1.1 Philosophies of science

The broad, theoretical and applied context of sustainability and resiliency in SES systems and the fundamental elements of CAS systems were discussed previously in Chapter 1. This chapter extends this discussion in more depth with details on different modeling approaches, assumptions, and mathematical/computational representations. A brief background on CAS modeling linked with a Systems Science and the Ecosystem perspective is first presented.

Ludwig von Bertalanffy is widely considered the founding father of the General Theory of Systems (*Allgemeine Systemlehre*) or General System Theory (GST). In his book, *General System Theory: Foundations, Development, Applications* published in 1968, von Bertalanffy provides a general framework of scientific inquiry that views the world in terms of irreducibly integrated systems and the need for a general science of organized complexity. It focuses attention on the whole, as well as on the complex interrelationships among its constituent parts. This way of seeing is not an alternative, but a complement, to the specialized way. It is more all-embracing and comprehensive, incorporating the specialized perspective as one aspect of a general conception (Laszlo and Krippner, 1998). Von Bertalanffy considered the systems approach as a critical prerequisite for: 1) envisaging, elucidating and accurately defining problems involving CAS without being over constrained by a given specialized field or scientific discipline, and a reductionist perspective, 2) obtaining realistic solutions, and, 3) enabling the development of new branches of knowledge through collective, interdisciplinary investigation and inquiry. In particular, von Bertalanffy identifies three distinguishing kinds or levels in the general description of natural phenomena, namely: analogies, homologies and explanation (Von Bertalanffy, 1968). Analogies are, "superficial similarities of phenomena, which correspond neither in their causal factors nor in their relevant laws" (e.g., plant versus animal organism growth and development). Homologies, on the other hand, have different causal factors, but formally identical laws (e.g., heat versus fluid flow and hydrodynamics). Explanation refers to, "specific conditions and laws that are valid for an individual object or class of objects."

The British mathematician and philosopher, Alfred Whitehead, together with Bertrand Russell co-authored *Principia Mathematica* (PM) (1910, 1927), one of the twentieth century's most important works (Whitehead and Russell, 1997). Their aim was to deduce all the fundamental propositions of logic and mathematics from a small number of logical premises and primitive ideas, establishing that mathematics is a development of logic. While their aim was to deduce the logical foundation of mathematics, their world-view was not static, but dynamic and termed "process philosophy" (or the ontology of becoming) that is based on the premise that being is dynamic and that the dynamic nature of being should be the primary focus of any comprehensive philosophical account of reality and our place within it[1]. This philosophy has been furthered by John B. Cobb Jr., a renowned American theologian, philosopher, environmentalist, and pioneer in environmental ethics who has coauthored many works; two of major significance include, *For the Common Good: Redirecting the Economy Toward Community, Environment, and a Sustainable Future*, and, *Is It Too Late? A Theology of Ecology* (Cobb, 1995; Daly and Cobb, 1994). Cobb emphasizes ecological interdependence, whereby every part of an ecosystem is reliant on all of its other parts, and argues that humanity's most urgent task is to preserve the world on which it lives and depends, linked with Whitehead's concept of "world-loyalty." Process philosophy argues that there is urgency to adopt and ap-

---

[1]Stanford Encyclopedia of Philosophy,
http://plato.stanford.edu/entries/process-philosophy/

ply a systems perspective to our world comprising a web of interrelated processes of which we are integral parts, so that all of our choices and actions have consequences for the world around us (Cobb, 1995; Mesle, 1993). Other relevant works that further discuss systems theory, process philosophy, and our sustainable future, include *Complexity Theory for a Sustainable Future* (Norberg and Cumming, 2008), *Philosophy of Complex Systems* (Gabbay et al., 2011), *Social Self-Organization* (Helbing, 2012), *Complex and Adaptive Dynamical Systems* (Gros, 2013), *Modeling Complex Systems* (Boccara, 2010) and *Multiscale Analysis of Complex Time Series: Integration of Chaos and Random Fractal Theory, and Beyond* (Gao et al., 2007).

While GST theory aims to provide a universal explanation of the structure and behavior of systems, through integrated learning, one must also recognize that there are fundamental limits to rationality, whereby no logical construct can represent everything, and no analytical or computer simulation model can address all scientific questions and needs (Hall, 2010). Kurt Gödel's *Incompleteness Theorems* (1931) shows that in any formal logical system consisting of axioms and inference rules, one can construct statements that are neither true nor false (i.e., paradoxical sets) and therefore, no consistent system can be used to prove its own consistency, nor can proof be proof of itself (Raatikainen, 2015). This means that there is a fundamental limit to proof and computation—one cannot derive all mathematics from purely logical axioms, and therefore either the system must be inconsistent, or there must in fact be some truths of mathematics which could not be deduced from them. Gödel's Incompleteness Theorems showed that Russell and Whitehead's proof was insufficient, and that no formal system, such as mathematics, is complete.

Hofkirchner (2005) discusses the impact of Bertalanffy's General Systems Theory (GST) *Weltanschauung* or "Worldview" in forging the merger of Systems Theory (STh) with Evolutionary theory into "Evolutionary Systems Theory" that applies to both biotic and abiotic aspects of ecosystems, alongside human socio-economic systems (SES). Specifically, GST theory consilience or the principle that evidence from independent, unrelated sources can "converge" to strong conclusions, can emerge across different lines of scientific inquiry leading to "unity through diversity" epistemologically, ontologically, and ethically. Consilience refers to the synthesis of knowledge from different specialized fields of human endeavor. Edward O. Wilson, advocate for sociobiology as the study of the genetic basis of the social behavior of all animals, including humans, in his book, *Consilience: The Unity of Knowledge*, like von Bertalanffy, identifies and discusses a wide range of methods that have been used to unite the sciences and might in the future unite them with the humanities (Wilson, 1998). He discussed many examples of consilience, including the unification of Darwin's theory of evolution with genetics, the unification of forces, and the Standard Model of particle physics.

Given an inevitable perpetual situation of multiple claims for truth, the essential requirement for advancing scientific knowledge is not necessarily that concepts, logical constructs and models are universally generalizable, but that scientific hypotheses and models must be falsifiable. Herein lies a key knowledge gap for achieving sustainability of ecosystems and SES when there is a limit to our rational understanding—how does one falsify scientific hypotheses when the number of plausible explanations continues to increase and they become more interrelated and intertwined, as part of integrated problem solving? At the same time, how do both theorists and practitioners avoid "self-referencing" a given dataset, hypothesis test result, and/or output of a model, by accepting their validity if they only reference themselves. Is the answer to just obtain as many as possible available datasets, test as many competing hypotheses that can be conceived as possible, and/or compare the outputs of as many different models as possible—or is there another way that is guided by integrated scientific learning and relies on fundamental statistical aspects inherent to real-world systems? As some answers are neither absolutely true nor absolutely false (strictly undecidable) then is sustainability in a modern society (i.e., comprising SES) even compatible? Such questions infer the importance or value of relative, rather than absolute measures, and the importance of predictability and predictive error measures, over associative ones. A problem or system is intractable if its complexity grows exponentially with the size of the input. As the complexity of most continuous multivariate problems is exponential in the number of variables and the only way to break their computational intractability is to replace their guaranteed, worst-case prediction error with their best-case stochastic assurance, using stochastic sampling methods (Traub, 2007). Traub (1997) further suggests that there is a closer linkage between intractability than undecidability and scientific questions.

In *Future Pasts : The Analytic Tradition in Twentieth-Century Philosophy: The Analytic Tradition in Twentieth-Century Philosophy*, the complex interaction between science and analytic philosophers (i.e., Friedrich Frege, Alfred Whitehead, Bertrand Russell, Ludwig Wittgenstein, Rudolf Carnap, Hans Reichenbach, Carl Hempel and Ernest Nagel) and multiple discontinuities and misunderstandings that have prevailed, are discussed with a historical context (Floyd and Shieh, 2001). Byron Jennings discusses a 21st century view of the nature of science, whereby, "...science advances through a sequence of models with progressively greater predictive power. The philosophical and metaphysical implications of the models change in unpredictable ways as the predictive power increases." Jennings articulates several philosophical arguments on the link between observations (sensory input), models and theories. A model is any theoretical construction used to describe or predict observations, and a theory is systematically organized knowledge applicable in a relatively wide variety of circumstances. Theories integrate a system of assumptions, accepted principles, and rules of procedure devised to analyze, predict, or otherwise explain the nature or behavior of a specified set of phenomena[2] (Jennings, 2007).

- It is impossible to deduce from a finite series of observations a generalization that applies to them all (David Hume)

- It is impossible to empirically establish any fact, and all empirical knowledge, scientific models, and scientific laws are only tentative (Karl Popper)

---

[2] *The American Heritage Dictionary of English Language,* Third Edition, Houghton Mifflin Company. Boston, 1992.

- Models cannot be verified as they depend on assumptions that cannot be rigorously proven, only falsified (Karl Popper)

- Empirical observations take on meaning within the context of models via induction (Roger Bacon)

- Scientific Paradigms are a set of interlocking assumptions and methodologies that define disciplines/fields of study, helping scientific communities to create avenues of inquiry (Thomas Kuhn)

- To be useful and informative in science, a model must be logically and internally consistent, but does not have to be reasonable. When there is a major change of models, within a paradigm shift, new models are, by definition, unreasonable. But simply because a model is unreasonable doesn't imply that it is correct.

- Although models cannot be proven correct, or strictly false, they may be shown to have a limited range of validity for explaining observations. Their predictions may be validated (i.e., judged and compared) against historical data (hind-casting), or near-real time and future scenario data (forecasting). Sufficient availability and quality of data used to train a model versus independently validating its predictions is crucial.

- Models must generate predictions that can be tested against future observations.

- If two models generate the same predictions for all observations, they are scientifically equivalent (despite possibly different sets of assumptions).

Ernest Nagel in, *The Structure of Science* argued that most scientific hypotheses can be tested only indirectly, and he explored causality and patterns in scientific explanation linked with four characteristic types of models: deductive, probabilistic, functional, and genetic types. These model types differ to the extent that they involve a direct reduction (deduction) versus indirect reduction (induction) of existing and competing theories. In essence, he argued that laws and theories that reduce possible explanations (i.e., congruence) to those that have highest predictive power should be based on a well-established methodology, not arbitrarily chosen (Suppes, 1994). Nagel also argued that there is significant potential for congruency in the scientific explanations of goal-directed processes. His inquiry is still very relevant today, and so is the role of systems theory in reconciling analytic and sociological understanding and approaches (Suppes, 1994; Nagel, 1979). In this way, the science, at its core, involves observational-constrained model-building (across different model types) involving the analysis and control of uncertainty/error in relation to predictive power, and neither induction or falsification. This suggests that as theories and models compete, convergence in science requires the analysis and control of uncertainty in relation to predictive power, not induction nor falsification. Furthermore, while the systems approach facilitates convergence, it must then be firmly grounded in statistical theory to reduce uncertainty and increase predictive power (Jennings, 2007).

3.1.2   New windows on the universe

An example of consilience has recently come to the forefront given both new experimental findings and theoretical insights in cosmological science. From the STh approach to scientific inquiry, this example serves to highlight the importance of understanding CAS dynamics in other domains such as ecosystem sustainability. The Standard Model (SM) theory is a theoretical framework used to explain how the basic building blocks of matter interact, involving four fundamental forces (i.e., gravity, electromagnetic, strong and weak forces) (Mann, 2010). The SM model has successfully explained almost all the experimental results and precisely predicted a wide variety of phenomena—including the mass of the Higgs boson that endows other particles with mass (aka the "God Particle"). An extension of the SM includes the hypothesized or theorized graviton particle, composite particles (e.g., including the existence of other particles predicted by supersymmetric theories (Amsler et al., 2008; Oerter, 2006). The one shortfall it has, despite being a complete theory of fundamental interactions, is that the full theory of gravitation (described by general relativity) is not incorporated into the model. It also does not account for a viable dark matter particle with all the required properties needed in observational cosmology, or neutrino oscillations and their non-zero masses. The largest and most recent test of the Standard Model was the search for the Higgs boson conducted at the CERN (Conseil Européen pour la Recherche Nucláire, or European Council for Nuclear Research, Geneva, Switzerland). Both experiments at the Large Hadron Collider (LHC) (i.e., ATLAS AND CMS experiments) in 2012 reported seeing this particle (or a new particle) with a mass of about 125 GeV/c^2 (CMS, 2012). The standard model also predicts that the Higgs field couples to fermions through a Yukawa interaction, giving rise to the masses of quarks and leptons, and further experimental findings reported in 2014 indicate that there is strong evidence for the direct coupling of the 125 GeV Higgs boson to down-type fermions (CMS, 2014). The existence of the Higgs boson could explain data obtained by CERN scientists using the LHC, but other particles could have created the data, suggesting there might be alternate explanations (e.g., techni-Higgs consisting of more fundamental techni-quarks that may bind together in various ways to form both techni-higgs particles, including other combinations that may form dark matter). Dark matter is considered to comprise 95% of all matter in the universe, leaving 5% that we directly observe, and dark energy explains why the universe is expanding at an ever-growing rate instead of collapsing under its own gravity—and they represent the missing link in cosmology. This theorizes the existence of a yet undetected force, called the technicolor force, that binds techni-quarks into particles (Belyaev et al., 2014). As successful as the SM is, it does not explain why neutrinos, the second most abundant particle in the universe after photons, have mass, and their quantum-mechanical transformation between three flavors of quantum states that combine all three neutrino masses, namely the electron, muon, and tau[3].

The renowned theoretical physicist, Stephen Hawking, in seeking to integrate gravity with the other fundamental forces, gravitational and quantum field theory

---

[3]See recent update by Boyle (2014), `http://arstechnica.com/`

(involved with the escape of energy and information from black holes, and the nature of dark matter and dark energy) has proposed that black holes don't in reality exist with "event horizons." Instead, the gravitational collapse of white dwarfs, neutron and other stars into black holes produce "apparent horizons." Apparent horizons are different than the theorized event horizons as nothing (not even light) that enters a black hole can get out or can be observed from outside the event horizon, and any radiation generated inside the horizon can never escape. Nonetheless, while black holes themselves may not radiate energy, electromagnetic radiation and matter particles may be radiated from just outside the event horizon via "Hawking radiation." Self-amplifying Hawking radiation, emitted from a black body (i.e., an object that absorbs all incident electromagnetic radiation, regardless of frequency or angle of incidence, and in thermal equilibrium), is predicted to be released by black holes, due to quantum effects near the event horizon that has been mimicked and observed in the laboratory, using a narrow, low density, very low temperature atomic Bose–Einstein condensate[4], containing an analogue black-hole horizon and an inner horizon, as in a charged black hole (Steinhauer, 2014). For a non-rotating black hole, the Schwarzschild radius delimits the event horizon of a spherical object, while rotating black holes have distorted, non-spherical, event horizons that only temporarily hold matter and energy. Hawking highlights how the space outside of an event horizon approaches a Kerr solution in gravitational collapse, but inside, the metric is chaotic, so that its approximation by a smoother Kerr metric is responsible for the information loss in gravitational collapse, such that collapsed objects would radiate chaotically, rather than deterministically. This is just like weather forecasting on Earth, where there is effective information loss such that one cannot reliably predict weather more than a few days in advance (Hawking, 2014). Such findings highlight the importance of stochastic behavior and CAS dynamics within collapsed stellar objects. In this sense, major advances in cosmological understanding may be complementary to understanding weather forecasting to unravel the limits to predictability, resiliency, and sustainability of SES and CAS, and the understanding of chaotic dynamics through Earth, Ocean and Atmospheric experimentation. New direct observational evidence of gravitational waves (i.e., ripples in the fabric of spacetime) from a binary black hole merger by The Laser Interferometer Gravitational-Wave Observatory (LIGO) Virgo Interferometer Collaborations, supports the existence of binary stellar-mass black hole systems. The discovery of gravitational waves opens a new window on our Universe and new opportunities for studying CAS dynamics via transient dynamics of gravitational waves, whether they are here on Earth or out in space[5].

Systems theory aims to unify science and interdisciplinary knowledge, and while structurally a system is considered a divisible whole, it functionally is both dy-

---

[4]A Bose–Einstein condensate (BEC) is a state of matter of a dilute gas of bosons (i.e., particles having integer spin, as opposed to fermions having half-integral spin) are cooled to temperatures very close to absolute zero

[5]B.P. Abbott *et al.* (LIGO Scientific Collaboration and Virgo Collaboration). "Observation of Gravitational Waves from a Binary Black Hole Merger." *Physical Review Letters*. 116, 061102 (11 February 2016).

namic (with emergent properties) and indivisible such that "the whole is greater than the sum of its parts." Such a system is a complex system composed of many interacting parts, or agents, that behave collectively such that the behavior of the entire system is not trivially but stochastically related to the behavior of its individual agents. Solving real-world problems involving CAS systems, according to Evolutionary Systems Theory (EST), Consilience Worldview and Process Philosophy, require scientific methods[6] to shift from being validated, adapted and applied not only for specific, domain-specific analysis but to aid in more general data and methodological synthesis. Methodology shifts ontologically, from theoretical constructions towards real-world system needs and applications. While sustainable development descriptions and guidelines shift ethically to integrated, scientifically defensible, socially and corporately accountable action plans. Troncale (2009) has identified general obstacles, areas of potential, and case studies for Systems Theory (STh) (Troncale, 2009). There is a great need to outline a strategic approach for STh investigation that links more directly with specific lines of inquiry, evidence and theory spanning scientific disciplines. For example, such an outline could start with chaos in black holes and cross-apply it to weather and ecosystem forecasting linked with the essential elements of CAS systems. Such interdisciplinary study could help understand leading statistical characteristics of pattern (i.e., stochasticity, self-organization, scaling) and process (asymmetry, interactions, heterogeneity) within CAS systems[7]. At the same time, more rigorous consensus on a consistent statistical architecture or framework for validating models and identifying dysfunctional elements will likely be required. A big step in this direction is being taken with the launch of "Future Earth"[8] that aims to integrate, "...contributions from systems science that links disciplines, knowledge systems and societal partners to support a more agile global innovation system." In line with systems theory and consilience, "Future Earth" is a global research platform designed to provide the knowledge needed to support transformations toward sustainability, and seeks to foster, build, and connect knowledge to increase the impact of research, to explore new development paths, and to find new ways to accelerate transitions to sustainable development.

The future development of systems theories will no doubt continue to evolve as knowledge is integrated at different extents and rates across diverse scientific fields, in addition to scientific inquiry conducted at the systems level itself. In *Facets of Systems Science*, George J. Klir (2001) discusses the principles, foundation and application of systems science (Klir, 2001). According to Kilr, "There is little doubt that, in the foreseeable future, systems science will continue to be the principal intellectual base for making advances into the territory of organized complexity, a territory that still remains, by and large, virtually unexplored." He highlights that the challenge is to how to deal with systems and associated problems whose complexities

---

[6]A wide variety or families of methods exist that include approaches that are qualitative and/or quantitative, hard, soft and hybrid/mixed analytic approaches, being grounded in logic, rational/critical thinking

[7]See Chapter 1.

[8]Future Earth, `www.futureearth.org/`

extend beyond predictability and information-processing limits, and that the answer likely lies in the fundamental relationship between complexity, uncertainty, credibility, and usefulness. For example, D-Wave, founded by Geordie Rose, is a company in British Columbia, Canada, that is tackling current information-processing limits, by developing the world's first commercially viable quantum computer that uses subatomic particles to process complex calculations[9]. In addition, reducing limits in predictability likely will require statistical science-based reasoning at the interface of Analytic and Process Philosophy (Shields, 1996). Laszlo and Krippner (1998) articulate how the relationship between STh and the study of perception, and how changing nature of human cognitive maps is increasing in its importance. "Cognitive Maps" are agents of cultural cognition[10] that "...serve as vehicles for societies to probe environments quickly and effectively are the means to their ability to keep pace with accelerating rates of change" (Laszlo and Krippner, 1998). A *cognitive map* can be defined as, "the mental image or representation, made by human individuals and groups, of their environment and their relationship to it, involving not only the rational aspects of attitudes and behaviors, but also the values and belief components that shape human perception." (Laszlo and Krippner, 1998). In this way, they are constructs of our underlying patterns of conception and perception amenable to empirical (i.e., scientific) investigation. As Laszlo and Krippner further explain, cognitive maps include cultural context (i.e., social values and preferences), and while culture is, "a property of the group or of the individual's relationship to the group," cognitive maps related both individuals and social groups or communities. Individualism (and communitarianism) are powerful predictors of whether individuals actually believe that climate change is a risk. Re-framing the climate issue in a way that favors solutions in line with the value systems of individuals, depolarizes an issue (Kahan et al., 2012)[11].

Alternative cognitive maps can be integrated to develop a general conceptual framework for operationalizing sustainability models, frameworks and tools. The integration of cognitive maps enables building frameworks without losing the essential features of a system, and recognize that addressing different problems presents different knowledge limitations and problem boundaries. Straussfogel and von Schilling (2009) elaborate on Systems theory and its co-evolution with information theory, game theory, cybernetics and chaos theory, theory of autopoiesis, complexity theory, and dynamic systems theory, involving diverse "soft" and "hard" applications and disciplines (e.g., engineering, biology, ecology, geography, sociology, psychiatry, and neurology). The "hard systems" approach is concerned mainly with controlling, managing, or constructing systems, and the "soft systems" approach focuses on seeking meaning and solutions from the system itself. Soft systems recognize that the dynamics of human societies and socio-economic systems are fundamentally

---

[9]Orton, T. *Business Vancouver*, January 29, 2015, `www.biv.com`

[10]A societal cognitive map describes the general orientation of a given culture at a given time.

[11]Mooney, C. "The climate debate is brutal and dysfunctional but there is still a way out", January 29, 2015, *Washington Post*, Energy And Environment,
`http://www.washingtonpost.com/news/energy-environment/wp/2015/01/29/`
`the-climate-debate-is-brutal-and-dysfunctional-but-theres-still-a-way-out/`

constrained by values, perspectives, interpretations and levels of knowledge, super-imposed on any limits of the predictability and uncertainty of the best available ana-lytical solutions to problems. Systems theory will likely become more relevant in the future, given its role in connecting diverse systems, actions, and uncertainties linked with international Sustainable Development Goals (SDG)s. Figure 3.1 provides an outline of the progressive development of a systems theory and its theoretical branches, along with an anticipated, strong convergence of existing theories into an emerging system for sustainable development involving human-machine symbiosis.

3.2   Complex, Adaptive Systems (CAS)

Complex adaptive systems (CAS) arise in many research fields, and practical ap-plications in the economy and financial markets, transportation networks, energy generation and distribution, ecosystems, health delivery and biomedical systems, safety and security systems, telecommunications, galaxy dynamics, condensed mat-ter systems, biological evolution, the brain, the immune system, granular materials, the collective motion of human societies and animal groups, to soil microbial com-munities (Gao et al., 2013; Newman, 2011). A CAS can be formally defined as, "A system that exhibits collective or emergent properties from small-scale interactions." When adaptation is included in systems behavior, complexity arises. Levin (2002) identifies the study of CAS as, "...from cells to societies, is study of the interplay among processes operating at diverse scales of space, time and organizational com-plexity" and "...is the study of systems limited in their predictability." *Emergence* can be defined as a process whereby larger entities, patterns, and regularities or stabili-ties arise through interactions among smaller or simpler entities that themselves do not exhibit such properties. Cadenasso et al. (2006) define ecosystem *complexity* as, "biocomplexity as the degree to which ecological systems comprising biological, social and physical components incorporate spatially-explicit heterogeneity, organizational connectivity, and historical contingency through time." In broader terms complexity is defined as "richness in structure or behavior." (Green, 2014) There are many types or forms of complexity—such as computational (e.g., algorithmic), dynamical (e.g., deterministic), and organizational (e.g., aggregate, hierarchical) (Manson, 2003; Re-itsma, 2003; Manson, 2001). In particular, computational complexity theory relates to the difficulty of performing certain tasks, such as calculating a particular num-ber or solving a quantitative problem, extending beyond the specific attributes of CAS alone. Like complexity, many other concepts also have a wide range of def-initions. Despite the notion of complexity being clear to many, it is still difficult to precisely or robustly define. There is no clear consensus on what constitutes a different definition, or metric of complexity. Complexity Theory focuses on complex, non-linear, open systems, whereas Chaos Theory deals with simple, deterministic, nonlinear, dynamic, closed systems that are extremely sensitive to initial conditions resulting in an unpredictable chaotic response to any minute initial difference or

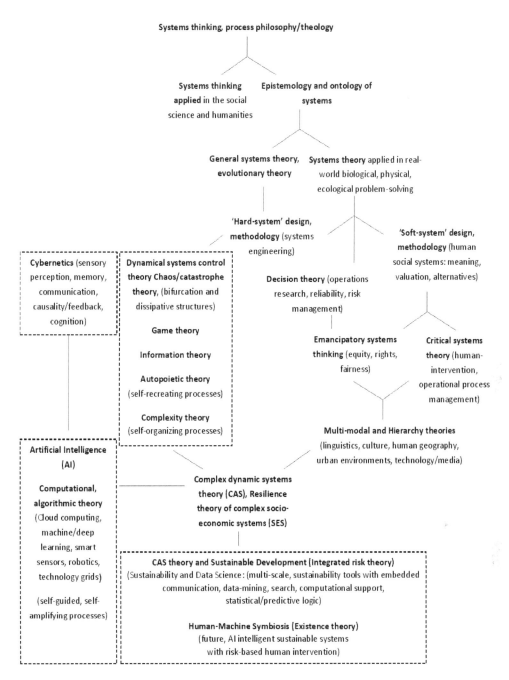

FIGURE 3.1: Outline of the progressive development of systems theory and its theoretical branches, and the convergence of theories into an emerging system for a sustainable development approach involving human-machine symbiosis (adapted from Laszlo and Krippner (1998)).

perturbation (Reitsma, 2003; Lorenz, 1995). Chaos is the inherent/controlled randomness within a CAS, defined as, "unpredictable order, whereby there is inherent unpredictability in the state and state dynamics/behavior of a CAS given its high sensitivity to initial conditions,' Chaos has, however, has the more common definition or interpretation of, "complete confusion and disorder a state in which behavior and events are not controlled by anything" (Manson, 2001; Lorenz, 1995). In addition to the types of complexity identified by Manson (2001), Reitsma (2003), from the perspective of geography, identifies 'statistical complexity', 'phase-transition', and 'chaos derivatives', 'system variability', and 'relative and subjective' types (Reitsma, 2003). O'Sullivan (2004) further discusses complexity science in the context of human geography, requiring renewed engagement between different disciplines to their perspectives (O'Sullivan, 2004). Despite the broad context of complexity, and the wide mix of complexity types, at its core is the realization that because one cannot describe, measure, or model all the variables in a complex system at once, theoretical insights and data from small-scale experiments need to be combined to better understand and predict large-scale patterns and processes.

In a survey of complex systems science, Newman provides a brief overview of its main themes and methods along with an annotated bibliography of a selected set of classic papers, recent books, and reviews (Newman, 2011). The study of CAS can encompass both top-down and bottom-up approaches (Straussfogel and von Schilling, 2009). The bottom-up approach involves devising and validating, "simplified mathematical models that, while they may not mimic the behavior of real systems exactly, try to abstract the most important qualitative elements into a solvable framework from which we can gain scientific insight." The top-down approach involves devising integrative (i.e., comprehensive and realistic) models, typically computer simulation models, which, "represent the interacting parts of a complex system, often down to minute details, and then to watch and measure the emergent behaviors that appear." (Newman, 2011) Modeling of CAS brings in new concepts of complexity (Cadenasso et al., 2006; Chu et al., 2003; Carlson and Doyle, 2002), self-organized criticality (SOC) (Bak et al., 1988), co-evolution (Rammel et al., 2007) and dual-phase evolution (Paperin et al., 2011), renormalization (Wilson, 1975), memory (Cadenasso et al., 2006), synchrony (Puetz et al., 2014), and distributed control (Choi et al., 2001).

3.2.1 Self-organized criticality (SOC) and power-law scaling

A large, interactive system evolves toward a self-organized critical state, whereby a minor event acts as a tipping point and can cause a cascading catastrophe or avalanche. In this state, the frequency and magnitude of events follow a "power-law" distribution (with a slope close to unity) that is both statistically stable and has no characteristic scale such that events are correlated across all scales, termed "scale invariance." Bak and coauthors defined this as Self-Organised Criticality (SOC), declaring it a universal, ubiquitous theory of CAS dynamics, linked with a simple model as an analogy of the dynamics of a sandpile (Bak, 1996; Bak and Chen, 1991; Bak et al., 1988). The sandpile (cellular automata (CA)) model consists of particles randomly dropped onto a square grid of boxes. When a box accumulates four particles they are redistributed to four adjacent boxes, or lost off the edge of the grid. Such redistributions can lead to further instabilities with the possibility of more particles being lost from the grid, contributing to the size of each avalanche. A power-law distribution (Solé and Bascompte, 2006), is given by:

$$Y(s) = kA(s)^{-\eta}, \eta > 1, s > s_{min} \tag{3.1}$$

where Y refers to a single or set of driving spatial, temporal, or coupled spatio-temporal ecological variables of interest; $A(s)$ is the area of a given unit size at the scale s; k is a constant, η is the *scaling exponent*, and s_{min} is a minimum threshold (i.e., non-zero) value for area, s, determined from observational data.

This sandpile model, along with several other cellular automata (CA) slider-block and forest-fire models, all exhibit self-organized critical behavior, and it has been argued that SOC can explain forest fires, earthquakes, landslides, wars and stock market crashes (Turcotte, 1999). But these models ignore a great deal of the information and detail in real-world systems, such that they may mimic similar behavior exhibited in these systems, but they are too simple and naive so cannot be used to reliably predict their dynamics. This is supported by work of Frigg (2003), who has sought clarity on SOC, by considering what SOC is and what it is not. His study concludes that SOC is a family of models (sandpile model etc.) that are related to each other via formal analogy having heuristic value, but that SOC models are gross oversimplifications of real-world systems, lacking any comprehensive, realistic description of any specific system.

Power-law scaling is ubiquitous in ecosystems, in body size, patterns of abundance, distribution and richness, the size and duration of epidemic events, and in trophic food webs (see Clauset et al. (2009); Newman (2005); Marquet et al. (2005) and references therein). For example, while there is an evolutionary trend toward larger body size in animals over time for increased mobility, fecundity, and survival potential, minimizing the predation risk of offspring for higher reproductive success (i.e., larger species tend to appear later in a phylogeny), there also exists an allometric scaling of physiological processes with body size (Freedman and Noakes, 2002). Under allometric scaling, the metabolic rate, stride length, or life span of an organism is related according to a power-law scaling of its body mass, under the assumption that an organism's net energy intake, i.e., its gross energy intake minus its metabolic costs, is an allometric function of its current body size, and that the same allometric exponent applies to these two processes (Pauly, 1981). Freedman and Noakes (2002) have further examined why the maximum size of teleosts (i.e., that include all the world's important sport and commercial fisheries, having a symmetrical homocercal tail and highly-evolved, bony fishes), is so much smaller than that of elasmobranchs and other marine animals (Freedman and Noakes, 2002). Intrinsic (genetic, physiological) and extrinsic/environmental (size of habitat, prey, temperature), intra- and interspecific competition for food and other resources between the organisms did not adequately explain the discrepancy in growth rates between teleosts and the other marine organisms. Instead, it is gill size (i.e., the surface area that limits oxygen intake via respiration) that limits both the metabolic and growth rates, whereby gill area increases with body weight according to a power-law scaling of $\eta = 0.2$, corrected for differences in movement modes and/or different caudal fin aspect ratios (Longhurst and Pauly, 1987; Pauly, 1981; Hughes and Morgan, 1973)[12]. In aquatic ecosystems (systems that are hierarchically structured by ecosystem levels, organized by branching based on trophic level, and predator-prey interaction), Salcido-Guevara et al. (2012) have found that "ecosystem metabolism" also follows a power-law scaling, alongside the allometric scaling of organism body size. They estimated the logarithm of biomass per production (B/P), and related it to the Trophic Level (TL) for 98 aquatic ecosystems, spanning different regions of the globe and geographic latitudes. When the scaling was corrected for different trophic transfer efficiencies, it was 0.726, very close to a theoretical value of 0.75 (or quarter-power law scaling) (Salcido-Guevara et al., 2012; Savage et al., 2004). Marquet et al. (2005) discuss how individual/organism level attributes involving power-law scaling explains and predicts, in part, patterns at the level of populations that can propagate at upper, population, and community levels of organization.

3.2.2   Self-similarity and fractional scaling

A fractal is a geometric shape that is *self-similar* and has a non-integer or fractional dimension, such that scale invariance is an exact form of self-similarity. Self-similarity is a property whereby a whole has the same shape as one or more of the parts (Mandelbrot, 1982). Anomalous diffusion i.e., diffusion under the combined action of a bias and a random noise (refer to Equation 1.36), differs from Gaussian diffusion, as it obeys a power-law scaling, such that the expected value of mean-squared displacement over time follows,

$$E(\Delta x) \propto K_\gamma t^\gamma \qquad (3.2)$$

whereby the case of Gaussian diffusion occurs when the anomalous diffusion coefficient, $\gamma = 1$, with subdiffusion and superdiffusion occurring when $0 < \gamma < 1$ and $\gamma > 1$, respectively; K ($cm^2 s^{-\gamma}$) is the anomalous diffusion constant. Anomalous material and information diffusion and long-tailed waiting time distributions cause slowly decaying memory effects and give rise to a *fractional* Fokker–Planck equation for CAS close to equilibrium (refer to Chapter 1, Equation 1.63 for non-fractional case). As is shown below, a fractal time series and power spectra coincide with solutions to a fractional derivative-based, stochastic differential equation exhibiting anomalous diffusion (Chen et al., 2010).

The *fractional* Fokker–Planck equation, for probability density, $p(x,t)$, is given by

$$\frac{\partial p(x,t)}{\partial t} = {}_0\widehat{D}_t^{1-\gamma} \left(\frac{\partial}{\partial x} \frac{\alpha(x,t)}{\eta_\gamma} p(x,t) + \frac{\partial^2}{\partial x^2} \sigma^2(x,t) p(x,t) \right). \qquad (3.3)$$

Here, η_γ denotes the generalized friction coefficient (constant) ($kg\ s^{\gamma-2}$). Identifying common factors and setting $\sigma^2(x,t) = \kappa_\gamma$ (a constant), in the above equation, where

---

[12]Fishbase, `www.fishbase.org/`, 272 records for 110 fish species

κ_γ is the anomalous diffusion coefficient $(m^2 s^{-\gamma})$ (note: the product of these two coefficients, $\eta_\gamma \kappa_\gamma = k_B T$, the Boltzmann constant) (Heinsalu et al., 2006), yields,

$$\frac{\partial p(x,t)}{\partial t} = {_0}\widehat{D}_t^{1-\gamma} \left(\frac{\partial}{\partial x} \frac{\alpha(x,t)}{\eta_\gamma} + \kappa_\gamma \frac{\partial^2}{\partial x^2} \right) p(x,t). \tag{3.4}$$

When $\gamma = 1$, the above equation reduces to the ordinary Fokker–Planck equation (i.e., Equation 1.63). ${_0}\widehat{D}_t^{1-\gamma}$ is the integro-differential operator of the *Riemann–Liouville* fractional derivative (with base point of zero), defined as (Heinsalu et al., 2006),

$$_0\widehat{D}_t^{1-\gamma} p(x,t) = \frac{1}{\Gamma(\gamma)} \frac{\partial}{\partial t} \int_0^t dt' \frac{p(x,t')}{(t-t')}^{1-\gamma} \tag{3.5}$$

where $0 < \gamma < 1$, and $\Gamma(\cdot)$ is the Gamma function. This convolution integral has a slowly decaying power-law kernel representing memory in CAS dynamics.

With the above theoretical motivation, we now consider observing or sampling anomalous diffusion dynamics. Let $X(t)$ represent the time evolution of a CAS as a regularly (i.e., equal interval) sampled process forming a time-series function containing a superposition of periodic (i.e., harmonic) contributions or components having different wavelengths/periodicity or frequencies. For a time-series over a specified time interval T, the mean and variance of the signal (denoted, $\hat{X}(t)$) and $\sigma^2(X(t))$, is defined as,

$$\hat{X}(t) = \frac{1}{T} \int_0^1 X(t) dt, \quad \sigma^2(X(t)) = \frac{1}{T} \int_0^1 [X(t) - \hat{X}(t)]^2 dt \tag{3.6}$$

In the frequency domain, the Fourier transform of the time-series, $X(t)$, provides the amplitude, $A(f,T)$ of a signal, where f denotes frequency (Hz), and T is period (s), for which its inverse involves integration over all amplitudes,

$$A(f,T) = \int_{-\infty}^{\infty} X(t) e^{2\pi i f t} dt, \quad X(t) = \int_{-\infty}^{\infty} A(f,T) e^{2\pi i f t} df. \tag{3.7}$$

With the Fast Fourier Transform (FFT) method for regular sampled time-series (or Lomb-Periodogram and other periodogram methods applying irregular sampling), within the finite time interval $0 < t < T$, the power spectral density (Y) of $X(t)$ is defined as,

$$Y(f) = \lim_{T \to \infty} \frac{1}{T} |A(f,T)|^2. \tag{3.8}$$

In the case of fractal time-series, the power, $S(f)$, scales across frequency in a self-invariant way (i.e., $kA(s) \to f$), such that (Cervantes-De la Torre et al., 2013),

$$Y(f) \propto f^{-\eta} \tag{3.9}$$

for uncorrelated white noise-like systems with a power spectrum (or power spectral density function, PSD) that is independent of frequency ($\eta = 0$), for moderately-correlated ($1/f^{\eta}$, $\eta > 0$) noise systems ($\eta = 1$), and for strongly-correlated (Brownian noise-like) systems ($\eta = 2$) (Cervantes-De la Torre et al., 2013). Yodzis and McCann (2007) clarify how chaotic solutions to autonomous discrete-time models of single populations tend to have blue spectral densities (i.e., blue signatures) where high-frequency fluctuations prevail, that may contrast with real time-series data that show red densities (red signatures) where low-frequency fluctuations prevail. They show how models of *interacting* populations (i.e., real ecosystems) tend to have resonant spectra, with fundamental harmonics of relatively low frequency, whereby in the absence of extremely long-time series (i.e., sufficient monitoring data record length), the spectra appear to be red.

Few observable, real-world patterns and processes (collectively termed phenomena) obey power laws over all the values of a variable of interest, s. More often the power law applies only for values greater than some minimum (i.e., defined here as s_{min}), such that extreme values or the "tail" of a distribution follows a power law (Clauset et al., 2009). More generally, given a shift between a scale s to αs, α, Y changes to $Y(s)$, and if Y is scale-invariant, then,

$$Y(s) = \alpha^{-\eta} Y(\alpha s), \eta > 1 \tag{3.10}$$

where $\alpha^{-\eta}$ is the *scaling translation factor* or *scaling dimension*.

A log-log plot can be used to identify power-law scaling behavior, but Pareto quantile-quantile or $Q - Q$ quantile-quantile plots and mean residual life plots can also be used. The signature of a power law is a linear relationship (i.e. with slope given by the spatial scaling factor α^{η}) in a log-log plot of Y(s) versus A(s). Power-law scaling with exponents less than unity cannot be normalized as its value diverges and does not normally occur in nature (Newman, 2005). As detailed by Newman (2005), after computing the normalization constant, the mean value of a power-law distributed variable, is given by,

$$\hat{Y} = \frac{k}{2 - \eta}[A^{-\eta+2}]_{s_{min}}^{\infty} \tag{3.11}$$

The scaling exponent typically lies in the range $2 < \eta < 3$. The mean $\hat{Y}$ becomes infinite if $\eta \leq 2$. Power laws with such low exponent values have no finite mean, and the mean would diverge if one had an infinite number of samples. In such cases, the mean is not a "well-defined" quantity or metric, because it can vary enormously from one measurement to the next, especially at different scales, s. When $\eta > 2$, the mean *is* well-defined and is given by,

$$\hat{Y} = \frac{\eta - 1}{\eta - 2} s_{min}, \eta > 2. \tag{3.12}$$

For a power-law distribution, the second distribution moment, or root-mean-square (RMS) is given by (Newman, 2005),

$$\widehat{Y^2} = \frac{\eta - 1}{\eta - 3} s_{min}^2, \eta > 3. \tag{3.13}$$

If $\eta \geq 3$, then the variance is well-defined and finite. Here we also see that under scale invariance conditions (i.e., power-law distributed), the mean is proportional to scale linearly, but its variance is proportional to scale quadratically, as expected. In summary, for a well-defined metric that is finite (i.e. enabling simulation-distribution re-sampling) and scale *invariant*, scaling exponents of the mean across scale, s, must satisfy $\eta > 2$, and its variance must satisfy, $\eta > 3$. Otherwise, it is scale-dependent with a scaling exponent that must be determined from cross-scale observational data, and used to spatially weight the variance between different subregions when upscaling or downscaling is undertaken. Otherwise, there is a high risk that the variance is underestimated when applying simple neighborhood averaging approaches, along with substantial bias in the mean, depending on the particular sampling method employed in measuring a given variable of interest, and model re-sampling of its observed distribution.

The scaling variance δV is defined as the difference between the variance of a response variable, $Y(s)$ and $Y(\alpha s)$, related by a given finite scale translation factor, α,

$$\delta V = \sigma^2_{Y(s)} - \sigma^2_{Y(\alpha s)} \tag{3.14}$$

where,

$$\sigma^2_{Y(s)} = \frac{1}{n} \sum_{i=1}^{n} (Y(s)_i - \bar{Y}(s))^2, \quad \sigma^2_{Y(\alpha s)} = \frac{1}{n} \sum_{i=1}^{n} (Y(\alpha s)_i - \bar{Y}(\alpha s))^2 \tag{3.15}$$

where n is the total number of statistical samples. The estimated difference in net-variance, δV, reduces to,

$$\delta V = (1 - \alpha^{2\eta}) \cdot \frac{1}{n} \sum_{i=1}^{n} (Y(s)_i - \bar{Y}(s))^2 \implies \delta V = (1 - \alpha^{2\eta}) \cdot \sigma^2_{Y(s)}, \alpha > 0.$$

When the scaling exponent (η) is fractional (i.e., non-integer), termed "fractal scaling," both the mean and variance may not exist, but a relationship between the autocorrelation function (ACF) of $X(t)$ (denoted r_{xx}), assuming that it is sufficiently smooth on $(0, \infty)$, is,

$$r_{xx}(\tau) = E[X(t)X(t+\tau)] = \int_{-\infty}^{\infty} X(t)X(t+\tau)p(x)dx, \tag{3.16}$$

where $p(x)$ is the probability distribution function (pdf) of $X(t)$, τ is time lag, and $E[\cdot]$ denotes the expectation of a random variable. It can be shown that r_{xx} converges for large $\tau \to \infty$ to (Li, 2010),

$$r_{xx}(\tau) \propto c|\tau|^{2H-2} \tag{3.17}$$

for a constant $c > 0$, where $0.5 < H < 1$ for H termed the "Hurst exponent" or "Hurst parameter," expressed equivalently, in terms of the fractal scaling or dimension exponent (η) as, $0 < \eta < 1$, is $0 < (2 - 2H) < 1$. This Hurst exponent (H) is a measure of the smoothness of fractal time series based on its asymptotic behavior.

Fractal processes can have strongly heavy-tailed PDFs, slowly decaying ACFs, and PSDs of fractional noise $(1/f)$, both long- and short-range lag dependence (LRD and SRD, respectively) and local or global self-similarity classified into Gaussian versus Non-Gaussian and LRD versus SRD-types (Li, 2010). The degree of lag dependence within fractal time-series is an important feature, whereby processes exhibit short-term and longer-term (extended) memory. In such cases, memory can provide a significant advantage in environmental or economic forecasting under high volatility and uncertainty, such as the state transition-fitted residual scale ratio (ST-FRSR) method for financial time-series (Richards, 2004), or the fuzzy time-series with fractal analysis (FTFA) method for environmental (soil, water, air) pollution (Chen and Wang, 2013). Also, when no scale translation ($\alpha = 0$) is involved, no spatial or temporal correction to estimated variance is required. However, when upscaling or downscaling when spatial dependence exists in a given process, and $\alpha > 0$, a correction factor that increases or decreases variance, respectively, is required.

Referring to Equation 3.10 for re-scaling or scale translation of a general time-series, a *fractal* time-series is self-similar in terms of its time-duration or periodicity T, such that,

$$Y(s) = kT^H Y(\alpha s), 0 \leq H \leq 1, \quad , < H >= \frac{log(Y(\alpha s)/Y(s))}{log(T)} \tag{3.18}$$

where k is a constant, T is a time-duration, the ratio $Y(\alpha s)/Y(s)$ is termed the re-scaled range, H is the Hurst exponent, and $< H >$ its estimate. Here, $H > 0.5$ indicates persistent, whereas $H < 0.5$ indicates antipersistence.

Rescaled Range Analysis (RRA) of stationary time-series with long record length and no gaps or trends can be applied that considers the difference between the maximum and the minimum cumulative values and the standard deviation of observed values (Kale and Butar, 2011). For noisy data in the presence of trends (i.e., long-term autocorrelation of non-stationary signals), De-trended Fluctuation Analysis (DFA) can be applied to determine scaling behavior (Golińska, 2012; Peng et al., 1995). Nonlinear forecasting that distinguishes inherent randomness from measurement error in time series, is also discussed by Sugihara and May (1990). While there are several alternative methods available for estimating the exponent of power-law distributions from empirical data, the maximum likelihood estimation reportedly outperforms other methods in both accuracy and precision.

Power laws, with the attribute of scale invariance, appear widely in physics, biology, earth and planetary sciences, economics and finance, computer science, demography and the social sciences (Newman, 2005). Solé and Bascompte (2006) further detail and discuss self-organization within complex ecosystems in relation to scaling. Clauset et al. (2009) provide a statistical framework for discerning and quantifying power-law behavior in empirical data, outlining how the scaling parameter, η, can be estimated from data, and the relative performance of alternative estimation methods,. Power-law scaling is seen in a wide range of real-world interplay—including the intensity of earthquakes (Saleur et al., 2012), the frequency of bird-species sightings, and the degree to which different cellular proteins interact. They specifically highlight that identifying and quantifying power-law distributions by the approximately linear behavior of a histogram on a log-log plot, while necessary, is not a sufficient condition of true power-law behavior. In a study of animal mobility, Cuddington and Yodzis (2002) simulated fractal environments and found that population dynamics become diffusion-limited, with the rate that individuals interact being a function of the fractal dimension of the environment, whereby the lowest rates of interaction are found in spatial environments with the lowest fractal dimension (Cuddington and Yodzis, 2002). Such findings indicate that *invariant* scaling within a terrestrial or marine/ocean environment reduces individual mobility and induces "diffusion-limited" population dynamics. For this reason, spatially-dependent landscape considerations (e.g., habitat-based "basins of attraction", or interaction/predatory-prey spatial arena, connectivity and life-history migratory routes) may be critically important for developing long-term, effective population protection or conservation measures, such as establishing multi-species, ecosystem-based marine conservation zones and marine protected areas (MPAs).

3.2.3 The hierarchical architecture of CAS

Although an ecological variable of interest follows a power-law distribution, this does not necessarily imply that a system is in a self-organized critical state. Also, power-law scaling or scale invariance does not necessarily require an underlying fractal geometry or hierarchical structure—a hierarchy may arise spontaneously from a heterogeneous system in a SOC state. Instead, self-organization can be viewed to drive hierarchical structuring of CAS, establishing power-law scaling relationships that, depending on the degree of chaos, can take on fractal properties. So, while a single, driving or control variable may be associated with such a critical state according to SOC theory, there is a competing theory, called "hierarchy theory," which considers that ecosystems do not function as sandpiles, but are instead hierarchically structured, evolving with complexity built upon complexity. In this way, hierarchy represents a central scheme of the "architecture of complexity, and often manifests itself in the form of modularity in nature." (Levin, 2000). Hierarchy theory is essentially the scaled-up version of SOC theory that applies to CAS. This combination of theory is necessary to explain multiple sets of variables interacting between ecosystem components (or modules), combining fast and slow processes occurring across spatio-temporal scale, while also explaining self-organizing behavior and statistical, power-law distributions.

Real-world ecosystems are often intermediately complex with a sufficient degree of both structural stability and dynamical or behavioral flexibility (e.g., degrees of freedom), such that, as suggested by Levin (2000), they exist in the middle between two ends of continuum from simple SOC to modular structure. Hierarchical structure can be nested, where all lower-level components are completely contained by the next higher-level, or non-nested. Recent evidence that astronomical, geological and biological cycles, spanning billion-year to decadal durations, exhibit fractal scaling behavior (i.e., period tripling, fractal scaling in full and half-cycles) is a strong indication that self-similar patterns develop with a degree of chaos, as the hierarchical structure of CAS adapts and evolves (e.g., complex hierarchies with both nested and non-nested levels). Yet, despite such chaos, there is also a surprising degree of synchrony (Puetz et al., 2014). Understanding these systems to the point of being able to predict their behavior in the face of perturbations requires different strategies that vary in both their approach and applied scale (Yodzis and McCann, 2007).

Wu and David (2002) discuss hierarchical theory and its application in modeling CAS. In particular, they discuss how hierarchy theory, itself, emerged from interdisciplinary scientific investigation, and has subsequently undergone refinement and expansion. An illustration of the major concepts of this theory is shown in Figure 3.2. An idealized depiction of a real-world, nested hierarchy, has both vertical and horizontal structure with hierarchical levels or "holons" having different characteristics (e.g., cycling and response rates) and interacting vertically in an asymmetric way—the higher levels in the hierarchy exert constraints on the lower levels (boundary conditions), whereas, the lower levels guide the higher levels by providing initial conditions (Wu and David, 2002). Also, higher levels are generally larger modules involving slower, low frequency processes, compared with lower levels that are smaller and evolve faster at higher characteristic frequencies. Within each level, different sub-systems or modules interact horizontally and vertically, in a more symmetric way. Interactions between components within a given module may be stronger and more frequent than across modules.

The degree that a CAS structure is nested or "nestedness" can have a different context and interpretation, depending on the specific system involved. For example, nestedness in mutualistic trophic networks can be defined as the tendency for ecological specialists to interact with a subset of species that also interact with more generalist species (James et al., 2012; Almeida-Neto et al., 2008). Furthermore, while nestedness is a significant aspect of the hierarchical structuring of CAS (i.e., a leading covariate), this does not preclude that simpler metrics may exist that could serve as better causal-based predictors of how various structures change over time (James et al., 2012). The metrics of entropy, Lyapunov exponents, and fractal dimensions are generally applicable in detecting and quantifying chaotic behavior. Yet, "Shannon's entropy" is just one of many statistical measures of uncertainty in a probabilistic process (Feldman and Crutchfield, 1997; Lòpez-Ruiz et al., 1995). Another *statistical* measure of complexity, termed LMC-complexity, quantifies the interplay between the information (H) stored in a system and its distance to the equipartition, called dis-equilibrium (D), given by (Lòpez-Ruiz et al., 1995),

$$C_{LMC}(\mathbf{X}) \equiv \Omega(\mathbf{X}) \cdot \Delta(\mathbf{X}) = - \left(K \sum_{i=1}^{N} p_i log_2 p_i \right) \sum_{i=1}^{N} \left(p_i - \frac{1}{N} \right)^2 \qquad (3.19)$$

where $(p_1, p_2, ..., p_N)$ are the probabilities of the N states $(X_1, X_2, ..., X_N)$ of a CAS, where $\mathbf{X}$ is a random variable, and probabilities satisfy the normalization condition $\sum_{i=1}^{N} p_i = 1$. K is a constant. For physical systems in equilibrium at temperature T, K would be equal to κ_B, associated with the probability p_i that they occupy a state i with energy E partitions that are weighted by the Boltzmann factor, $exp(-E_s/\kappa_B T)$, with units of J/K or entropy units, of statistical physics. Here, Ω is the Shannon

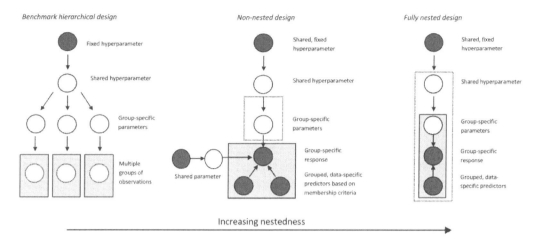

FIGURE 3.2: An idealized depiction of a real-world ecosystem with (a) non-nested, (b) partially nested, and (c) fully nested hierarchical structure.

information/entropy, defined by,

$$\Omega(\mathbf{X}) = -\sum_{x} p(x) log_2 p(x), \tag{3.20}$$

and Δ is a measure of a probabilistic hierarchy among the states or departure of $p(x)$ from uniformity (Feldman and Crutchfield, 1997), defined as,

$$\Delta(\mathbf{X}) = \sum_{x} \left(p(x) - \frac{1}{N} \right)^2 \tag{3.21}$$

Shannon information (Ω) is the thermodynamic entropy, S, for a physical system at equilibrium (i.e., a canonical ensemble of equiprobable microscopic states). For perfect order and perfect disorder (maximal randomness), C_{LMC} is zero. In the case of perfect order, $\Omega = 0$, $C_{LMC} = 0$, and perfect disorder, $H = log_2 N$, $p(x) = 1/N$.

Feldman and Crutchfield (1997) have explored how a statistical complexity changes as hierarchical characteristics or parameters change, such as interaction strength and degree of nonlinearity, and the behavior of the C_{LMC} statistical complexity metric in the thermodynamic limit $(N \to \infty)$, and provide a modified C_{LMC} measure, given by,

$$C'(x) \equiv \Omega(x) \cdot \Delta(p(x)|\widehat{p(x)}) \tag{3.22}$$

where,

$$\Delta(p(x)|\widehat{p(x)}) \equiv \sum_{x} p(x) log_2 \frac{p(x)}{\widehat{p(x)}} \tag{3.23}$$

is the *relative information gain* or *Kullback–Leibler information metric* between two independent distributions $p(x)$ and $\hat{p}(x)$ (note between *distributions*, not *states*) (Feldman and Crutchfield, 1997).

Hierarchical structuring is important for how systems learn, re-design themselves and adapt in response to driving (control) variables through direct interactions, in response to indirect interactions (co-evolution) and environmental uncertainty. Applying a co-evolutionary perspective to natural resource management leads to, SES structured as hierarchically-arranged mosaics of co-evolving social, technological and environmental processes or elements. The coevolutionary perspective, in particular, represents our dependency on various technologies as part of social, economic and environmental development and sustainability (i.e., human-machine symbiosis) (Rammel et al., 2007). Bale et al. (2015) discuss new ways forward in applying complexity science in understanding energy systems and system change. CAS theory can, in part, help to explain how best to transition to sustainable energy systems on the basis of experiments and modeling insights from hierarchical ecological networks (Jorgensen and Nielsen, 2015).

External controls and internal constraints impose negative feedback, reducing the dimensionality or degrees of freedom within a CAS. This makes CAS more nested. Positive feedback is introduced when controls and constraints are reduced, and greater autonomy is enabled between variables and modules in a hierarchical structure. With a combination of negative and positive feedback loops, and nonlinearity between variables and processes, it is difficult to represent, map, and translate hierarchical structure across different systems—especially when such models are used in prediction (i.e., hind-casting and forecasting). In ecosystems - competitive or predation trophic interactions are negative feedbacks, whereas mutualism between two or more species introduces positive feedback. Cross-validation of hierarchical models used for prediction also faces challenges, because correlations (as a measure of association) can be easily misrepresented or misinterpreted as inferring *causality* between variables, modules and levels in a hierarchical structure—termed in ecology as *ecological fallacy*, whereby group-level correlations are attributed to individual-level causes (Freedman, 2001). The interplay between control and emergence within dynamic energy supply chains and networks involving regulators, firms, multiple agents with different business strategies, levels of cooperation, available technologies is an area of increasing relevance and interest in terms of both devising models with high prediction capabilities and enabling their validation against historical or near-real time (NRT) measurement data and expert knowledge (Li et al., 2009; Choi et al., 2001; Funtowicz and Ravetz, 1994).

3.3   Alternative Approaches, Models, and Uses

A wide variety of alternative theoretical approaches, and models are available, but they rely on different assumptions that may or may not be overly restrictive and realistic. The usefulness of different approaches is linked to their different levels of flexibility for making the best use of available spatio-temporal data and its quantity, quality, and scale. Popular methods that provide a broad range of flexibility to explore these aspects include Bayesian hierarchical modeling, machine-learning approaches (e.g., Bayesian belief networks (BNs), causal models (CMs), Artificial Neural Network (ANN)s), knowledge-based models (KBMs), individual or agent-based models (ABMs), game-theoretic models (GTMs), and other hybrid-based, coupled-component models (CCMs) or system dynamic models (SDMs). Hybrid approaches include model-fusion, and crowdsourcing. Hybrid approaches often involve geospatial (i.e., spatially and temporally-explicit) interactions, social or group human and machine-based learning, scenario planning dynamics and optimization—all taking place within a simulated virtual geospatial environment. Ratzé et al. (2007) outlines key concepts of scale, level of organization, holon, constraint, and interactions, within hierarchy theory. They outline existing CAS simulation modeling methods and formalisms that includes ordinary and partial differential equation models, compartmental models, discrete event and CA models, individual and multi-agent based models, and coupled and hybrid approaches. Kelly et al. (2013) also provides an updated review of the top five state-of-the-art, commonly applied approaches or model types for integrated environmental assessment and management and a decision tree for selecting the most appropriate approach under standard applications. An extended decision tree for CAS ecosystem modeling that includes existing modeling approaches, including Bayesian hierarchical modeling, is provided in Figure 3.3. In the next sections, Bayesian hierarchical, agent-based, and game-theoretic modeling approaches are discussed in greater detail. Network modeling approaches that include Artificial Neural Networks and Bayesian Belief Networks are detailed further in the next chapter, linked with a discussion on AI and Machine Learning (ML).

3.3.1 Bayesian hierarchical modeling

Two key statistical questions involved in hierarchical modeling, are whether a frequentist or Bayesian approach should be used or a combination of both, and how many levels should be specified in a given hierarchical model (Cressie et al., 2009; Lele and Dennis, 2009). A frequentist approach considers $P(D|H)$, the probability of data randomly sampled from a given population distribution, given a single or fixed set of null hypotheses, which are mutually exclusive and either true or false. The Bayesian approach, considers instead, $P(H|D)$, the probability of a hypothesis, given the data, and applies the Bayes theorem, treating prior data as fixed, and hypotheses as random with a given probability of being true or false.

Hierarchical (also known as multilevel or mixture) modeling can be applied in data reduction, estimation, prediction and causal inference. While this method can outperform classical regression in terms of model predictive accuracy, it may be useful also to guide model and variable selection and explore how model and data uncertainty propagates in space and time. Its aim is to achieve more flexible models and analysis methods for the analysis of environmental space-time data—termed Hierarchical Dynamical Spatio-Temporal Models (HD-STMs) (Cressie and Wilke, 2011). It provides a stochastic modeling framework for testing multiple models against multiple measurement datasets. It also enables the propagating of uncertainty, involving hard measurement/instrumental uncertainties and/or softer, more subjective data uncertainties linked with expert judgment and problem-specific rea-

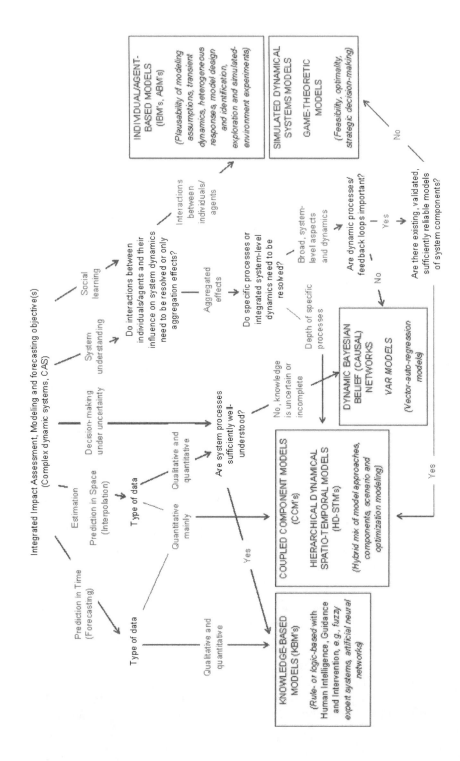

FIGURE 3.3: Decision tree for integrated modeling approach under standard application (modified/extended from Kelly et al. (2013)).

soning. Uncertainty quantification and propagation is critical for selecting between competing models or competing measurement datasets. Hierarchical modeling is a generalization of linear and generalized linear modeling (GLM), whereby a model is prescribed for the regression coefficients with parameters estimated from data (Wang and Gelman, 2014; McCullagh and Nelder, 1989; Banarjee et al., 2005). This can be generalized further in terms of decomposing a complicated joint probability distribution into a simpler representation of conditional models and factors of a given spatio-temporal process. Wilke et al. (1998) and Wilke (2003) present a BHM

TABLE 3.1: Levels or stages of Bayesian representations of hierarchical CAS (i.e., dynamic spatio-temporal) models (adapted from Wilke et al. (1998)).

Stage	Variables	Model	Sub-model
1	Data model	$[D|Y,\theta_1]$	
2	Process model	$[Y|\mu,\beta,u']$	
3	Spatial scales	$[\mu,\beta,u'|\theta_3]$	
	Spatial priors		$[\mu|\theta_\mu]$, $[\beta|\theta_\beta]$
	Space-time dynamics	$[u'|\theta_{u'}]$	
4	Process parameters	$[\theta_1,\theta_2,\theta_3|\theta_4]$	
	Data error variances		$[\theta_1|\theta_4(1)]$
	Process error variances		$[\theta_2|\theta_4(2)],[\theta_\mu|\theta_4(\mu)], [\theta_\beta|\theta_4(\beta)]$
	Dynamic parameters		$[\theta_{u'}|\theta_4(u')]$
5	Hyperparameters	$[\theta_4]$	

modeling framework and discuss the primary stages involved. These include: 1) modeling the measurement/data sampling process (*data modeling*), 2) representing large versus small features and slow versus fast dynamics of underlying process dynamics (*process modeling*), and 3) capturing spatial and temporal dependencies and inter-dependencies (i.e., space-time dynamics includes stationary/nonstationary spatial covariance, temporal autocorrelation, and space-time coupling interactions, and dynamic parameters including higher-order, non-local and extended memory (i.e., spatial and temporal lag effects), 4) specification of model parameter distributions (*priors*) and their conditional independence, and 5) specification of model parameter distributions that are interdependent (*hyperpriors*). These stages are summarized in Table 3.1, embellished from Table 1 of Wilke et al. (1998). In this Table, the listed variables are denoted, $\theta_1 = \sigma_{x,t}^2$, $\theta_2 = \sigma_Y^2(x_i)$, $i = (1, ..., n)$, $\theta_3 = (\theta_\mu, \theta_\beta, \theta_{u'})$, and the set of hyperparameters and their hyperpriors are, $\theta_4 = (\theta_4(1), \theta_4(2), \theta_4(3))$.

Given a joint probability distribution of parameters, $\boldsymbol{\theta}$ and data, $\mathbf{D}$, that is the product of a prior distribution, $p(\boldsymbol{\theta})$ and data sampling distribution, $p(\mathbf{D}|\boldsymbol{\theta})$,

$$p(\boldsymbol{\theta}, \mathbf{D}) = p(\boldsymbol{\theta})p(\mathbf{D}|\boldsymbol{\theta}). \tag{3.24}$$

Bayesian inference uses the Bayes rule of conditional probability to condition parameters to known data, $\mathbf{D}$, to update and estimate them as additional evidence is learned, and to predict unobserved data based on observed data. The *unnormalized* posterior probability distribution (or density), conditioned on known data (Gelman et al., 2013), is given by,

$$p(\boldsymbol{\theta}|\mathbf{D}) \propto p(\mathbf{D}|\boldsymbol{\theta})p(\boldsymbol{\theta}) \tag{3.25}$$

where $\boldsymbol{\theta}$ are parameters whose probability is influenced by data or observational evidence $\mathbf{D}$, $p(\boldsymbol{\theta})$ is the *prior* probability (probability of $\boldsymbol{\theta}$ before $\mathbf{D}$ is observed), $p(\boldsymbol{\theta}|\mathbf{D})$ is the *posterior* probability of $\boldsymbol{\theta}$ after $\mathbf{D}$ is observed, $p(\mathbf{D}|\boldsymbol{\theta})$ is the probability of observing $\mathbf{D}$ given $\boldsymbol{\theta}$, also called the *likelihood*, where $p(\mathbf{D})$ is the marginal likelihood or evidence. Bayes rule therefore states that the "posterior probability is proportional to the prior times likelihood," or similarly, "posterior is prior times likelihood over evidence." The posterior is the result of updating our prior information with data (i.e., learning). Since the denominator does not depend on any $\boldsymbol{\theta}$, it acts as a proportionality or normalization constant, and is obtained by averaging the "prior times likelihood" (i.e., numerator) over all possible parameters, $\boldsymbol{\theta}$, such that,

$$p(\boldsymbol{\theta}|\mathbf{D}) = \frac{p(\boldsymbol{\theta},\mathbf{D})}{p(\mathbf{D})} = \frac{p(\mathbf{D}|\boldsymbol{\theta})p(\boldsymbol{\theta})}{p(\mathbf{D})}, \quad p(\mathbf{D}) = \int_{\boldsymbol{\theta}} p(\mathbf{D},\boldsymbol{\theta})p(\boldsymbol{\theta})d\boldsymbol{\theta}, \qquad (3.26)$$

where $\boldsymbol{\theta}$ is assumed to be continuous ($p(\mathbf{D})$ becomes a summation if $\boldsymbol{\theta}$ is assumed discrete). Bayesian hierarchical modeling (hereafter, BHM) is based on the Bayes rule for obtaining the joint or posterior probability distribution of a space-time process, $\mathbf{Y}$, with model parameters, $\boldsymbol{\theta}$. If additional latent or hidden hierarchical structure is assumed for a process, $\mathbf{Y}$ (i.e., $\mathbf{Y}$ is a vector of parameters for a 2-stage, 3-stage or higher, multi-level hierarchical model with hyperparameters), then the normalized relation given by Equation 3.25 can be re-expressed as,

$$p(\boldsymbol{\theta},\mathbf{Y}|\mathbf{D}) \propto p(\mathbf{D}|\mathbf{Y},\boldsymbol{\theta})p(\mathbf{Y},\boldsymbol{\theta})p(\boldsymbol{\theta}) \qquad (3.27)$$

and when normalized becomes,

$$p(\boldsymbol{\theta},\mathbf{Y}|\mathbf{D}) = \frac{p(\mathbf{D}|\mathbf{Y},\boldsymbol{\theta})p(\mathbf{Y},\boldsymbol{\theta})p(\boldsymbol{\theta})}{\int_{\boldsymbol{\theta}} p(\mathbf{D}|\mathbf{Y},\boldsymbol{\theta})p(\mathbf{Y}|\boldsymbol{\theta})p(\boldsymbol{\theta})d\boldsymbol{\theta}}. \qquad (3.28)$$

Bayesian estimation generally assumes *exchangeability* whereby there is no ordering or grouping of the parameters and there is symmetry between prior distribution parameters. This means that uncertainty as a joint probability distribution is invariant to permutations of its indices, and can apply to each level of a hierarchical model (Gelman et al., 2013). It is also typically assumed that any random variables or a sequence of them, are *Independent and Identically Distributed (IID)*, whereby if two random variables share the same probability distribution then the distributions are mutually independent. Exchangeability is a broader assumption than IID, because any sequence that is IID is also exchangeable, but the converse does not hold (i.e., exchangeable sequences. while being identically distributed are not always independent, so are not necessarily IID).

Forecasting/prediction of the prior probability distributions involves marginalizing over parameters, and for the posterior probability distribution, involves the assumption of conditional independence between observed data, $\mathbf{D}$ and unobserved (or new) data, $\tilde{\mathbf{D}}$, marginalized over the posterior. The *prior predictive* distribution is given by,

$$p(\mathbf{D}) = \int_{\boldsymbol{\theta}} p(\mathbf{D},\boldsymbol{\theta})d\boldsymbol{\theta} = \int_{\boldsymbol{\theta}} p(\mathbf{D}|\boldsymbol{\theta})p(\boldsymbol{\theta})d\boldsymbol{\theta} \qquad (3.29)$$

and the *posterior predictive* distribution is,

$$p(\tilde{\mathbf{D}}|\mathbf{D}) = \int_{\theta} p(\tilde{\mathbf{D}}, \theta|\mathbf{D})d\theta = \int_{\theta} p(\tilde{\mathbf{D}}|\theta, \mathbf{D})p(\theta|\mathbf{D})d\theta = \int_{\theta} p(\tilde{\mathbf{D}}|\theta)p(\theta|\mathbf{D})d\theta. \quad (3.30)$$

In the above equations, the spatio-temporal process model, $\mathbf{Y}$, with parameters θ, is assumed to consist of a single level associated with a vector of parameters θ. They can, however, be further decomposed, for example, into a spatio-temporal process having a spatial, site-specific mean μ, parameters β associated with a large-scale temporal component and additional parameters associated with spatial-temporal dynamics, $\mathbf{u}'$. Also, $p(\mathbf{D}|\mathbf{Y}, \theta)$ can also be expanded or decomposed into multiple conditional probabilities associated with data from multiple sources assumed to be independent (i.e., with any dependence between them attributable to the process $\mathbf{Y}$) (Wilke, 2003). In Chapter 4, Bayesian prediction of a spatio-temporal process is further discussed, not with multiple sets of the same type of data, but instead multiple, different data types (e.g., leading and auxiliary data).

The measurement process or data model associated with a given dataset, $\mathbf{Z} = Z(x, t) : (x, t) \in D, \mathbb{D} \in \mathbb{R}^2$ then $\mathbf{Z}$ comprises a space-time lattice or grid comprising a total of m spatial locations, x, and n time points, t, i.e., $Z(x_i, t_j); i = (1, 2, .., m); j = (1, 2, ..., n)$. Previously in Equation 1.34, the stochastic equation for a time-varying process was defined. Here, we extend this equation to include spatial variability, where $\mathbf{x}$ can denote a point, site, or area centroid location of a neighborhood in space (i.e., spatial location) and t, time. In general, a spatio-temporal process model for $Y(s, t)$ is given by,

$$Y(x, t) = f(x, t) + u(x, t) + h(x, t), \quad (3.31)$$

where $f(\cdot)$ is a function of resolved variables, $u(\cdot)$ is a function of unresolved drift variables, and $h(x, t)$ is a Wiener process. This is a continuous, stochastic process in time (i.e., standard Brownian motion with zero mean and unit variance). It is a random-walk process having stationary, independent (i.e., Gaussian-distributed) increments. If one assumes that $f(\cdot)$ is a spatial mean of a process $\mu(x)$ (i.e., resolved in space and independent of time), and $u(\cdot)$ can be separated into two contributions—one that is stationary at/over large spatial-scale and slower time-scales (denoted $u_s(x, t)$), in relation to another (denoted $u_f(x, t)$) that is stationary over small spatial-scales and fast/short time-scales, then,

$$Y(x, t) = \mu(x) + u_s(x, t) + u_f(x, t) + h(x, t). \quad (3.32)$$

For example, $u_s(x, t)$ could be resolved as an inter-annual, seasonal or monthly spatial mean, in relation to $u' = u_f(x, t)$, as a daily-scale mean. If a hyperparameter is also introduced comprising a set of time-integrated, spatial indicator variables, $\beta(x)$, along with $h(x, t)$ as Brownian, uncorrelated white-noise, independent of space and time, modeled as a Gaussian distribution with mean zero and unit variance, denoted ϵ, then one obtains the reduced stochastic, hierarchical space-time model,

$$Y(x, t) = \mu(x) + u_s(\beta(x), t) + u'(x, t) + \epsilon, \quad (3.33)$$

where $Y(s,t)$ is a conditionally independent random variable. Because the necessary integration to obtain the posterior distribution is analytically intractable, Markov-Chain Monte Carlo Simulation (MCMC) using the Metropolis–Hastings algorithm or the Integrated Nested Laplace Approximation (INLA) method can be applied to obtain a sample from the posterior distribution (Rue et al., 2009).

3.3.2   Spatio-temporal processes

Covariance is a measure of the strength of the correlation between two or more sets of random variates. It is a fundamental metric for spatio-temporal model-based inference (i.e., estimation and prediction). This is because reliable inference requires representing and understanding uncertainty in both spatial and temporal variation in the covariance of ecosystem and other environmental processes (Sampson et al., 2001). Moreover, the behavior of a covariance function near the origin has large consequences on predictive distributions and uncertainty (Stein, 2005). A spatio-temporal process can, in general, possess a range of properties, such as: 1) symmetric or asymmetric, 2) non-separable or separable in space and time, 3) non-stationary or stationary, and 4) isotropic/homogenous or anisotropic/heterogeneous. These conditions can be expressed for the space-time covariance function (see Porcu et al. (2006); Stein (2005); Gneiting (2002) and references therein). For reliable estimation and prediction, it is crucial to recognize that many statistical properties may provide compact/non-compact support in space and time. All or just one of the above conditions may exist within a given CAS, although there is a tendency for many of these conditions to characterize real-world processes (Paciorek and Schervish, 2006; Porcu et al., 2006; Stein, 2005). The definition of these properties, and various methods for building models with covariance assumptions built in (i.e., not just mean or variance assumptions) are presented.

Let $Z(\mathbf{u}) = [Z_1(u), Z_2(u), ..., Z_p(u)]^{\mathsf{T}}$, be a vector of p, *Univariate* Spatio-Temporal Random Functions (STRF) associated with a set of spatio-temporal variables of interest (i.e., coregionalized variables), defined on the domain $S \times T \subseteq \mathbb{R}^{d+1}$, where $d \leq 3$.

Let $Z(\mathbf{u}), \mathbf{u} = (\mathbf{s},t) \in S \times T \subseteq \mathbb{R}^{d+1}$, represents a Multivariate Spatio-Temporal Random Function (MSTRF) with components, $Z_i, i = (1, ..., p)$ where $\mathbf{s} = (s_1, s_2, ..., s_n)$ are spatial coordinates within the domain $S \subseteq \mathbb{R}^d$ and $\mathbf{t} = (t_1, t_2, ..., t_m)$ are temporal coordinates within the domain $T \subseteq \mathbb{R}$.

Let observational or measurement data of the p variables, Z_i at points $\mathbf{u}_\alpha \in S \times T$, as finite realizations of a MSTRF process, denoted as $Z_i(\mathbf{u}_\alpha)$, where $i = (1,,,.,p)$ and $\alpha = (1, ..., n_i)$ (Genton and Kleiber, 2015; de Iaco et al., 2012). The cross-covariance, C_{ij}, is then given by,

$$C_{ij} = Cov(Z_i(\mathbf{u}), Z_j(\mathbf{u}')) = E\left((Z_i(\mathbf{u}) - m_i(\mathbf{u}))(Z_j(\mathbf{u}') - m_j(\mathbf{u}'))\right), i \neq j \quad (3.34)$$

and the direct covariance (hereafter termed just covariance), C_{ii}, is,

$$C_{ii} = Cov(Z_i(\mathbf{u}), Z_i(\mathbf{u}')) = E\left((Z_i(\mathbf{u}) - m_i(\mathbf{u}))(Z_i(\mathbf{u}') - m_j(\mathbf{u}'))\right) \qquad (3.35)$$

The *space-time* cross-correlation is given by,

$$\rho_{ij}(\mathbf{u}, \mathbf{u}') = C_{ij}(\mathbf{u}, \mathbf{u}')/\sqrt{C_{ii}(\mathbf{u}, \mathbf{u}')C_{jj}(\mathbf{u}, \mathbf{u}')}$$
$$i, j = (1, ..., p), \mathbf{u}, \mathbf{u}' \in S \times T.$$

Furthermore, if $Z(\mathbf{u})$ is a Gaussian multivariate random field (e.g., multivariate normal distribution or MVN) then it is fully represented by a mean vector, $m_i = E[Z_i(\mathbf{u})]$ and cross-covariance $C_{ij}(\mathbf{u}, \mathbf{u}')$, $i, j = (1, ..., p)$. The cross-covariance and direct covariance of a MSTRF function, to be *permissible* or *admissible* (i.e., symmetric and real-valued), must be positive semi-definite (or non-negative definite) and satisfy (de Iaco et al., 2012; Christakos, 2000),

$$\sum_{i=1}^{p}\sum_{j=1}^{p}\sum_{\alpha=1}^{n}\sum_{\beta=1}^{n} \lambda_{\alpha i}\lambda_{\beta j}C_{ij}(\mathbf{u}_\alpha - \mathbf{u}_\beta) \geq \mathbf{0}, \forall \mathbf{u}_\alpha, \lambda_{\alpha \mathbf{i}}, \qquad (3.36)$$

involving p components and $\mathbf{u}=(u_1, ..., u_n)$ points in a spatio-temporal domain, $S \times T$ (i.e., $u_1=(x_i, t_1)$, $u_2=(x_2, t_2)$,...,$u_n=(x_n, t_n)$).

Continuous space-time covariance functions are equivalent to positive semi-definite (non-negative) finite measures iff (Cressie and Wilke, 2011; Gregori et al., 2008; Bochner, 1933),

$$C(\mathbf{h_s}, h_t) = \int_{\mathbb{R}^d}\int_{\mathbb{R}^d} e^{-\imath \omega \mathbf{h_s} + th_t \omega} dF(\omega, \tau) \qquad (3.37)$$

where F is a distribution function on $\mathbb{R}^d \otimes \mathbb{R}$.

A MSTRF is *second-order stationarity* (SOS) (also termed weakly stationary), if for any STRF Z_i, $E[Z_i(\mathbf{u})] = m_i$, $var[Z_i]=\sigma_Z^2 < \infty$ and for any pair of STRF's Z_i and Z_j, the cross-covariance C_{ij} only depends on the spatio-temporal separation or lag vector, defined as $\mathbf{h} = (\mathbf{h_s}, h_t)$ between the points $\mathbf{u}$ and $(\mathbf{u} + \mathbf{h})$ (i.e., separation vector with spatial component, $\mathbf{h_s} = (\mathbf{s}_i - \mathbf{s}_j)$, and temporal component, $h_t = (t_i - t_j)$ between two space-time points),

$$C_{ij}(\mathbf{h}) = E\left(Z_i(\mathbf{u} + \mathbf{h}) - m_i)(Z_j(\mathbf{u}) - m_j)\right) = E\left(Z_i(\mathbf{u} + \mathbf{h})Z_j(\mathbf{u})\right) - m_i m_j \quad (3.38)$$

such that, when $i = j$, the STRF covariance function is obtained (de Iaco et al., 2012). A MSTRF that is SOS, is also partially *separable*, when its cross-covariance satisfies

$$C_{ij}(\mathbf{h}) = \rho(\mathbf{h})a_{ij} \qquad (3.39)$$

and is *fully separable* when,

$$C_{ij}(\mathbf{h_s}, h_t) = \rho_S(\mathbf{h_s})\rho_T(h_t)a_{ij} \qquad (3.40)$$

where $\mathbf{h} = (\mathbf{h_s}, h_t) \in D \times T$, $i, j = (1, ..., p)$ and a_{ij} are the elements of a $p \times p$ positive definite matrix, ρ_S is a spatial correlation function, and ρ_T is a temporal correlation function. *Separability* means that the spatio-temporal covariance structure can be decomposed or factored into a product of a purely spatial lag $\mathbf{h_s}$ and a purely temporal lag h_t component, whereby,

$$C_{ij}(\mathbf{h_s}, h_t) = \frac{C(\mathbf{h_s}, 0)C(\mathbf{0}, h_t)}{C(\mathbf{0}, 0)}, \tag{3.41}$$

which allows for computationally efficient estimation and inference by reducing the number of parameters in the covariance matrix especially for large complex datasets (Genton and Kleiber, 2015; Gregori et al., 2008; Porcu et al., 2006). However, for separable covariance functions, small changes in the locations of observations can lead to large changes in the correlations between linear combinations of the observations that arise as a discontinuity (i.e., lack of mean-square differentiability in either space or time), because they are not smoother away from the origin than at the origin (Stein, 2005). *Symmetry* exists when the cross-covariance satisfies,

$$C_{ij}(\mathbf{h}) = \rho(\mathbf{h})a_{ij} \tag{3.42}$$

and it is *fully symmetric* when,

$$C_{ij}(\mathbf{h_s}, h_t) = C_{ij}(-\mathbf{h_s}, h_t) = C_{ij}(\mathbf{h_s}, -h_t) = C_{ij}(-\mathbf{h_s}, -h_t) \tag{3.43}$$

where $h = (\mathbf{h_s}, h_t) \in D \times T, i, j = (1, ..., p)$. Asymmetry is observed when the response of one variable affects another variable delayed in time. Apanosovich and Genton (2010) identify two main types of asymmetry—a Type 1 that occurs when there is a time lag in a correlation between variables that are individually fully symmetric, and a Type 2 that occurs when there are velocities (Apanosovich and Genton, 2010),

For a separable process, where there is no spatial-temporal interaction, the *spatio-temporal* covariance matrix (also termed a dispersion or variance-covariance matrix) is then given by,

$$\mathbf{\Sigma}_{S,T} = \mathbf{\Sigma}_T \otimes \mathbf{\Sigma}_S \tag{3.44}$$

where $\mathbf{\Sigma}_T$ is a $m \times m$ purely temporal covariance matrix with elements $(C_{ij})_{m \times m}$ and $\mathbf{\Sigma}_S$ is a $n \times n$ purely spatial covariance matrix with elements $\mathbf{\Sigma}_S = (C'_{ij})_{n \times n}$, where $\otimes$ denotes the Kronecker matrix product. $\mathbf{\Sigma}_{S,T}$ is then the $mn \times mn$ block matrix,

$$\mathbf{\Sigma}_{S,T} = \begin{pmatrix} C_{11}(s_1, s_1)\mathbf{\Sigma}_T & C_{12}(s_1, s_2)\mathbf{\Sigma}_T & \dots & C_{12}(s_1, s_n)\mathbf{\Sigma}_T \\ C_{21}(s_2, s_1)\mathbf{\Sigma}_T & C_{22}(s_2, s_2)\mathbf{\Sigma}_T & \dots & C_{2n}(s_2, s_n)\mathbf{\Sigma}_T \\ \vdots & \vdots & \ddots & \vdots \\ C_{n1}(s_2, s_1)\mathbf{\Sigma}_T & C_{n2}(s_2, s_n)\mathbf{\Sigma}_T & \dots & C_{n2}(s_n, s_n)\mathbf{\Sigma}_T \end{pmatrix}_{mn \times mn}$$

Its matrix inverse as $(\mathbf{\Sigma}_{S,T})^{-1}$ is then just the Kronecker product of the inverse matrix of each component i.e., $(\mathbf{\Sigma}_T)^{-1} \otimes (\mathbf{\Sigma}_S)^{-1}$, and similarly for its determinant, $|\mathbf{\Sigma}_{S,T}|$.

Environmental, atmospheric, and geophysical processes are typically influenced by constant and variable advection (not just diffusion) such as prevailing winds, ocean currents and pressure gradients, where the assumption of full symmetry is not realistic or reliable. As Gregori et al. (2008) note, typically covariance models that are always positive are specified, but real-world problems involving CAS stochastic dynamics, require covariance models that allow for negative values or oscillations from positive to negative values (Gregori et al., 2008). They propose an easy-to-implement class of models that admits this type of covariances, and they test it on biometrical data. The so-called *Generalized Product–Sum (GPS)* model and its mixed forms combine spatio-temporal covariances through weighted sums and products of valid marginal covariance functions C_s and C_t, where the weights must be non-negative (Gregori et al., 2008), whereby,

$$C_{ij}(\mathbf{h_s}, h_t) = k_1 C_s(\mathbf{h_s}) C_t(h_t) + k_2 C_s + k_3 C_t \qquad (3.45)$$

and $k_1 > 0$, $k_2 \geq 0$ and $k_3 \geq 0$ is sufficient for positive definiteness. In the linear model coregionalization (LMC) method, this approach is further generalized to include non-stationarity in either time or space. Ma (2008) discusses recent developments in building spatio-temporal covariance models, and Sampson et al. (2001) provide a review of alternative methods for estimating nonstationary spatial covariance for environmental applications.

The main approaches for building *stationary* cross-covariance models are the linear model coregionalization (LMC), the kernel and covariance convolution, the latent dimensions, and the multivariate Matérn (Genton and Kleiber, 2015). While higher-dimensional stationary representations for nonstationary MSTRFs exist, in many applied situations, neither a nonstationary variogram, nor an analytic mapping to higher dimensions, is known (Bornn et al., 2012). Multiresolution bases for expanding the covariance can be especially useful when spatial data is sparse, discontinuous, or varies in its degree of smoothness. They are also useful when one seeks an approximation that requires the thresholding of matrix elements within the covariance matrix, Σ to ensure matrix sparsity (Nychka et al., 2002). An approach for constructing and applying *nonstationary* models is, for example, moving window kriging based on stationary covariance (Tadic et al., 2014; Chen et al., 2006; Haas, 1990), nonlinear deformation or "image warping" of the geographic coordinates (Sampson et al., 2001), nonparametric latent dimensions (Bornn et al., 2012), variable convolution of a stationary process (Genton and Kleiber, 2015; Apanosovich and Genton, 2010; Ver Hoef and Barry, 1998), and change of support models (Nychka et al., 2002). Wood (2010) also discusses statistical inference and robust covariance estimation (that discounts the extreme-values or tails in the covariance distribution) when modeling noisy nonlinear processes.

The *linear model coregionalization (LMC)* method avoids specifying the cross-covariance of a MSTRF explicitly. Instead, it is assumed to be stationary and anisotropic over relatively small, local spatial regions, and thus is decomposed into a linear combination of r independent univariate STRFs, such that (Apanosovich and Genton, 2010),

$$C_{ij}(\mathbf{h_s}) = \sum_{k=1}^{r} a_{ik}a_{jk}\rho_k(h), \tag{3.46}$$

where $1 \leq r \leq p$ and $\rho_k(\cdot)$ are valid stationary correlation functions, and $A = a_{ij}$ is a $p \times r$ full rank matrix. When $r = 1$, the cross-covariance function $C_{ij}(\mathbf{h})$ is separable. Further, if one associates A with a principal component (i.e., eigenvalue/eigenvector) transformation (i.e., a set of orthogonal local stationary processes across k subregions or sub-grids), then a space-time process can be assumed to be stationary in time, but nonstationary in space, such that,

$$Z(\mathbf{s}, t) = \sum_{i=1}^{p} k_i(\mathbf{s} - s_i)Z_i(\mathbf{s}, t) \tag{3.47}$$

where Z_i is locally-stationary and $k_i(\mathbf{s} - s_i)$ is a spatial weighting function centered on s_i. Similarly, if a space-time process is assumed to be stationary in space, but nonstationary in time, then,

$$Z(\mathbf{s}, t) = \sum_{i=1}^{p} k_i(t - t_i)Z_i(\mathbf{s}, t) \tag{3.48}$$

involving the time-weighting function $k_i(t - t_i)$ is a weight function centered at t_i.

Furthermore, one can obtain an eigenvalue/eigenvector decomposition of covariance for any choice of discrete basis, ζ, and varying dimensionality, d, under the assumption that $\mathbf{Z}$ is a Gaussian, zero-mean process, such that,

$$\boldsymbol{\Sigma} = \zeta \mathbf{D} \zeta^{\mathsf{T}} = \zeta \mathbf{H}^2 \zeta^{\mathsf{T}} \tag{3.49}$$

where $\zeta\zeta^{\mathsf{T}} = \mathbb{I}$, where $\mathbb{I}$ denotes the identity matrix and $\mathbf{D} = \zeta^{-1}\Sigma(\zeta^{\mathsf{T}})^{-1}$ and $\mathbf{H} = \mathbf{D}^{1/2}$, where $\mathbf{D}$ and $\mathbf{H}$ are "sparse" matrices (i.e., a matrix that is close to diagonal having non-zero diagonal elements and many/most off-diagonal elements that are zero), such that ζ may not be orthogonal and $\mathbf{H}^2$ may not be diagonal.

Nychka (2002) outlines how to obtain sample-based estimates of $\hat{\boldsymbol{\Sigma}}$ for a data model, $\mathbf{Z}$. By numerically solving the matrix equations below, one can obtain the matrix $\hat{\mathbf{H}}$ via singular-value decomposition (SVD) of $\zeta^{-1}\mathbf{Z}$ and basis functions ζ to yield the matrix $\hat{\mathbf{D}}$. Next, by applying smoothing between spatially-adjacent elements in $\boldsymbol{\Sigma}$, and shrinkage to a stationary model, an estimate for the sample covariance matrix, $\hat{\boldsymbol{\Sigma}}$ can be obtained, for T time points (Cline and Dhillon, 2006; Nychka et al., 2002),

$$\hat{\boldsymbol{\Sigma}} = \frac{1}{T}(\mathbf{ZZ^{\mathsf{T}}})$$

$$\hat{\mathbf{D}} = \frac{1}{T}(\zeta^{-1}\mathbf{Z})(\zeta^{-1}\mathbf{Z})^{\mathsf{T}}.$$

$$\hat{\mathbf{D}} = \hat{\mathbf{H}}^2$$

The *kernel convolution* method assumes that all processes $Z_i(\mathbf{u})$ are generated by the same underlying spatio-temporal process which imposes strong dependence that can be highly restrictive (Genton and Kleiber, 2015; Apanosovich and Genton, 2010; Ver Hoef and Barry, 1998) and defines,

$$C_{ij}(\mathbf{h_s}) = \int_{\mathbb{R}^d} \int_{\mathbb{R}^d} k_i(h_{\tau_1}) k_i(h_{\tau_2}) \rho(h_{\tau_1} - h_{\tau_2} + \mathbf{h_s}) dh_{\tau_1} dh_{\tau_2} \qquad (3.50)$$

where $\mathbf{h_s}$ is a spatial lag/separation distance, k_i are square-integrable kernel functions and $\rho_k(\cdot)$ is a valid stationary correlation function. In the case of the covariance convolution method,

$$C_{ij}(\mathbf{h_s}) = \int_{\mathbb{R}^d} C_i(\mathbf{h_s} - k) C_j(k) dk, \qquad (3.51)$$

where C_i are the chosen stationary covariance functions that are square-integrable whereby some closed form solutions exist (e.g., a Matérn function).

The *latent dimensions* method creates additional latent dimensions representing variables to be modeled proposed by Apanosovich and Genton (2010) (Apanosovich and Genton, 2010). The MSTRF $\mathbf{Z}$, or each of its component, Z_i, are represented as a point, $\boldsymbol{\zeta}_i = (\zeta_{i1}, \zeta_{i2}, .., \zeta_{ik})^\mathsf{T}$ in a new dimension, $\boldsymbol{\zeta} \in \mathbb{R}^k$, $i = (1, ..., p)$ for $k \in \mathbb{Z}$, $1 \le k \le p$.

$$C_{ij}(s_i, s_j) = C'((s_i, \zeta_i), (s_j, \zeta_j)), s_i, s_j \in \mathbb{R}^d \qquad (3.52)$$

where $\mathbf{C}'$ is a valid STRF on $\mathbb{R}^{d+k}$. Bornn et al. (2012) provide a non-parametric method for learning of latent dimensions (i.e., dimensional expansion), especially useful when dealing with high-dimensional nonstationary covariance modeling—with thin-plate splines as the mapping function. They test this approach on solar radiation and air pollution data. To "learn" the latent dimensions, $\boldsymbol{\zeta}_i$, they propose to minimize the spatial lag or separation distance between non-latent spatial coordinates $\mathbf{s}$ and augmented coordinates $\boldsymbol{\zeta}$, according to,

$$\hat{\phi}, \boldsymbol{\zeta} = \underset{\phi, \boldsymbol{\zeta}'}{\arg\max} \sum_{i<j} \left(\nu_{i,j}^* - \gamma_\phi \left(d_{ij}([\mathbf{s}, \boldsymbol{\zeta}']) \right) \right)^2 + \lambda_1 \sum_{l=1}^{k} ||\boldsymbol{\zeta}'_{.,l}||, \qquad (3.53)$$

which considers the inter-site dispersion (sites i and j), ν_{ij}^*, defined as,

$$\nu_{ij}^* = \frac{1}{|\tau|} \sum_{\tau} |Y(s_i) - Y(s_j)|^2, \tau > 1. \qquad (3.54)$$

where the augmented dimension vector $\zeta \in \mathbb{R}^n \times \mathbb{R}^k$, involves n spatial coordinates and k is optimal (or near-optimal) under the above minimization. A function can be constructed, whereby $F(s) \approx \zeta$. $d_{ij}([\mathbf{s}, \boldsymbol{\zeta}'])$ is the ith, j^{th} element of the distance matrix of augmented locations $[\mathbf{s}, \boldsymbol{\zeta}']$, $\zeta'_{.,1}$ is the lth column of the augmented dimension vector, and λ_1 is a shrinkage parameter linked with a group Least Absolute Shrinkage and Selection Operator Penalty (LASSO) term that allows predefined groups of covariates to be selected or not selected (see Hastie et al. (2008); Yuan and Lin (2006)), §3.4.1, pg. 68). In the case of rare, extreme, or low sample of events, model overfitting with shrinkage of regression coefficients can be achieved with penalized shrinkage and LASSO regression methods, which can improve risk prediction model accuracy and interpretability (Pavlou et al., 2015; Tibshirani, 1996).

A stationary covariance function is *isotropic* if it is invariant under translation and rotation (Porcu et al., 2006), such that,

$$C_{ij}(\mathbf{h_s}, h_t) = C(\mathbf{h'_s}, h'_t), \tag{3.55}$$

iff (if and only if) $||\mathbf{h_s}|| = ||\mathbf{h'_s}||, |h_t| = |h'_t|$, where $||\cdot||$ is the Euclidean norm on $\mathbb{R}^d$. Another approach to modeling nonstationary covariance of STRFs is to specify the Matérn class that is stationary and isotropic with an anisotropic metric. This can be extended to MSTRFs, as Gneiting (2002) has demonstrated. A multivariate Matérn model (termed the Gneiting class) can be constructed, whereby both correlation and cross-correlations are Matérn functions.

$$C(s) = \sigma^2 \frac{1}{\Gamma(\nu)} 2^{\nu-1} \left(2\sqrt{\nu} \frac{s}{\rho}\right)^\nu K_\nu \left(2\sqrt{\nu} \frac{s}{\rho}\right), \tag{3.56}$$

where s is distance, ρ is the spatial range parameter and $K_\nu(\cdot)$ is the modified Bessel-function of the second-kind and order ν, and $\rho, \nu > 0$. It is $\nu - 1$ times differentiable, compared to the case of a Gaussian covariance that is infinitely differentiable, and as $\nu \to \infty$, this function approaches the Gaussian covariance function with an exponential form, and $\nu = 0.5$ results in a nonstationary, exponential covariance function.

Specifying the Mahalanobis distance metric (Mahalanobis, 1936), anisotropy, whereby a MSTRF or STRF has directional or angular dependence, can be represented in this Matérn (i.e., K-Bessel class) function, defined as,

$$S'(\mathbf{s_i}, \mathbf{s_j}) = \sqrt{(\mathbf{s_i} - \mathbf{s_j})^\mathsf{T} \mathbf{A^{-1}} (\mathbf{s_i} - \mathbf{s_j})}, \tag{3.57}$$

where $\mathbf{A}$ is an arbitrary positive definite matrix. Substituting this metric S' into Equation 3.56 yields a new class of nonstationary covariance functions, C' given by (Paciorek and Schervish, 2006),

$$C'(\mathbf{s_i}, \mathbf{s_j}) = \sigma \frac{1}{\Gamma(\nu) 2^{\nu-1}} |\mathbf{\Sigma}_i|^{\frac{1}{4}} |\mathbf{\Sigma}_j|^{\frac{1}{4}} |\frac{\mathbf{\Sigma}_i + \mathbf{\Sigma}_j}{2}|^{-\frac{1}{2}} (2\sqrt{\nu Q_{ij}})^\nu K_\nu(2\sqrt{\nu Q_{ij}}) \tag{3.58}$$

where $\mathbf{\Sigma}_i = \mathbf{\Sigma}(\mathbf{s_i})$ is the stationary, isotropic covariance matrix of the Gaussian kernel, centered at $\mathbf{s_i}$. Q_{ij} is the quadratic form given by,

$$Q_{ij} = (\mathbf{s_i} - \mathbf{s_j})^\mathsf{T} \left(\frac{\mathbf{\Sigma}_i + \mathbf{\Sigma}_j}{2}\right)^{-1} (\mathbf{s_i} - \mathbf{s_j}). \tag{3.59}$$

In the case where a covariance function is both positive semi-definite *and* stationary (refer to Equation 3.37), the cross-covariance matrix can be mapped from the time- to the frequency-domain forming a cross-spectral density function, f_{ij}, whose elements are given by,

$$f_{ij}(\omega) = \frac{1}{(2\pi)^d} \int_{\mathbb{R}^d} C_{ij}(\mathbf{h_s}) e^{-i\mathbf{h_s^\mathsf{T}}\omega} d\mathbf{h_s} \tag{3.60}$$

for $i, j = (1, ..., p)$ and where $\mathbf{h_s} = (s_1 - s_2)$. Here the superscript "T" denotes the transpose of the spatial separation or lag vector $\mathbf{h_s}$ and $\imath = \sqrt{-1}$ is the imaginary unit of an imaginary/complex number. Angular frequency is $\omega = 2\pi f$, $\omega \in \mathbb{R}^d$.

$$f(\omega) = \begin{cases} \sigma^2 e^{-\omega^2/4\nu} & (1) \\ \sigma^2(\omega^2 + \eta^2)^{-1} & (2) \\ \sigma^2(\omega^2 + \eta^2)^{(-\nu-1/2)} & (3) \end{cases} \tag{3.61}$$

The above spectral power equations are for: 1) Gaussian covariance, 2) exponential covariance and 3) Matérn (K-Bessel) covariance. A new class of nonstationary, non-separable covariances has been proposed by Fuentes et al. (2008) and Stein (2005) and includes dependency on the time-lag, τ, given by,

$$f(\omega, \tau) = \gamma(\alpha^2\beta^2 + \beta^2|\omega|^2 + \alpha^2\tau^2 + \epsilon|\omega|^2\tau^2)^{-\nu}(4). \tag{3.62}$$

Recent applications include: stationary covariance modeling of daily mean air pollution concentrations of particular matter up to 10 micrometers in size (PM_{10}) likely responsible for serious adverse health effects, particularly in urban ecosystems, because of their ability to reach the lower regions of the respiratory tract, with effects on breathing and respiratory systems, damage to lung tissue, cancer, and premature death[13] (Cocchi et al., 2007), modeling spatially varying cross-covariances of soil nutrients (Guhaniyogi et al., 2013), exploring regional variation in spatial covariance of ecosystem services (Anderson et al., 2009), modeling of space-time covariance for improved environmental health assessment, and epidemic propagation (Kolovos et al., 2013; Christakos, 2000).

Further methodological advancement has also been made recently with the testing of a new model whereby non-separability arises from temporal non-stationarity on tropospheric ozone data (Bruno et al., 2009). Tensor decompositions of spatio-temporal covariance are also being explored for reproducing and understanding wind data (Suryawanshi and Ghosh, 2015). Anderson et al. (2009) have studied the regional variation in spatial covariance in 100 km^2 regions and 4 km^2 sub-regions for three ecosystem services (carbon storage, agriculture value, and recreation) and biodiversity (priority species for conservation). They find a consistent, mixed pattern of either none, or mixed positive and negative relationships illustrating that one can arrive at diametrically opposing conclusions about relationships between ecosystem services and biodiversity by studying the same question within different regions. Kolovos et al. (2013) have introduced new space-time nonstationary covariance functions and they integrate them into epidemic propagation models informed by existing/historical flu epidemic datasets. Their approach may help to identify possible causal mechanisms associated with anticipating epidemic outbursts and identifying hotspots, controlling the speed of epidemic spread, and identifying interactions and relevant explanatory variables.

Bannister (2008) has reviewed how covariance estimation is used in generating forecast error surfaces in spatio-temporal atmospheric variational assimilation-based

---

[13]Environmental Protection Agency (EPA), `www.epa.gov/airtrends/aqtrnd95/pm10.html`

forecast models. This approach involves a background model error covariance matrix, **B**, and an observation/measurement error covariance matrix, **R**. The review finds that major issues of non-Gaussian behavior have been left largely unaddressed in operational Earth-system forecasting systems, highlighting the need to improve the representation of model variable and error covariance for improved forecasting of atmospheric (climate and weather), air pollution and human health, ecosystem state and dynamics and environmental health indicators. Furthermore, when forecasts (assimilated by integrating historical and near-real time (NRT) measurement data) produce states that are far from true states, the error covariance matrix is often "inflated" by a fixed scaling factor. Zheng et al. (2013) apply an adaptive approach to the ensemble Kalman filtering-based forecast method, estimating the inflation factor for forecast error covariance matrices for selected spatially-correlated data, further improving their accuracy and reliability.

There are many published university textbooks, popular readers targeting environmental scientists and other natural resource professionals, and reference books available spanning spatial statistics, hierarchical modeling, geostatistical analysis and geospatial mapping. Several excellent texts that provide further statistical details and discussion are: *Statistical Analysis of Environmental Spatio-Temporal Processes* by Nhu Li and James Zidek (2006), *Hierarchical Modeling for the Environmental Sciences: Statistical Methods and Applications* edited by James Clark and Alan Gelfand (2006), *Hierarchical Modeling and Analysis for Spatial Data* by Sudipto Banerjee, Bradley Carlin, Alan Gelfand (2nd Edition, 2014) *Statistics for Spatio-Temporal Data* by Noal Cressie and Christopher Wilks (2nd Edition, 2013) , and *Spatial Statistics and Spatio-Temporal Data* by Michael Sherman, and *Geostatistics - Modeling Spatial Uncertainty* by Jean-Paul Chilès and Pierre Delfiner (2nd Edition, 2012), *Multiple-point Geostatistics: Stochastic Modeling with Training Images* by Gregoire Mariethoz and Jef Caers (2014) (Mariethoz and Caers, 2014; Chilès and Delfiner, 2012; Cressie and Wilke, 2011; Sherman, 2011; Clark and Gefland, 2006; Li and Zidek, 2006; Banarjee et al., 2005).

3.3.3 Agent-based models (ABMs)

Agent-based models (ABMs) are computational, rule-based models that consist of agents (i.e., entities, computational "objects" or real-world organisms or organizational units) that sense, respond, interact, and adapt within a spatio-temporal environment generating self-organizing, emergent dynamics. ABM models were first introduced within the spatial sciences as cellular automata (CA) (i.e., cell-based automata or computable media for connecting information and processing) to model urbanization and land-use and land-use change, but have rapidly expanded in their application in the social, economic and environmental sciences (Torrens, 2010). Macal and North (2010) provide a review of this relatively new modeling approach that is growing in terms of capability, relevance, and interdisciplinary problem solving application. Farmer and Foley (2009), of the Sante Fe Institute, argue that the economy needs AGM modeling because current models assume that "...humans have perfect access to information and adapt instantly and rationally to new situations, maximizing their long-run personal advantage. Of course real people often act on the basis of overconfidence, fear and peer pressure." Furthermore, they argue that adding such behavior to equilibrium-based models introduces too much nonlinearity and complexity for them to handle, such that "The policy predictions of the models that are in use are not wrong, they are simply non-existent." Their vision is, "To make agent-based modeling useful we must proceed systematically, avoiding arbitrary assumptions, carefully grounding and testing each piece of the model against reality and introducing additional complexity only when it is needed. Done right, the agent-based method can provide an unprecedented understanding of the emergent properties of interacting parts in complex circumstances where intuition fails."

Agents may be *autonomous*, whereby they make their own choices and interact within their environment outside the influence of a leader or global plan, *non-autonomous* where they follow a leader or global plan, or *semi-autonomous* where they may follow a leader or act independently, depending on their behavioral state and/or spatio-temporal conditions. The ability of an agent to sense its environment, and to search, take actions, and learn from its interactions with others and its environment can vary. The attributes of an environment in terms of its size, topology, and changing conditions affect how uncertain agents are in relation to limitations, and in their abilities to detect cues and behavioral responses. Agents may be identical and/or non-identical (heterogeneous), having different attributes or state variables with different physiological and/or behavioral traits. Rules on how agents interact can be deterministic or stochastic or both, involving a varying degree of chance or probabilistic event-based outcomes. The number of agents can range from a few to millions, and their rules can vary from simple to very complex, resulting in very different event-based statistical distributions. Rules also govern the interdependencies and feedbacks of how agents interact with each other and their environment, and can be prescribed at the level of individuals, aggregations, or at the system-level. ABMs are especially useful for exploring the outcome of non-Gaussian, long-tailed and power-law distributional assumptions, non-stationary patterns, non-equilibrium dynamics and the identification of extreme and SOC states. They enable a bottom-up exploration of CAS process dynamics and observable spatio-temporal patterns and their output can be tested empirically. While ABMs offer the benefits of being very flexible models that capture great reality (i.e., a more natural description of a system), these same features pose additional challenges. For example, even though ABMs can generate distinct or testable statistical event-based distributions and spatio-temporal patterns, their numerical or computer simulation output may be difficult to describe analytically, because their degrees of freedom and level of heterogeneity of rules and state- and event-dependent interaction dynamics can make them analytically intractable (Brännström and Sumpter, 2005).

Integrating insights and findings generated from different applications can prove to be a challenge, as there are so many different forms of ABMs, depending on the CAS modeling objective/s and their mix of theoretical/heuristic/inductive or empirical/deductive-based assumptions. They can be characterized in general terms by the type of agents, their attributes, how groups or collections of agents are configured, whether they exist in a geospatial environment, how they interact (i.e., locally as geographic neighbors, via state-mediated or state-dependent contact or interaction rules, or non-locally according to a network topology), as well as the degree that agents learn. For example, ecologists build ABMs as spatially-explicit individual-based models (SIBMs) constructed with mobile agents as identifiable individual organisms or social groups in modeling animal populations, or non-mobile agents such as individual trees in modeling forest succession. In addition, for ABMs in geography or social science, agents may include not only human individuals, and social groups, but also institutions as an additional hierarchy as complex mixes of social groups (DeAngelis and Grimm, 2014; Helbing, 2012; Morozov and Poggiale, 2012; Bonabeau, 2002; Newlands, 2002). DeAngelis and Grimm (2014) distinguish ABMs in two main categories: pragmatic and paradigmatic. The first involves the "simulation of specific populations, communities, or ecosystems, usually with a management goal in mind," while the second involves gaining a "better understanding of the underlying causes of ecological phenomena" and more general ecological questions (DeAngelis and Grimm, 2014). A standard protocol (the 'Overview, Design concepts, Details' (ODD) general protocol[14,15,16]) has been devised for IBMs and ABMs to facilitate communication, a shared understanding of their characteristics, attributes, and potential, and encouraging their broader application (Grimm et al., 2010, 2006; Grimm and Railsback, 2005). As highlighted by Grimm et al. (2010) unlike object-oriented programming (OOP) that integrates the state variables of model entities (agents) and agent behavioral processes into a single unit, the ODD protocol separates properties and methods.

North (2014) have recently demonstrated that ABMs are "computationally optimal" for a common class of problems and that no other technique can solve the same problem computationally using less asymptotic time or space. Also, they are universal models such that any computational model can be expressed as an ABM. ABMs can be further characterized by their number of agents (N), the scheduler (that specifies and handles the time step and scheduling of random/nonrandom, fixed/variable agent activation in a simulation), the interaction space size, resolution, dimensionality and topology (e.g. one or more dimensional finite lattices, grids, network), and the emulator (i.e., referenced probability distributions for initialization and statistical learning linked with agent behaviors, responses etc.). An emulator is a stochastic process that serves as a representation of an ABM (or any simulator in general), which incorporates full probabilistic specification based on beliefs and knowledge, and the use of an efficient emulator can considerably speed up ABM simulations, especially as their run-times increase with respect to real-time (Wilke, 2014). In addition, ABMs differ in their tracking or logging scheme for recording simulation output for post-analysis, and level of interactive/non-interactivity with a graphical user-interface (GUI). A general theoretical formalism is needed for ABMs to more generally determine the runtime and memory space requirements for executing, testing, and validating ABMs of different size and algorithmic complexity.

---

[14]ODD Standard, `www.openabm.org/page/standards`

[15]Open Agent-based Modeling Consortium, `https://www.openabm.org/`

[16]Network for Computational Modeling for Socio-Ecological Science (CoMSES)

TABLE 3.2: Elements of the Overview, Design concepts, Details (ODD) general protocol for ABMs (Grimm et al., 2010).

---

1. Purpose
Problem definition, hypotheses
2. Entities, state variables, and scales
Types, variables, behavioral rules/strategies, spatio-temporal resolution/s
3. Process overview and scheduling
Updating of discrete/continuous state variables, ordering of changing conditions, event interactions
4. Design concepts
Basic principles, emergence, adaptation, objectives, learning, prediction, sensing, . interaction, stochasticity, collectives, observation
5. Initialization
Initial values of state variables—deterministic or stochastic
6. Input data
Process dynamics of variables driven by external and internal forcing, linked with observed data or statistical distribution information
7. Sub-models
Component/modular membership, model parameters, units/dimensions, reference values/settings, model identification/design, variable selection, cross-validation/independent validation support

---

This is especially true as many real-world problems do not have corresponding analytical solutions (Nisan et al., 2007). Bayesian Agents (BA), Probabilistic cellular automata (P-CA), Discrete Event System Specification (DEVS) and game-theoretic (GT) are several developing theoretical formalisms for ABMs (Ahmad and Yu, 2014; North, 2014; Merrick and Shafi, 2013; Hooten and Wilke, 2010; Ziegler et al., 2000). A framework for multi-scale ABM biomedical and systems biological modeling called SPARK (Simple Platform for Agent-based Representation of Knowledge) is specifically tailored for application to problems in this domain[17] (Solovyev et al., 2010). Integrated Development Environments (IDEs) and software packages with varying levels of functionality and specificity that are available to implement ABMs as "modular imperative architectures," include Swarm, Repast, MASON, NetLogo, and Starlogo. ABMs have also been coupled to Geographical Information Systems (GIS) and geospatial computing environments (Torrens, 2010).

How ABMs should be optimally configured and optimally selected for a given real-world problem relies on an understanding of the initial and boundary condition specification, statistical information criterion, suitable validation procedure, robust sensitivity analysis, scalability, and reproducibility—all important aspects requiring more investigation (DeAngelis and Grimm, 2014; Wilke, 2014; Haase and Schwarz, 2009; Piou et al., 2009; Bousquet and Le Page, 2004). There is also considerable opportunity for improving mean-field approximations, upscaling/downscaling and integrating structural covariance matching, causal-based tracking and optimization schemes into ABMs (Filatova et al., 2013; Helbing, 2012; Morozov and Poggiale,

---

[17]SPARK ABM, `http://code.google.com/p/spark-abm/`

2012). The distributed-based computing, large data, and complex statistical learning aspects of designing, testing, and applying ABMs to solve real-world problems has led to a human-machine symbiosis, collective mind approach called crowdsourcing of geospatial information (Fursin, 2013; Hammon and Hippner, 2012; Heipke, 2010).

Crowdsourcing can be defined as, "The acquisition of data by large and diverse groups of people using web technology that is transferred to and stored within an integrated database, computing environment and/or machine learning architecture for automated processing and analytics." GeoWiki is an example of a geospatial crowdsourcing approach for global land-cover mapping (Fritz et al., 2009). The reliance on expert and non-expert agents with varying knowledge and information contexts is one contentious aspect of this approach. An empirical-automated statistical learning paradigm (or "automated statistician"), which represents another human-machine symbiosis approach, aims to integrate AI/machine learning and data science[18]. According to Kevin Murphy (a Senior Research Scientist at Google)[19], this initiative aims to address two core statistical challenges, "...the first that current machine learning methods require considerable human expertise in devising appropriate features and models, and the second that their output is often difficult to understand and trust." He further explains that the "automatic statistician project from Cambridge aims to address both problems, by using Bayesian model selection strategies to automatically choose good models and features, and to interpret the resulting fit in easy-to-understand ways, in terms of human readable, automatically generated reports." In guiding future ABM modeling and integration work, Filatova et al. (2013) identify methodological challenges, societal issues and future prospects from the application of ABMs to real-world socio-ecological systems. The validation of ABMs needs to move beyond simpler comparisons between model predictions and observations, and incorporate more fully the sensitivity analysis of agent behaviors and other processes, rules and mechanisms. Broader empirical testing of such assumptions also needs to be conducted, which would aid in attributing uncertainty and error to specific components, while also identifying critical knowledge gaps to improve how ABMs model real-world CAS dynamics (Goldstein and Coco, 2015).

Action selection by both humans and other organisms/animals is the task of resolving conflicts between competing behavioral alternatives (Seth et al., 2011; Prescott et al., 2007). A greater understanding is needed of why humans (and animals) act irrationally, and how different suites of behaviors, their selection, and interrelated actions taken at the level of individuals drives and/or co-varies in space and time with broader, integrated impacts groups, organizations and societies. ABM models are able to explore such uncertain, complex realms. Recently, Yu and Huang (2014) have investigated human decision-making strategies under conditions of uncertainty using an ABM model whereby the consequence of different assumptions on statistical learning and decision-making are tested. This work is motivated by the cognitive science question as to whether humans match or maximize when faced with different reward levels under conditions of uncertainty (Yu and Huang, 2014).

---

[18] Automated Statistician, http://www.automaticstatistician.com/
[19] www.automaticstatistician.com

They raise the interrelated question as to, "why the brain would build sophisticated representations of environmental statistics, only then to adopt a heuristic decision policy that fails to take full advantage of that information." They find that humans maximize on a fine, trial-by-trial time-scale. This process continuously updates internal beliefs about the spatial-temporal distribution of potential visual cues and target locations, indicating that human visual search is neither random nor heuristic, but instead fluctuates stochastically, closely tied with the beliefs about stimulus statistics.

In general, a multi-objective optimization problem can be defined as (Caramia and Dell-Olmo, 2008),

$$\min[y^{(1)}(\mathbf{x}), y^{(2)}(\mathbf{x}), .., y^{(q)}(\mathbf{x})], q > 1, \mathbf{x} \in \mathbf{S} \tag{3.63}$$

where $\mathbf{y}$ is a multidimensional function and $\mathbf{S}$ is a set of constraints defined as,

$$\mathbf{S} = \mathbf{x} \in \mathbb{R}^m : h(\mathbf{x}) = 0, g(\mathbf{x}) \geq 0. \tag{3.64}$$

The space in which the objective vector belongs is called the *objective space*, and the image of the feasible set is termed the feasible or *attained set*, defined as $C = y \in \mathbb{R}^n : y(\mathbf{x}), \mathbf{x} \in S$. A vector $\mathbf{x}^* \in S$ is termed *Pareto-optimal* if all other vectors $\mathbf{x} \in \mathbf{S}$ have a higher value for at least one of the objective functions, $y^{(k)}(\mathbf{x})$, ($k = 1, ..., q$) and have the same value for all the objective functions. A given solution or optimum point $\mathbf{x}^*$ can be defined as weak or strict, whereby this point

- is a *weak Pareto* optimum, iff there exists no $\mathbf{x} \in S$ such that $y^{(k)}(\mathbf{x}) < y^{(k)}(\mathbf{x}^*)$ for all $k = (1, .., q)$

- is a *strict Pareto* optimum, iff there exists no $\mathbf{x} \in S$ such that $y^{(k)}(\mathbf{x}) \leq y^{(k)}(\mathbf{x}^*)$ for all $k = (1, .., q)$, with at least one strict inequality

Based on the above definition of Pareto optimums, a point $\mathbf{x}$ *dominates* another point, $\mathbf{x}^*$ if $y^{(k)}(\mathbf{x}) < y^{(k)}(\mathbf{x}^*)$ for all $k = (1, .., q)$. The Pareto front or surface is the set of all optimal solutions belonging to the feasible or attained set, C, as depicted in Figure 3.4 for $n = 2$. Caramia and Dell-Olmo (2008) provide a more in-depth discussion of recent work that applies tabu search, simulated annealing, and genetic algorithms to obtain Pareto-optimal solutions for multi-objective optimization problems. Stepwise reduction in uncertainty in multi-objective optimization problems using Gaussian process emulators (GPEs) have also been applied to reduce of the volume of the excursion sets below the current best solutions (Pareto set) via a stepwise reduction scheme (Picheny, 2014).

From the equation for a general stochastic process (refer to Equation 3.31), where trend functions, $f(\mathbf{x}, t)$ are decomposed into linear terms with fixed coefficients, $\boldsymbol{\beta}$ the *emulation* of a single computer response Y, can be modeled according to,

$$\mathbf{Y}(\cdot) = \mathbf{f}(\cdot)^\mathsf{T} \boldsymbol{\beta} + \mathbf{Z}(\cdot) \tag{3.65}$$

where $\mathbf{f}(\cdot)^\mathsf{T} = (f_1(\cdot), ..., f_p(\cdot))$ is a vector comprising the trend functions, $\boldsymbol{\beta}$ is a

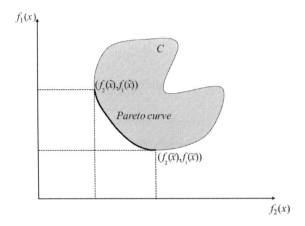

FIGURE 3.4: Depiction of the Pareto front of the attainable set of feasible solutions, C, for a two-dimensional, multi-objective optimization problem.

vector of unknown coefficients and $\mathbf{Z}$ is a Gaussian process (GP) with zero mean with a known *covariance* kernel distribution. Now assume that GP models are fitted to each of the objectives, $\mathbf{y}$ and that n experiments have been performed, yielding,

$$\begin{aligned} \mathbf{X}_n &= (\mathbf{x}_1, \mathbf{x}_2, ..., \mathbf{x}_n) \\ \mathbf{Y}_n &= (\mathbf{y}(\mathbf{x}_1), ..., \mathbf{y}(\mathbf{x}_n)) \end{aligned}$$

and $\mathbf{A_n}$ as a set of corresponding measures for this set of multi-objective events. Let $\mathbf{X}_n^*$ as a current approximation or current Pareto set $(m \leq n)$ in relation to an actual Pareto set consisting of non-dominated points. $\mathbf{Y}_n^*$ is then the *current Pareto front*. The GPE method decomposes a multi-objective space using a tessellation Ω_i, $i \in (1, ..., I)$, with cells of size $I = (m+1)^q$, where $\cup_{i \in I} \Omega_i = \mathbb{R}^q$ and $\cap + i \in I \Omega_i = \emptyset$. Here, $\emptyset$ denotes an empty set (Picheny, 2014). Each cell is a hyperrectangle, defined as,

$$\Omega_i = (\mathbf{y} \in \mathbb{R}^q | y_{i-}^{(k)} \leq y^{(k)} \leq y_{i+}^{(k)}), \quad k = (1, ..., q). \tag{3.66}$$

Thus, a cell Ω_i *dominates* another cell Ω_j if *any* point in Ω_i dominates *any* point in Ω_j. After n steps or iterations, assuming pairwise independence of $Y^{(1)}, ..., Y^{(q)}$, the probability that a cell Ω_i contains $\mathbf{Y}(\mathbf{x})$ is,

$$\begin{aligned} p_n^i(\mathbf{x}) &= P[\mathbf{Y}(\mathbf{x}) \in \Omega_i] \\ &= \prod_{k=1}^{q} p_n^{i(k)} = \prod_{k=1}^{q} \left[\Phi\left(\frac{y_{i+}^{(k)} - m_n^{(k)}(\mathbf{x})}{s_n^{(k)}(\mathbf{x})} \right) - \Phi\left(\frac{y_{i-}^{(k)} - m_n^{(k)}(\mathbf{x})}{s_n^{(k)}(\mathbf{x})} \right) \right], \end{aligned}$$

where $\Phi(\cdot)$ is the cumulative distribution function (CDF) of the Gaussian distribution. Siirola et al. (2004a) discuss the single objective, ϵ-Constraint method, weighted-metric and goal programming, and evolutionary algorithm methods to compute Pareto fronts. They provide an ABM, multi-agent approach to computing the Pareto front that demonstrates dramatic effects in terms of algorithmic speedup and scaling efficiency obtained by allowing a mixture of such computational algorithms to *collaborate* to find a Pareto front solution. They define two new ABM evaluation metrics. The first, is the *normalized, non-dominated hypervolume error* as a measure of the closeness that a set of non-dominated solutions represents the actual Pareto-optimal front, sensitive to the distance between a reference and solution set and diversity of solutions across the front. The second metric is the *fraction of clusters identified* that provides a measure of the extent to which a solution set identifies real-world, physically meaningful solutions across the Pareto surface. These metrics are particularly useful for characterizing the performance of an ABM in different real-world applications, and extending the set of existing statistical metrics on the distance, proximity or closeness to a Pareto front, the spacing, deviation and spread of solutions along the front, or both (e.g., hyperarea) (Giannakis et al., 2012; Koutsourelakis, 2008; Siirola et al., 2004a).

Heuristic methods for multi-objective optimization for large combinatorial and distributed-control problems include simulated annealing (SA), tabu search, swarm intelligence (e.g., particle, ant-colony and hybrid optimization algorithms), genetic algorithms (GAs), and ABM methods for computing Pareto optimal fronts (Oremland, 2011; Panigrahi et al., 2011; Siirola et al., 2004b). Pareto optimization is a multi-objective optimization method that focuses on not just a single objective solution, but a suite of solutions that lie along what is termed the Pareto frontier, defined as the boundary whereby the solutions cannot be further improved upon in terms of a single objective, without sacrificing another (Oremland and Laubenbacher, 2014). Both cognitive and statistical theory are very relevant to the design and interpretation of the ABM model simulation output because of the reliance on behavioral strategy and action selection, learning, and other aspects of computational neuroscience in these models (Epstein, 2007; Prescott et al., 2007).

The optimization of action selection in ABM models, within the context of multi-objectives, employing heuristic methods holds great importance and relevance for sustainable development (Oremland and Laubenbacher, 2014). Arneth et al. (2014) in a study of global models of human decision-making for land-based mitigation and adaptation assessment under climate-change uncertainty, propose that ABMs should be integrated with dynamic global vegetation models (DGVMs) to better represent land-use change (LUC) processes and understand emergent patterns. This is seen as crucial because current global-scale modeling tools need to better account for social/societal learning and human behavioral responses to environmental change.

Ng et al. (2011) have developed an ABM of BMP involving crop production costs, water, nutrient inputs and other decisions, coupling it to a watershed scale hydrological-agronomic model to investigate the potential impacts of agent decision-making on stream nitrate load, and the driving influence of conventional commodity markets and crop trading/pricing, inclusion of second-generation biofuel crops and carbon pricing and allowances. Their ABM compares a deterministic farmer with rational behavior and perfect information with a stochastic farmer, where pricing, costs, yields, and weather are stochastic and known in the context of underlying probability distributions and simulations that mix both these types of farmer "agents." They find that, generally, the farmer agents who adopt new practices are those who interact more with other farmers and who are less risk averse (i.e. more adaptable and less cautious), and are able to efficiently adjust their expectations and have higher confidence in crop yield forecasts (i.e., slower to reduce their forecast confidence). Valbuena et al. (2008) highlight that land use/cover change is the results of cumulative individual agent farmer decisions and the definition of agents is not often based on real data, ignoring the inherent diversity of farmers and farm characteristics in rural landscapes. They provide an empirical approach to defining the farmer agents.

A further example of an ABM that has been developed to better understand agricultural economics is a multi-agent and cellular automata type that uses heterogeneous farm-household models to capture social and spatial interactions in an explicit way—it provides greater resolution on local decisions, such as the introduction of improved land use practices as a farm investment decision to cope with natural resource degradation (Berger, 2001). Rather than treating farmland investment just as an exogenous process of technological change, which could mislead policy recommendations and decisions, use of this ABM model enables one to determine the degree to which it should be an endogenous process and be considered in policy decisions—particularly relevant when land-use trade-offs between competing uses and transitions (agricultural, forestry, conservation) or when it is converted to urban use. Spatially-explicit (or simply spatial) ABMs have also been applied to analyze the expansion of urban areas into farmland and forests using the "Evoland" model (Guzy et al., 2008). Likewise, a spatial ABM to understand the impact of animal search behavior, in addition to stocking density, in determining the spatial structure of grazing systems, has implications for sustainable grazing management (Swain et al., 2007).

3.3.4 Game-theoretic (GTM) models

Game theory is the study of strategic decision-making (Merrick and Shafi (2013) and references therein). Simple games with a few players and a few possible strategies can quickly become more complex as more players participate; they exhibit both rational and irrational behaviors and the space of possible strategies they can take, expands (Galla and Farmer, 2013). Many games have at least one stable outcome or "Nash equilibrium." Optimal outcomes, payoffs or "utility" of individuals in simple games may not mimic reality, and maximizing individual benefits may not match an evolutionary stable strategy (ESS) involving resource use and rate of depletion. In human and animal societies, mixed strategies involving trickiness (e.g., mistakes, misunderstandings, iterated processes, changing payoffs) and cooperation in real-world CAS can emerge under certain conditions and circumstances as individuals learn and adapt in respond to perceived risks or threats to avoid collective failure (i.e., Tragedy of the Commons) (Santos and Pacheco, 2011). So, for example, avoiding the effects of climate change may be framed as a public goods dilemma, in which there is significant risk of future losses. Moreover, in real-world games, the utility of switching to a new state or strategy in the future given may change according to an individual's current state (Akiyama and Kaneko, 2000).

Game-theoretic models (GTMs) often rely on equilibrium and stationary assumptions designed to produce dynamical equilibria, and while substantial attention is given to the robustness of equilibria to changes in model parameters (i.e., sensitivity analysis), rarely are such models independently validated for their robustness to different behavioral assumptions (de Marchi and Page, 2014). This contrasts with ABMs, where optimality, equilibrium and rationality are not assumed, but instead are replaced by non-equilibrium and heuristic assumptions (Hadzikadic et al., 2010). Both ABMs and game-theoretic models may specify optimal goal, objective functions as payoff-maximizing, survival or evolutionary "fitness" functions and impose constraints on agent rules, and initial and boundary conditions. Also, "agents" in ABMs are termed "players" in GTMs. In ABMs, an emphasis on initial or starting conditions being set as realistic as possible is made, based on prior probability distribution functions and expert knowledge. For example, logical rules can be identified through participatory democratic decision-making, educational gaming, role-playing and intelligent tutoring approaches involving relevant real actors within a real CAS (van Bilsen et al., 2010).

Complex, dynamic games with mixed strategies are in a higher complexity class that are solved in so-called non-deterministic time complexity or "polynomial time" (NP-complete complexity class) (Nisan et al., 2007; North, 2014). An algorithm is said to be of polynomial time if its running time is upper bounded by a polynomial equation in the size of the input for the algorithm (i.e., $T(n) = O(nk)$ for some constant k), and it is associated with several complexity classes as the time bound of computational devices (Turing machines) (i.e., P, NP, ZPP, RP, BPP, BQP)[20],[21].

Game-theoretic modeling may provide insights that we can mimic in promoting international environmental cooperation and establishing strong consensus and agreement following not a top-down approach at the international and national scale, but instead a bottom-up, polycentric or integrated approach that involves multiple institutions, parties/groups and individual actors, self-organized on a local and regional-scale. Game-theory changes ones perspective that all players must co-operate all the time, but instead, it shows that cooperation can emerge from both cooperative and non-cooperative approaches, albeit the time-scale, possible actions, and the costs that are available may be dependent on which strategy dominates over time. In this way, a bottom-up approach can re-engineer institutional pathways, build adaptation capacity, creating strong local institutions that help with environmental enforcement, while facilitating the emergence of wider cooperation, better communication, and equality in sustainability decision-making, especially when uncertainty is high (Vasconcelos et al., 2013; Tavoni et al., 2011).

Nonetheless, as Tavoni et al. (2011) highlight—inequality is a key constraint on meaningful cooperation, as often those who are deprived and poor are not willing to compensate for inaction and the lack of a strong commitment by the privileged and rich and a willingness to take up a sizable share of a collective burden. Instead, success may hinge on social groups and key promoting institutions to eliminate

---

[20]Complexity classes, `http://en.wikipedia.org/wiki/Complexity_class`
[21]NP-complete problems, `http://en.wikipedia.org/wiki/NP-complete`

inequality over the course of a public goods game, whereby the rich players signal willingness to redistribute early on. In addition, there are different types of sanctioning to discourage and deter non-cooperative behavior in climate agreements that can emerge (rather than being imposed) under the bottom-up approach, involving multiple, local/regional-based organizations and institutions that follow an integrated approach defined through a broader national and international consensus. Such a shared development pathway may be the best one to discover feasible solutions and establish the most optimal ones, especially as there are many competing and inter-related, real-world complex, global dilemmas (Finus and Caparroós, 2015; Tavoni, 2013; Vasconcelos et al., 2013).

4

Real-World Application of Integrated Methods

CONTENTS

> "It is baffling, I must say, that in our modern world we have such blind trust in science and technology that we all accept what science tells us about everything - until, that is, it comes to climate science....So, thank goodness for our young entrepreneurs here this evening, who have the far-sightedness and confidence in what they know is happening to ignore the headless chicken brigade and do something practical to help....We have spent the best part of the past century enthusiastically testing the world to utter destruction; not looking closely enough at the long-term impact our actions will have."

H.R.H. Prince Charles, Prince of Wales, A speech by HRH The Prince of Wales at the Unilever Sustainable Living Young Entrepreneurs Awards, Buckingham Palace (30 January, 2014)

> "During the course of Hubble's working life such grand ambitions were thrown into doubt when the inability of the telescope to function properly turned it into the most ridiculed object on Earth, with one Senator anointing it with the derogatory tag of a 'techno turkey'....The scientific mega-project of Hubble captured a complexity of organizational life, where management constantly endured the spectre of failure."

Mark Egan, Leicester University, Hubble, trouble, toil and space rubble: The management history of an object in space (Egan, 2009)

4.1   Perspective

Some perspective and context on integrative research studies and the design of ecosystem experiments in needed before further discussing real-world applications of integrated methods. Knowledge domains and lines of inquiry/fields do not advance at the same pace, with some developing much more slowly due to a myriad of constraints, while others accelerate. Inquiry may provide an incremental step toward a solution to a defined problem, may solve it, or may solve a different undefined problem. In turn, while new knowledge may generate new insights, they may not be immediately useful, in terms of being "actionable" or "operationalizable" in society. Furthermore, even if it is deemed scientifically or technologically useful, it may be judged or evaluated not based on its scientific merits, such as whether competing theories are considered, testable hypotheses, realistic assumptions, with sufficient robust support from data and model-based inference etc. Instead, scientific and technological (and other) solutions may be evaluated relying on a broader SES context and relevance driven by social attitudes, perceptions, and socio-political optics, which may not be logical or rational, and may even frame and introduce hidden biases of a research study or design from the onset of scientific inquiry.

A broader integration of knowledge for accelerating scientific and technological advancement in addressing global problems facing both ecosystems and humanity is urgently needed. Such integration needs to include alternative perspectives, addresses multiple (complementary and competing) knowledge domain/field constraints and uncertainties, and eliminate biases. Mowery et al. (2010) argue that scientific and technological solutions to global climate change must differ in the traditional design of "big-push" centralized programs exemplified by the Manhattan Project, Apollo space, or Hubble telescope programs. Instead they must involve decentralized development and deployment. Such operative programs have broad requirements, such as predictability and reliability, realizable societal or social benefits, cost-effectiveness, deployment ease, and operational flexibility. This latter requirement recognizes and anticipates that scientific and technological solutions will continue to evolve and improve in their effectiveness and reliability, and that social, economic, and environmental knowledge may become distributed or fragmented. They may also develop between a diverse set of knowledge stakeholders and potential users and interest groups within the public and private sector. They also need to be integrated and shared to ensure that both supply- and demand-side public policies are well-informed by existing knowledge and are not only developed, deployed and adopted when and where they are needed, but are also responsive and adaptable as well.

Real-world problem solving recognizes the inherent interconnectivity of human-ecosystem interactions and the resilience of ecosystems and human societies to transform, adapt, and respond across a hierarchy of spatio-temporal scales and processes, supported by CAS theory. It confronts uncertainty by acknowledging that some events and/or trends may be directly/indirectly controllable and measurable, while others are not. It also confronts fundamentals in our approach to scientific research inquiry by acknowledging the usefulness of different research study philosophies, interdisciplinary scientific and non-scientific qualitative and quantitative knowledge and perspectives. As Creswell (2013) further details, scientific research involves making claims about what is knowledge (ontology), how we know it (epistemology), what values go into it (axiology), how we write about it (rhetoric), and the processes for studying it (methodology). This leads to the four main schools of thought about knowledge claims being postpositivism, constructivism, advocacy/participatory, and pragmatism.

Postpositivism is a deterministic, reductionist perspective grounded in the scientific method but that challenges the absolute truth (i.e., positivism) of knowledge. It reflects on issues affecting experiments and causes that influence outcomes through analytical metrics, observer selection, and behavior. *Constructivism* considers that subjective meanings are socially constructed based on background sets of experiences, social interactions and through negotiated, collective discussion. Historical context and cultural norms dictate how individuals construct their intents, interpretations and meanings, whereby meaning is varied, multiplied and made more complex.

The *advocacy/participatory perspective* is practical and collaborative and challenges the structure of laws and theories of postpositivism. It frames problems strongly in terms of their constraints, and considers research inquiry as intertwined with politics and a political agenda linked with a power or influence-seeking groups. Knowledge is integrated based on social constructions, but is strongly influenced by decisions on future outlook, current priorities, social influences and issues (e.g., inequality, marginalized groups), social-based metrics and media (e.g., democratic consensus, trust, transparency, networked communication and information technology) and social motivations advancing an action agenda for change. Knowledge and truth is what works at a given instance or application setting and within a reality that is not independent of the mind.

Pragmatism views knowledge arising, "out of actions, situations, and consequences rather than antecedent conditions (as in postpositivism). There is a concern with applications - "what works"—and solutions to problems. Instead of methods being important, the problem is most important, and researchers use all approaches to understand the problem." In this way, pragmatism is not committed to any one system or perspective, and encourages both the selection and integration of quantitative and qualitative methods, techniques, and procedures.

Many other factors come into play in real-world ST problem solving such as monitoring and experimental cost, logistics, computational efficiency, infrastructure and management support. For example, a recent report released in 2015 by the US National Academy of Science reviews past achievements and future ocean science priorities for 2015–2025. This review was a part of a decadal survey to find a new strategy for optimizing investments to advance knowledge in the most critical and/or opportune areas of investigation, to provide guidance on the most effective portfolio of investments realistically achievable to address the most significant priorities, accompanying realized reductions in ocean monitoring infrastructure (e.g., research vessel fleets, ocean drilling, global moorings, coastal arrays, regional cabled ocean observatories) and Oceanographic Technology and Interdisciplinary Coordination (OTIC) support. Similar real-world challenges exist in many other countries and ecosystems (i.e., terrestrial, urban), in addition to ocean ecosystems, that critically need novel integrated solutions to enable cost-effective operational support involving integrated sensing, geospatial intelligence and planning, geocomputation and predictive analytical capabilities (NAP, 2010).

Real-world research and development environments typically involve interdisciplinary teams with a broad set of knowledge and expertise with the capacity to grapple with as realistic assumptions as possible to obtain robust solutions with minimal risks, realizing that often inquiry involves learning and risks, whereby unanticipated intermediate steps and solutions are required. They often seek to benefit from latest ST advances to address their objectives, while attempting to address challenges and provide useful solutions that can be staged, improved and refined, and packaged into products or tools with maximal anticipated societal impact or other audience influence, gain (e.g., profit) and life-time. Alongside such pursuits, real-world problem solving integrates a broad array and hierarchy of limitations and constraints linked with dedicated effort/labor and time, data quality and quantity, theoretical and practical understanding and recommendations, existing methods and models, and their estimation and predictive power, operational requirements such as the need to adhere to established laws, regulatory limits across multiple jurisdictions, international protocols, guidelines, industrial process designs, data collection and sharing protocols, institutional information technology and knowledge, risk management, and business reporting requirements.

Real-world problem solving involves knowledge inquiry that is a mixture of the postpositivism, constructivism, participatory and pragmatism perspectives. These mixtures involve sequential, concurrent or transformative strategies of research inquiry. They may, for example, couple existing or pre-determined approaches with newer emerging methodology. They may address both open- and closed-ended questions, assimilate multiple forms of data, and explore either a subset or a more complete set of considerations/possibilities. An area that needs strengthening is guidance for integrating new statistical insights, and the knowledge gaps for each of these research strategies. This would greatly aid the integration of knowledge across disciplines in a more robust, consistent and congruent way, enabling better a wide array of operational needs, such as rapid impact appraisals, medium-term outlooks and/or long-term scenarios, to better inform decision making and guide government policy.

The mixture of data, methods, approaches and research perspectives are closely aligned with the challenges of the applied statistician—in understanding a system in its breadth and depth sufficiently to devise new models and ST approaches, and identifying and testing assumptions (e.g., open boundaries, non-Gaussian, non-stationarity). In addition, statistical work is needed to develop and adapt knowledge frameworks, models, tools and techniques integrating interdisciplinary knowledge content and context effectively, to translate interdisciplinary knowledge across teams, and communicate with stakeholders and practitioners who may have less statistical or other expertise and knowledge, or time to learn it.

4.2 Ecosystem Observation, Experimentation, and Modeling

A guiding or prescriptive, integrated framework for ecosystem research studies of real-world problems involving SES considers all of the essential statistical elements of pattern (i.e., heterogeneity, interaction, scaling) and process (i.e., stochasticity, asymmetry, adaptation) (refer to Figure 1.6 in Chapter 1). Because it mixes theory, assumptions, methods, data, and research perspectives, it may require single and/or multiple, and short and/or Long-Term Experiments (LTEs), each potentially involving relative extents and degrees of observation, intervention, and manipulation. Cook et al. (2004) reviews the importance of "learning to roll with the punches" involving adaptive experimentation in human-dominated systems and the importance of measuring and manipulating feedbacks between ecological processes and activities of resident humans, particularly within urban environments. They provide a summary of major advantages and disadvantages they identify associated with widely-applied experimental approaches (i.e., small-plot, large-plot/landscape, watershed-scale, chronological sequences and natural, historical ecological experiments, simulation and modeling, social surveys, adaptive management and experimentation) (Table 4.1). Carl Walters discusses how the management of resources, in itself, is an adaptive process, in which regulatory, conservation and enhancement actions can be treated as deliberate experiments having uncertain outcomes (Walters, 2002). Trade-offs in control and replication, spatial scale, resources required, and human activities and perceptions are considered. For example, small-plot experiments exclude the consideration of many ecosystem and management control processes, but permit extensive replication, strict controls and factorial treatments.

Large-scale experiments involving the pairwise comparison (i.e., nested sampling) between catchment units within ecosystems, whole-lake or broad monitoring of open ocean responses to nutrient enrichment and pollution, may contribute significantly in the future to the scientific basis of ecosystem management (Carpenter et al., 1995). Nonetheless, while large-scale ocean or terrestrial catchment/landscape experiments consider ecosystem processes more broadly, the ability to manipulate, control, and extensively replicate to increase statistical confidence may be impractical. To address such needs, adaptive experimentation via controlled experimentation allows for replication to better gauge causality and uncertainty, but requires *a priori* correlation analysis modeling. The different approaches thus rely on the integration of data and models to different extents when generating new insights and knowledge. As discussed by Eberhardt and Thomas (1991), ecological and environmental research experiments usually do not meet the stringent criteria for modern, controlled experimental design, whereby sub-sampling is often mistakenly substituted for true replication, and sample sizes are too small for adequate power in tests of significance, and argues for hypothesis testing involving a mixture of sampling procedures. In real-world applications, not just one, but several of these experimental approaches may be utilized and integrated to better measure and understand how ecological and human impacts vary over spatial and temporal scale.

A conceptual model of the functioning of the Earth system in time scales of

TABLE 4.1: Advantages and disadvantages of experimental approaches that trade-off control and replication, spatial scale, resources required, and a consideration of human activities and perceptions (see Cook et al. (2004) for further examples and references therein).

Approach	Advantages	Disadvantages
Small-plot experiments	Allows extensive replication, strict controls, and factorial treatments involving several predictor variables	Small spatial scale excludes consideration of many processes. Human element cannot be simulated easily.
Large-plot, landscape experiments	Controls and some replication possible. Allows consideration of processes operating at scales greater than a few metres	Extensive replication much more difficult. Manipulation may be impractical or unethical.
The watershed approach	Allows study of whole ecosystem functioning. Before-after comparisons and untreated 'reference' watershed useful.	Replication often not possible. Strict controls difficult. Manipulations may be impractical or unethical.
Chronological sequences and natural experiments	Opportunism. Observation of different stages of sequence at once. Active manipulation can be avoided. Good for social-anthropological studies	Controls generally lacking. Confounding factors. Key events may gave occurred in past and critical information missing.
Historical ecology	Allows consideration of many factors and large spatial and temporal scales. Human and non-human feedbacks clearly elucidated.	Post hoc explanations with limited prediction ability. Limited ability to replicate.
Simulations and modeling	Allows consideration of many factors and large spatial scales. Can require limited field-based infrastructure.	May not include key factors. Possible lack of "realism." Human behavior is difficult to predict.
Social surveys	High ability to replicate and comparability. Can include large spatial and temporal scales. High correlation potential.	Low potential to identify causality. Generally lacks non-human feedbacks.
Adaptive management	Embraces uncertainty and human/non-human feedbacks. Policy and management benefits.	Mostly limited to resource management questions and intervention by institutions. Difficult to generalize findings.
Adaptive experimentation	Embraces uncertainty and human/non-human feedbacks. Controlled experimentation elucidates causality and allows for higher rates of replication. Social, policy, and management benefits.	Requires *a priori* correlation analysis or modeling. Spatial scale limited by logistic constraints. Ethical considerations.

decades to centuries used in framing many global change programs of research, along-side an overview of Earth system processes operating across spatial and temporal scale, is provided in Figure 4.1. This illustrates the real-world diversity of processes operating at different spatial and temporal scales and the importance of integrating an understanding of scaling when developing integrated research frameworks, conducting experimental ecosystem studies, and linking data to models. Physical and ecological phenomena tend to line up, approximately, along the diagonal direction in such space-time scale diagrams, although variations may sometimes be large (Innes, 1998). Ecosystem research studies using an adaptive design and comprising a mixture or set of controlled, quasi-control, and uncontrolled experiments, can potentially yield a broad set of quantitative and qualitative data, knowledge and insights (see Figure 4.2). Complete randomized block designs (RCBs) investigate the effects of a single factor, with test subjects grouped into blocks and all treatment levels applied to each block in random order to eliminate interaction effects via unmeasured confounding factors. Replication in such experiments enables quantifying main, interaction and random effects. When there is more than one factor in an experiment, test subjects are randomly assigned to all combinations of factors and all treatment levels—more specifically, latin-square (LS) sampling designs (i.e., with two sources of variation), factorial experiments involving more than one (i.e., multiple) factors, and split-plot sampling designs when an observer cannot control or adjust one factor as often as another are utilized. For example, for a factorial experiment with 3 factors (f), each with 4 levels (L), there are 64 possible combinations (i.e., L^f), whereas for one that has a combination of a two-, three- and four-level factor has 24 combinations (i.e., $2 \cdot 3 \cdot 4$).

Fractional factorial designs, unlike full factorial designs, consider only the most important factors for studies with a large number of factors/treatments where the number of combinations in a full factorial design is too high to be logistically feasible and/or where it is necessary to account for interaction-effects between sampling units or blocks and reduce confounding between factors. Here, response surface modeling (RSM) is often used by assuming *a priori* functions that describe the possible response of main or interaction effects (e.g., linear, quadratic, cubic) at different treatment levels. Nested sampling involves sampling at two different locations to account for location-specific conditions or treatments. For experiments involving more treatments than blocks, with and without sub-sampling, and despite whether the blocks are fixed or random, *in-complete* randomized block designs are typically used. While *randomized-controlled* trials (RCTs) and split-plot experiments aim to maintain spatial and temporal homogeneity in sampling, *uncontrolled* experiments adapt to different levels of observed spatial and temporal heterogeneity. Split-plot experiments are a type of nested-factorial experiment, whereby the levels of one factor are assigned at random to large plots, and the large plots are then divided into smaller "sub-plots" or "split plots," with the levels of a second factor randomly assigned to these small plots within the large plots.

LTEs first began in agriculture (1843–1856 in Rothamsted UK) at the field scale, to better understand crop rotation, fertilizer and meteorological effects on

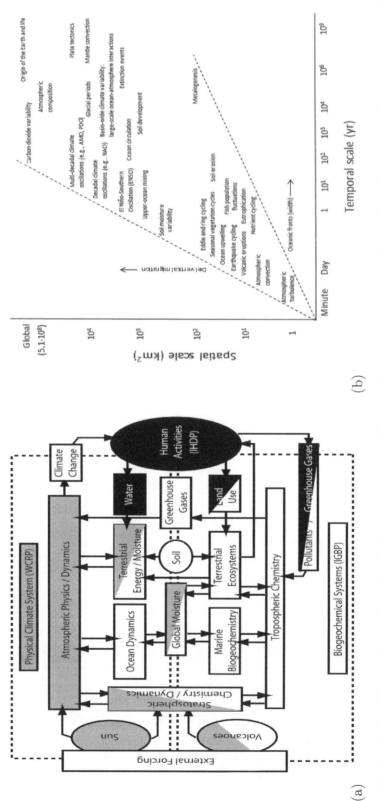

FIGURE 4.1: a) A conceptual model of the functioning of the Earth system in time scales of decades to centuries (Mooney et al., 2013; NASA, 1988), b) Earth system processes operating across spatial and temporal scale (adapted from Edwards et al. (2010)).

crop yield and production (Silvertown et al., 2010)[1], included within a UK's long-term environmental observation framework activities (UKEOF)[2]. Such controlled experiments rely on measuring differences between treatments and control, and such analyses can be greatly aided by previous studies on design sensitivity—such as analyses that identify leading characteristics (e.g., spatial covariation via cross-sectional surveys) or leading temporal variation (e.g., time-sequences using cohort-longitudinal prospective studies; causal reasoning using case-control retrospective studies). Nonetheless, when randomized experimentation is infeasible or unethical, such that there is no intervention, observational studies are necessary (Rosenbaum, 2010). While events and treatments are not under the control of observers or investigators in observational studies or experiments, more broadly natural experiments are those that involve anticipated impacts or outcomes that are not under an observer's control. Whole-ecosystem experiments that have been undertaken are discussed further by Fayle et al. (2010) in the context of tropical forests involving an intensive use of resources to manipulate ecosystems to generate an ecosystem-wide response, or impacts that can be reasonably extrapolated to the entire ecosystem. While such experiments may offer considerable advances in understanding interactions, feedbacks, and synergistic effects within ecosystems, there is a need for a consistent, shared yet flexible experimental design/sampling plan for such ecosystem-scale experiments. Zidek et al. (2014) have investigated the effects of preferential sampling (i.e., involving response-biased sampling) linked with case-control experiments associated with the selection of site locations within environmental monitoring networks (e.g., urban air pollution monitoring sites for detecting noncompliance with air quality standards and informing human health public policy). They model site selection bias associated with the black smoke (BS) particulate matter measure data (624–193 sites, 1970–1996) of the UK National Air Quality Information Archive[3], with results that show how selection bias leads to an overestimation in the number of sites exceeding regulatory concentration limits. They highlight that selection bias can accumulate over time and can require increasingly greater adjustments, especially as forcing attainment of stringent environmental standards can entail large economic costs.

In general, when the random allocation of treatments is not possible, it is assumed that treatment variables and covariates are not affected by how treatments are allocated, and to reduce selection bias due to unmeasured factors or misclassification, stratification, matching, covariate and weighting adjustments are used in assigning treatments, and unmatched treatment allocations are discarded. For example, propensity score matching is a statistical technique in which a treatment case is pairwise matched with one or more control cases based on each case's propensity score. For binary response variables, logistic parametric regression is often used. Alternatively, other statistical techniques, such as the non-parametric Classification and Regression Tree (CART) method can be used.

---

[1]UK Environmental Change Network (ECN),
`www.rothamsted.ac.uk/environmental-change-network`
[2]UK Environmental Observation Framework (UKEOF), `www.ukeof.org.uk`
[3]UK AIR, `http://data.gov.uk/publisher/uk-national-air-quality-archive`

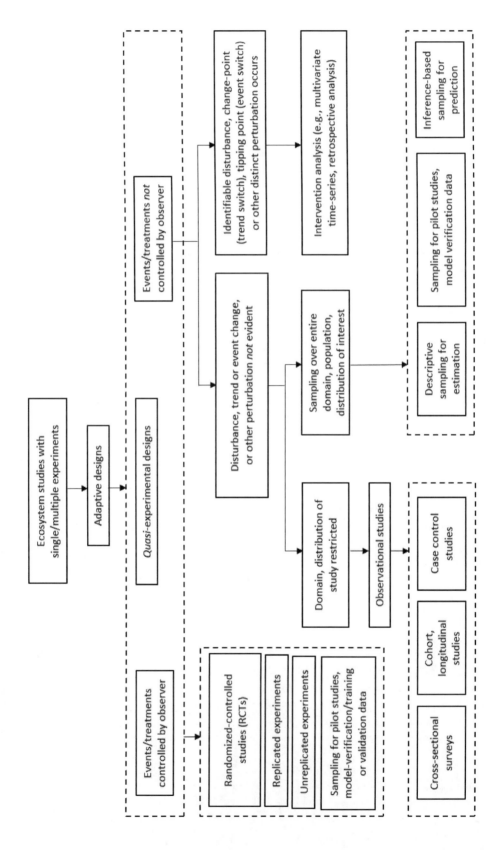

FIGURE 4.2: Classification of ecosystem and environmental research studies for assessment, modeling and forecasting.

The *propensity score* is a statistical measure with a central role in observational studies that investigate causal effects (Rosenbaum and Rubin, 1983). It is a function of the conditional probability of an exposure given a treatment (i.e., survival probability) in association with a given set of observed (continuous-valued) covariates, $p(s,t) = P(Z = 1|s,t)$, defined in terms of the observed spatial (s) and temporal (t) covariates following a Bernoulli (i.e., binary) probability distribution function. With the logit transformation (i.e., the logistic link function for selection bias as a binary response variable $[0,1]$, or alternatively the quantile function of the logistic distribution), $r(s,t) = log(\frac{p(s,t)}{1-p(s,t)})$, the *propensity score*, representing the log of the odds ratio of an event. Such a response can be represented as a linear, additive function of the space-time variables with observational error,

$$log\left(\frac{p_i(s,t)}{1 - p_i(s,t)}\right) = \zeta_o + \sum_{j=1}^{n} \zeta_j Z_{ji} + \sum_{j=1}^{n} \sum_{k=1, k \neq j}^{m} \zeta_j \zeta_k Z_{ji} Z_{ki} + \epsilon_{i,t} \quad (4.1)$$

for the ith individual, based on constant linear coefficients, ζ, and comprising contributions due to n main ($\zeta_j Z_{ji}$) effects, mn interaction ($\zeta_j \zeta_k Z_{ji} Z_{ki}$) effects, and random error effects (ϵ_i), where $\epsilon_i \sim N(0, \sigma_i^2)$ independently and σ_i^2 are the additional model parameters to be estimated. Assuming time-invariant error ($\epsilon_{i,t} = \epsilon_i$) and a pooled-error variance ($\sigma_i^2 = \sigma_o^2$, $\forall i$) greatly simplifies the model. Alternatively, it can be extended to include age-structure ($Z_{.,l}$ for age l), time-varying covariates ($\zeta_{.,t}$) and time-varying random-error effects ($\epsilon_{.,t}$). A multinomial (i.e., rather than the current binomial propensity score) can also be considered, whereby event or probability outcomes can have three or more possible event types or impact risk levels (Hosmer and Lemeshow, 2000). Matching on a single binary response variable tends to produce treated and control groups that, in aggregate, are balanced with respect to observed covariates; however, individual pairs that are close on the propensity score may differ widely with respect to specific covariates. The logit model is also frequently used to test hypotheses in statistical ecology that address animal mark-recapture, occupancy, distance sampling, count, geospatial movement tracking/telemetry, and relative abundance population estimation issues and conservation challenges. King (2014) discusses how continuous covariate values are significantly more difficult to deal with than are the discrete-valued covariates, due to the observed data likelihood that is an analytically intractable integral over all possible values of covariate variables (i.e., rather than a discrete sum of discrete values). A Bayesian modeling approach for defining an underlying model of time-varying covariates in fully modeling the data likelihood has also been applied to investigate complexity in the survival probability of Soay sheep (i.e., that reside on the human-inhabited island of Hirta in the St. Kilda archipelago west of mainland Scotland), associated with covariation due to time, age, sex, individual phenotypes, genotypes, resource abundance, climate, and population size (King et al., 2006). Here, independent normally-distributed priors on the ζ coefficients along with an inverse gamma prior on random-effects was assumed. In addition to different forms of the logit model linked with its use in different epidemiological and ecological applications, there are different score matching techniques that also exist (e.g., exact, coarsened-exact, sub-classification, nearest-neighbor, optimal, genetic) that utilize this model form (Randolph et al., 2014). While randomization can be a more powerful approach for balancing covariates than matching on an estimate of the propensity score, it assumes unobserved covariates are *balanced*, or that both treated ($Z = 1$) and control ($Z = 0$) subjects having the same propensity score, $r(s,t)$, also have the same distribution of observed covariates. In comparison, while matching balances observed covariates via scoring, randomization balances not only observed covariates, but also unobserved covariates, and other potential responses.

While the difference between treatment, and control groups is modeled as a dependent variable in matching when treatments cannot be randomly allocated, while in randomization experiments, it is the independent coefficients ζ that measure the difference between treatment and control groups across blocks with randomly allocated treatments. Such RCB experiments are solved by standard analysis of variance (ANOVA) with statistical significance assessed by p-values, as the probability of rejecting the null hypothesis when it is true (type I or "false positive" error).

The probability that the observed statistic occurs by chance alone is determined by setting the p-value< 0.05 (i.e., $\alpha = (1 - 0.95) = 0.05$) (i.e., at a 95% significance level, the threshold value that p-values are measured against). The null hypothesis is then rejected in favor of the alternate that there is a significant difference between treatment and control. *P*-values and significance measure the probability of a hypothesis given the data. Also, while a hypothesis can be highly significant, with replication and reproduction, the same hypotheses can be determined to not be very probable. In a Bayesian approach, this can be verified by ensuring that a high $p(D|H)$ occurs only when $p(H|D)$ is high, such that an experiment is properly or well-structured. Furthermore, while the statistical significance based on p-values can be strong, they can be associated with correlations that have negligible actual effects, and minimal significance in the real world.

As Nuzzo (2014) clarifies, "When the UK statistician Ronald Fisher introduced the P value in the 1920s, he did not mean it to be a definitive test. He intended it simply as an informal way to judge whether evidence was significant in the P values." This is because for any statistical test, there is a trade off between *specificity* (i.e., "hit" or correctly rejected rate) as the proportion of negatives which are correctly identified as negatives (($1 - \alpha$), true negatives, TN), and *sensitivity* (i.e., "power," "recall" or the correctly identified rate) as the measure of the proportion of positives which are correctly identified as positives (($1 - \beta$), true positives, TP). These measures compare with the false alarm/type I error as the incorrect rejection of a true null hypothesis (α probability of false positive, FP), and the miss/type II as incorrect acceptance of a false null hypothesis (β probability of false negative, FN). Figure 4.4 provides an illustration of inferential hypothesis testing relative to a given value of a statistical criterion, such as a critical value of a test statistic, or threshold value of an environmental indicator. This illustration identifies the α (*FP*), ($1 - \alpha$) (*TN*), β (*FN*), ($1 - \beta$) (*TP*) probabilities associated with areas under probability distributions (normally-distributed) for a null H_o hypothesis, $p(H_o|D)$ and alternative hypothesis, H_a, $p(H_a|D)$, given evidence or data, D. Type I and II errors change depending on the degree of overlap between these probability

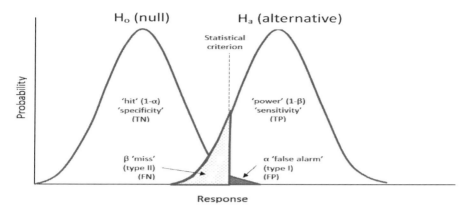

FIGURE 4.3: Illustration of inferential hypothesis testing relative to a given value of a statistical criterion, such as a critical value of a test statistic, or threshold value of an environmental indicator, identifying the α (FP), $(1-\alpha)$ (TN), β (FN), $(1-\beta)$ (TP) probabilities as associated with areas under probability distributions (normally-distributed) for a null H_o hypothesis, $p(H_o|D)$ and alternative hypothesis, H_a, $p(H_a|D)$, given evidence or data, D. Type I and II errors change depending on the degree of overlap between these probability distributions and their bias and skew that can change differently in response to environmental noise.

distributions and their bias, and skew that can change differently in response to environmental noise. To obtain more informative, predictive analytics, a *confusion* matrix for any statistical test or experiment can be specified, with associated precision, accuracy, sensitivity, specificity, discovery, prediction, omission, odds, and the likelihood statistical measures (Figure 4.4).

As highlighted by Rosenbaum (2010), the design of observational studies is crucial to ensure the quality and reliability of evidence, as statistical or other analytical analysis and modeling methods, whereby "design anticipates analysis." It may not be possible to salvage data if a sampling has been poorly designed or design aspects have been overlooked. Such aspects may pertain to how scientific hypotheses have been framed, the context of a study and whether assumptions are invalid, decisions on what data is collected and potential measurement gaps, where and how data is collected, biases in measured covariates, strategies, and tactics to limit uncertainty caused by covariates not being measured. David Powers and colleagues showcase the importance of using unbiased statistical measures, and encourage making greater use of confidence intervals over point-values of significance. This aim requires research studies to move beyond relying on traditional significance testing measures (e.g., p-value, sensitivity, specificity) for evaluating significance. They indicate the importance of using fully independent, repetitive sampling data from real-world experiments, where possible, for cross-validating, rather than relying heavily on statistical bootstrapping, Monte Carlo (MC)-simulated data. In particular, significance and confidence testing of machine-learning algorithms should involve cross-validation with data from real-world AI experiments (Powers and Atyabi, 2012; Powers, 2011).

Total population (P)	Condition positive (CP)	Condition negative (CN)	Prevalence = Σ CP/ Σ P	
Test outcome Positive (TOP)	True positive (TP)	False positive (FP) (type I)	Positive predictive value (PPV), Precision = Σ TP / Σ TOP	False discovery rate (FDR) = Σ FP / Σ TOP
Test outcome Negative (TON)	False negative (FN) (type II)	True negative (TN)	False omission rate = Σ FN / Σ TON	Negative predictive value (NPV) = Σ TN / Σ TON
Accuracy (AAC) = (Σ TP + Σ TN) / Σ P	True positive rate (TPR) (recall, sensitivity)= Σ TP / Σ CP	False positive rate (FPR) (fall out) = Σ FP / Σ CN	Positive likelihood ratio (LR+) = TPR / FPR	Diagnostic odds ratio (DOR) = LR+ / LR-
	False negative rate (FNR) = Σ FN / Σ CP	True negative rate (TNR) (specificity) = Σ TN / Σ CN	Negative likelihood ratio (LR-) = FNR / TNR	

FIGURE 4.4: The confusion matrix (or precision-recall chart) and its associated statistical derived measures of precision, accuracy, sensitivity, specificity, discovery, prediction, omission, odds and likelihood, applicable to problems involving a null with multiple-competing alternative hypotheses (Powers, 2011).

Large complex datasets obtained from LTEs, and monitoring observations at ecological sites and across ecological networks in both marine and terrestrial environments help to, "...assess the rate and direction of change, to distinguish trends from short-term variability, and to determine effects of infrequent, yet extreme events and time lags in response" (Edwards et al., 2010; Peters, 2010). These approaches can jointly lead to improved hypotheses, testing, knowledge and insights regarding processes and their interaction that underly observable patterns to ensure society does not suffer from environmental myopia in framing conservation, restoration, and manipulation decisions within ecosystems (Silvertown et al., 2010). Nonetheless, integrative methods are crucially needed to reliably identify the single or multiple-interacting processes giving rise to such patterns, to process, blend and contrast different datasets against alternative theories, and/or to determine spurious associations with no causative relationship between pattern and process. Moreover, integrative methods may contribute significantly to reducing data, model, and knowledge dimensionality to ensure experiments and monitoring networks can be maintained over the longer term with less cost and provide knowledge that is more robust, reliable (causal-based not just association-based), accessible, informative, and useful to a wide range of interdisciplinary user communities (McDonald-Madden et al., 2010; Peters, 2010).

Often, models used in many scientific disciplines are selected based on familiarity and convenience; model design choices can be fairly arbitrarily related to the choice of explanatory variables, functional forms, error/uncertainty assumptions. Furthermore, the search for significance across a range of possibilities (i.e., distinguishing real-world effects from the effects of chance alone) can be very minimal, with effects weak and inconsistent; important confounding variables and other sources of bias are often outside of a research study, assumed negligible or simply ignored. One is often more concerned with loss of statistical power than distortion of a p-value (Freedman, 2009). In turn, often the difference between model precision and accuracy is overlooked or misinterpreted—*precision* is exactness of detail, the degree of refinement with which an operation is performed or a measurement taken, while *accuracy* is correctness, the degree of conformity of a measure to a true value or standard (i.e., it includes *biases*).

There is also a great need for wider awareness of what constitutes good model-development practice, particularly when building integrative models linking qualitative and quantitative knowledge across a broad range of disciplines with different extents of model-building expertise (Jakeman et al., 2006). One often wants to explore more factors than one can control in field or laboratory experiments, and such factors are varied on a wider scale often in numerical experiments than in real-world experiments. Increasing the number of factors and their ranges results in more and larger interactions. For a model of f factors, the total number of terms in a decomposition of variance is as high as $2^f - 1$, termed the "curse of dimensionality" and requires an integrated approach to developing and evaluating models. This requires also one to apply efficient input-output model representations via global sensitivity analysis (Cariboni et al., 2007; Saltelli et al., 2000; Rabitz et al., 1999). According to King (2014), "Integrated models are designed to analyze multiple data

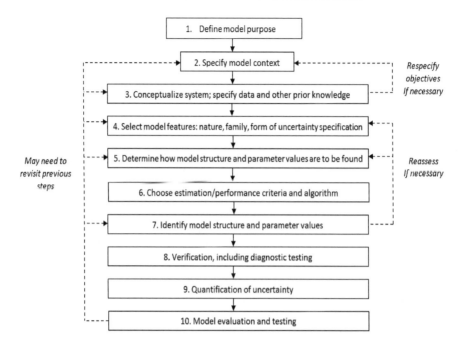

FIGURE 4.5: Iterative steps for developing and evaluating environmental models (Jakeman et al., 2006).

sources simultaneously within a single robust analysis, permitting information to be borrowed (or pooled) across datasets. Thus, integrated models can be viewed as multivariate observation processes that correspond to the different forms of data collected, with the system process describing the different underlying biological processes. The different observation processes may observe either the same underlying states or different states that evolve simultaneously over time but that depend on the same demographic parameters."

This leads us to ask what is a useful model and how can they be efficiently and reliably constructed? Jakeman et al. (2006) outline ten iterative steps for developing and evaluating models of environmental space-time processes and systems, shown in Figure 4.5. They argue that each of these modeling steps should be open to critical review and revision linked with the three key feedback loops (i.e., re-specify, re-assess and re-visit), in consort with end-users, to ensure that models are purposeful, credible, and that they sufficiently integrate perspectives, available data, and prior knowledge. Special focus needs to be given to these important feedbacks in the model development and in tuning research studies to adapt to changing purposes and contexts against consistent baselines provided by longer-term experimentation and observational study. Integrated models rely very strongly on causal-based reasoning and assumptions, yet in many cases, sufficient and reliable statistical knowledge on the interrelationship between key system variables is not available.

In general terms, a useful ecosystem model is then one that embodies the following critical feedback steps and can be re-specified, re-assessed and re-visited,

- Contends with both space and time
- Integrates interdisciplinary perspectives and knowledge
- Is supported by a measurement (i.e., experimental or observational) design methodology
- Contends with *non-stationary* spatial covariance
- Can incorporate multi-variate *responses*
- Handles missing, mismatched, systematic uncertainties in data
- Can be reliably upscaled/downscaled to systems of different size and functional/structural complexity
- Uses open and transparent assumptions, and their consequences for model design are easily identifiable
- Can generate insights, not previously known or measured
- Can be shared, enhanced, adapted, re-engineered for changing purposes and contexts.

4.3   Geospatial Intelligence and Planning

The global geopolitical landscape is changing dramatically as our natural resources are depleted. As communities and societies continue to become more complex (i.e., closely networked, with interdependent development and decision making across regional and national borders and boundaries), so too has the need for greater foresight, planning, and intervention in transboundary economic and environmental issues (Dühr et al., 2010). This is, in part, due to more integrated risks and uncertainties on the benefits and drawbacks of cooperation and competition, within the broader context of depleting resources, increasing needs for ecosystem services and increasing integrated environmental and socio-economic risks due to climate and global change. *Geography* involves the study of human use and interaction with the Earth and the identification of spatial-temporal variation in natural and human processes, applying both natural and social science principles to analyze and interpret change in the environment, such as globalization and cultural diversity, resource management, climate change and environmental hazards. *Geospatial intelligence and planning* specifically refers to the ability to describe, understand, and interpret processes (data, information), products (knowledge), and organizational aspects (functions, decision requirements), *to anticipate or predict human impacts of an event or action within a spatio-temporal environment. Intelligence* can be considered a core, integrative function that operates comprehension, invention, direction and censorship/criticism (Sternburg, 1982; Varon, 1936); however, Susan Epstein, in a recent position paper highlights the urgent need for understanding "collaborative intelligence" (CI) and its application in guiding and enabling a continued integration of human and artificial (machine) intelligence, and their collaboration in solving complex tasks in our sustainable future. CI aims to combine human and machine sensory modalities, communication/dialog streams, analytical, reasoning, critical judgment methodologies and abilities. Accordingly, in the future, "Successful CI could establish a synergy between people and computers to accomplish human goals" (Epstein, 2015).

Geostatistics is a branch of applied statistics that focuses on providing a quantitative description of observations of geological and ecological variables that are distributed in space, or in time and space. It refers to a specific set of models and techniques stemming from its early development and application in mining, meteorology and the theory of stochastic processes and random fields. It offers the practitioner, a probabilistic methodology for quantifying spatio-temporal uncertainty (Chilès and Delfiner, 2012). While still a relatively new term, *geocomputation*, refers to a critical component of geospatial intelligence and planning involving the complex, integrated geospatial techniques that combine geographical information science (GIS), spatio-temporal modeling, ML and AI computer techniques and are closely linked with geography and its integration across the social sciences (Paegelow and Olmedo, 2008). In particular, Sundaresan et al. (2014) in *Geospatial Technologies and Climate Change* showcase the practical application of geocomputation and associated geospatial technologies in providing more rapid monitoring and assessment, improved disaster planning and reliable early-warning systems in tackling regional climate mitigation and adaptation problems in different countries—from landscape flooding, landslides, dust storms, coral reef health, fishing zone advisories, and coastal vulnerability due to sea level rise. Geospatial data, models and technologies embedded within ICT will likely be, in the future, a major driver of research, innovation, growth and social change, and it offers huge potential for more widespread improvements in environmental and socio-economic sustainability performance under climate change, especially when accompanied by systemic changes in human social behavior (OECD, 010a). However, especially in developing countries, the use and adoption of ICT (e.g., smallholder farms) while aided by access to micro finance loans, membership in local community and industry associations, government awareness campaigns, and wealth, is strongly hindered when individuals and communities are poorer and farther away from major urban centers (Kiiza and Pederson, 2012).

New research directions outlined by the US National Geospatial-Intelligence Agency in 2010 are aligned with cross-cutting themes, which aim: 1) *to go beyond fusion with aggregation, integration, and conflation* with the goal of removing the effects of data measurement systems and facilitating spatial analysis and synthesis across information sources, 2) *to advance forecasting as an operational technique* to anticipate outcomes, trends, or expected future behavior of systems, integrating improved models of space-time dynamics and statistics, to enable its use in decision making, 3) *to create new paradigms for conveying certainty (and uncertainty) linked with new operational technologies* that enable modeling, representation, simulation, and the anticipation of behaviors and activities (i.e., of both individuals and the social networks to which they belong) and an integrated knowledge of evolving risks, 4) *to advance participatory sensing and the understanding of "human terrain" behavior* through the use of scalable sensor networks (fixed and mobile devices), so as to enable both experts and resource stakeholder communities to jointly participate

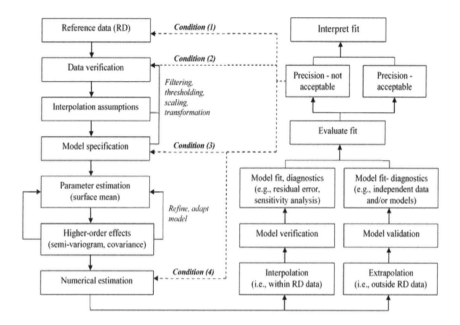

FIGURE 4.6: Adaptive interpolation and refinement of geospatial data with four feedback loops to ensure acceptable precision is achieved when verifying and validating geostatistical models (Newlands et al., 2011). RD refers to reference or training data.

in gathering, analyzing, sharing, and visualizing local knowledge and observations, and, 5) *to advance visual analytics and human-in-the-loop simulation capabilities* to enable integrated analytical reasoning aided by interactive visual interfaces for synthesizing information and insights from "massive, dynamic, ambiguous, and often conflicting data." (NAP, 2010)

4.3.1 Geospatial verification and validation

An adaptive methodology for the spatio-temporal model-based interpolation is shown in Figure 4.6, distinguishing model verification versus validation, and it includes several, critical feedback loops for modeling real-world, spatio-temporal phenomena (Newlands et al., 2011). *Verification* involves testing the assumptions of a model (e.g., normality, stationarity, isotropy) using the best quality data available, whereas *validation* involves assessing how well a model describes data, employing cross-validation comprising leave-one-out (LOOCV) procedures or using independent validation data (Haining, 1990). Validation typically involves more extrapolative assumptions and testing outside of a training dataset, while verification involves interpolative assumptions and testing specific to a training dataset. Interpolation and extrapolation may change depending on the spatial and temporal scale of available data. Also, by inter-comparing competing geostatistical models, one can estimate model structural uncertainty, in addition to parameter uncertainty.

After selecting a subset of reference or training data (RD), geospatial data is verified alongside any interpolative or extrapolative assumptions. An appropriate model is identified or specified. Parameters associated with main interaction and any higher-order effects are estimated. Interpolative and extrapolative assumptions are next distinguished, to guide verification and validation testing in evaluating and interpreting how well a model describes the data. When precision is not acceptable, re-verification of the reference dataset needs to be undertaken to consider both its quantity and quality, alongside its spatial and temporal scales, and how it has been selected. This involves also reconsidering how filtering, thresholding, scaling and transformation algorithms have been applied. Based on this, a model may need to be re-specified and this may involve the use of a different numerical solver based on a consideration of local versus global extremum in a problem's solution or parameter space.

While the iterative and feedback steps in modeling may be seen as rudimentary, clear, and familiar to many from an idealized perspective, from the real-world perspective assumptions change radically. There are often many unforeseen barriers to data and knowledge sharing and set requirements on what needs to be collected, evaluated and compared to address a specific or set of priority, data and/or knowledge gaps. Real-world research studies, typically involving both data collection and modeling components, include only parts of these iterative steps due to project management constraints (i.e., time, funding, personnel/human resource expertise). Crucial modeling feedback loops may not be accomplished, even in well-funded, multi-year research studies, due to inherent complexities coupled with project management constraints that are involved. The collection of new data is often given higher importance due to its real, tangible benefit by many individuals, studies, and funding institutions; however, modeling, that aims to verify, validate, and generate new insights by exploring a broad, integrated range of possibilities within the context of available data, with its abstraction, complexity, and inference is less directly applicable or "industrializable." Often, completing the modeling cycle (i.e., 10 iterative steps with feedback loops) is left for other studies to pick up where the previous ones have left off, with time lapses involved, lack of continuity in research data, knowledge and tools. New studies often have no access to the data used previously or the interdisciplinary expertise that existed in former teams. Alternatively, feedback loops may be simply overlooked based on justifications of lack of sufficient data or a lack of sufficient scientific support for renewed or continued investigation.

It is crucial to determine not only when model assumptions no longer hold and/or predictive power and accuracy of a model breaks down. It is also crucial to determine when assumptions lack verification or validation support, especially important when they are re-applied to address real-world problems having different contexts and accuracy needs. However, far too often, research studies tend to emphasize critical insights regarding model success, rather than how and when a model fails or loses its ability to inform. Too often, in the past, it has been difficult, if not impossible, to directly verify new findings against existing, peer-reviewed, published modeling results because their raw and manipulated data and code is not made available/shared. There are also many gaps in understanding and expertise that arise, due, in part, to a lack of global sharing protocols (i.e., data interoperability) between individuals, teams, institutions and countries. However, this situation is changing, aided by geographic and geospatial knowledge and context, as many scientific communities are now creating new standards for research, ensuring that data products, analyses and insights are accessible and workable openly across data and software-based distributed architectures.

The Global Gridded Crop Model Inter-comparison (GGCMI), a sub-project of AGMIP, is an excellent recent example of such innovation in research study that integrates data and modeling components, being framed to deliver its important scientific outcomes and deliverables so that they can evolve and develop during and well beyond its three-year project lifetime (Elliott et al., 2015). Analogous to long-term experimentation, to undertake integrated modeling in the real-world context, great continuity (i.e., not necessarily consistency in what is collected, or in analytical methodology) across data and modeling steps is needed. This would enable the completion of crucial steps in model development involving verification, validation and inter-comparison of datasets and model methodologies. It would also enable completion of crucial feedback loops necessary to improve the reliability of models, as data becomes available, and modeling methodologies and supporting technologies advance over time.

4.3.2 Geospatial technologies

A broad range of geospatial technologies now exist to support geographic research, geostatistical modeling and geocomputation (Table 4.2). These technologies can be combined into "integrated sensing" solutions so as to integrate data and model forecasts of field-scale soil moisture variability across global cropland areas for improved agricultural decision-support (Phillips et al., 2014). The Global Terrestrial Monitoring Network is an example of a sensing network used to globally monitor and better understand ecosystem processes that regulate the uptake and release of CO_2. This network also verifies and integrates data across major biomes from a network of ground-based, station/tower measurements, and satellite remote-sensing observations to profile and clarify larger spatial and temporal patterns and validate ecosystem models (Running et al., 1999). In addition to airborne, earth observation (EO) satellite and satellite-constellation remote-sensing platforms for global to regional monitoring and surveying the ocean and terrestrial ecosystems, many new opportunities are emerging that integrate new technology into geospatial assessment frameworks, for integrated sensing, and the use of geostatistical models and geocomputation algorithms to provide geospatial intelligence and planning in making resource use and management decisions. Examples of such technologies are 1) semi-automated, mobile surveys conducted by humans using backpacks of integrated smart sensor systems (e.g., thermal, laser, or spectral) and mobile/cell phones, or animals employing an integration of airborne surveying, shipboard tracking, and satellite-based archival tagging technology[4], 2) fully-automated surveys by intelligent robots (i.e., consistent observers) to conduct automated *in-situ* sensing,

---

[4]For example, in ecosystem-based management of highly-migratory fish populations, like Atlantic bluefin tuna, integrating shipboard tracking, satellite archival tagging and airborne surveys

and 3) unmanned aerial systems (UAS), Autonomous Unmanned Aerial Vehicles (or AUAV) (UAV) or drones, remotely piloted vehicles (RPV), remotely-operated aircraft (ROA), remotely-controlled helicopters, and hoverbikes[5]—all requiring an improved knowledge of intelligent sensor design and calibration, data integration and new integrated methodologies for geospatial operational planning, multi-tasking, task and goal changes, or adaptation.

Optical sensing systems measure reflectance from objects from the visible and infrared portions of the electromagnetic spectrum and have been applied in regional mapping of land-use and crop types, with increasing application to the finer field-scale (e.g., in precision agriculture involving tracking soil and leaf wetness, vegetative greenness, crop yield, disease presence, as indicators of crop development and health). Optical sensing, however, is of low quality in the presence of clouds and haze. While generally applicable to irrigated cropping systems, within arid-regions they are less useful, being of low image quality. This is because such regions often have heavy rainfall during the crop growing season and fewer cloud-free days. For this reason, an integration of optical with active and passive radar sensing are currently being explored (Forkuor et al., 2014) Such integration aims to benefit from UASs offering a practical, inexpensive substitute and/or complementary technology for rapidly assessing field conditions, impacts and potential risks. High-resolution satellite remote-sensing imagery is more costly and has critical impediments due to revisit times in image acquisition, the presence of clouds, and the frequency of storms and rain events affecting image reliability during crop growing seasons (Kouadio and Newlands, 2014; Zhang and Kovacs, 2012). UAS data can, however, be costly, unlike freely available, open data archives of satellite data. Moreover, UASs generate large amounts of data (e.g., 250 mosaics stitched together for a single agricultural field).

Future technological improvements in satellite sensors may compete at spatial and spectral resolutions comparable to UASs. For example, remote sensing (RS) Synthetic Aperture Radar (SAR) technology enables operational monitoring of soil moisture and its variability more rapidly, across large spatial regions around the world, with repeated sampling at different spatial scales and resolutions.. Satellite instruments can detect and measure the properties of emitted microwaves that are reflected back from surfaces and objects (i.e., "backscattered"). Imaging radars can have different polarization configurations, where polarization describes the orientation of the electric field component of an electromagnetic wave. Microwave remote sensing can be passive or active. The active type receives backscattering reflected from a transmitted microwave incident from the surfaces or objects on the ground, whereas the passive type just receives microwave radiation emitted from the surface and objects on the ground. Active microwave data, therefore, depends on several natural surface parameters such as the dielectric constant and the surface roughness. The complex dielectric constant is a measure of the electric properties of surface materials that are highly dependent on soil moisture content. Soil roughness is a relative concept depending upon wavelength and incidence angle. We refer in-

---

(Newlands and Porcelli, 2008; Gutenkunst et al., 2007; Newlands et al., 2007, 2006, 2004; Newlands, 2002; Newlands and Lutcavage, 2001).

[5]Hoverbikes, `www.hover-bike.com/MA/`

terested readers further to a comprehensive background of RS principals and radar polarimetry (Henderson and Lewis, 1998).

Sensors are autonomously powered devices capable of sensing signals from their surrounding environment. Sensing, communication (transceiver), computation (processing) and power are the four basic functions that a sensor network requires. Sensor networks provide observations at discrete point locations. Zecha et al. (2013) provide a general taxonomy for classifying sensor experimental platforms in terms of the area of mobility and application, data analysis method and off-line/online processing, level of data fusion, software architecture, passive/active sensing activity, transmission method, platform size, propulsion and the degree of automation. Often the most distinguishing aspect is the degree to which a sensor network engages in reactive (real-time or near-real-time (NRT)) and predictive (geospatial-based) sensing. Sensor measurement or "source" observational nodes are routed to "sink" nodes that gather the data from source nodes. *Wireless* sensors are limited in their power storage, computing, and communication abilities (power efficiency is of major importance), but enable otherwise impossible sensor applications, such as monitoring dangerous, hazardous, or remote areas and locations, reducing maintenance complexity and costs. Peer-to-peer *mesh networks* allow all the nodes in the network to have a routing capability and enable additional/new autonomous nodes to self-assemble (Wang et al., 2006). Communication between source and sink nodes is transmitted via a wireless medium (i.e., radio, infrared or optical). Wireless sensors can use different communication technologies that differ in their communication medium, operating frequency and bandwidth: IrDA (infrared light over short-range, point-to-point communications), wireless personal area network (WPAN) such as Wi-Fi, Bluetooth, ZigBee (short range, point-to multi-point communications), multi-hop wireless local area network (WLAN) (mid-range) to GSM/GPRS technology and CDMA cellular phone networks (long-distance communication) (Rehman et al., 2014; Wang et al., 2006).

Sensor-based agriculture is typically termed precision agriculture (PA) , "Smart Agriculture," Variable Rate Technology (VRT), Precision Farming, Global Positioning System (GPS) Agriculture, Farming-by-Inch, Information-Intensive Agriculture, and/or Site-Specific Crop Management depending on its specific goals and emphasis (Rehman et al., 2014). In general terms, it is "the application of geospatial techniques and sensors to identify variations within agricultural fields and to deal with them using alternative strategies" and it integrates sensing technologies to assess the current and future, potential condition of crops (i.e., stress, nutrient, yield potential), automated soil, water and air samplers that sense environmental conditions (i.e., soil moisture, soil compaction, nutrients, crop disease), geospatial guidance for seeding, fertilization, irrigation and spraying, weed control, and harvesting based on quantity and quality (Zhang and Kovacs, 2012). Sensor networks are also applicable in managing forestry and rangeland ecosystems, for example, in detecting and monitoring forest wildfires, pathogen detection, monitoring changes in canopy cover and differentiating rangeland vegetation (Salamí et al., 2014).

As McBratney et al. (2005) discuss, the development and adoption of PA is struggling, within developed and developing countries, due to the lack of proper sustain-

TABLE 4.2: Technological advancements for improved integrated sensing (adapted from Phillips et al. (2014)).

Technology	Advantages	Issues	Description
Smart sensors	Anomaly alerts, power saving methods, smart monitoring.	Alerts sent when problematic issues are detected, when observations cross specific thresholds.	Event alerts or triggers to power up/down certain sensors, precision monitoring.
Fully wireless	Better potential sensor distribution in the field.	Sampling not constrained by cables. Less exposed cables.	Autonomous nodes that communicate through a central data station. Nodes small/cheap enough to minimize damage or the cost to replace.
Mobile (UAV) sensing	Less human labor and sensing networks.	Adaptable sampling scheme based on local site and changing environmental conditions.	Robot teams with the ability to adaptively sample an area, controlled by a single field operator.
Operational remote-sensing	Daily regional and weekly local monitoring.	Daily regional-scale monitoring to identify hotspots. Ability to conduct frequent fine-resolution monitoring in an area experiencing extreme conditions.	Regional monitoring for frequent monitoring of hotspots and extreme conditions using satellites.
Integrated analysis platforms	Assimilate data from multiple sensors for analysis and modeling.	Integration of multiple sensor datasets for analysis. Standards to ensure data interoperability.	Integration and processing environment to process soil moisture and crop water consumption data.
Sustainability decision-support tools	Policy and best management recommendations.	Further integration of soil moisture datasets with ecosystem, crop and climate models.	Scenario modeling to investigate future crop adaptation, food production forecasting and environmental change.

ability tools—decision-support systems (DSSs) for integrating data from sensing technologies within models and environmental assessment frameworks for implementing sustainability decisions and actions, especially at the farm-scale. Several critical issues are identified, in order of importance, as: 1) a lack of appropriate criteria for economic assessment, 2) an insufficient recognition of temporal variation in the context of the collection and delivery of images in a timely manner, data resolution, image interpretation and data extraction issues, 3) a lack of a whole-farm ecosystem approach whereby all fields and components (e.g., soil, crop, livestock, water, energy, air etc.) on a farm are managed in a precise way, 4) the need for methods to assess crop *quality* (e.g., fiber length, thickness, strength, color, acidity/pH, protein, oil, and sugar content), 5) to expand product tracking and traceability involving electronic labeling as part of food product supply-chains and sustainability information for end consumers, and 6) the need to establish data protocols for PA technologies to enable farmers to comply with more rigorous environmental regulations and its potential use in environmental audits. In addition, they indicate the need for a stronger, cooperative intelligence approach that restores trust as a basis for the interaction between governments, farmers, and consumers in focusing on, "the means to achieve environmental objectives," not just, "the environmental goals to be achieved."

Reliable protocols and standards for the production of on-farm geospatial data (e.g., yield potential, yield risk, yield loss) maps, assessing environmental condition, damages and risks beyond geolocation position system (GPS) guidance (e.g., yield gaps, soil moisture levels, soil erosion, and the nutrient leaching losses, management zone delineations) are needed. Protocols must be sufficiently robust while also meeting concurrent, specific data needs and information requirements of an increasingly broadening variety of on-farm experiments, analyses and impact assessment methodologies (Zhang and Kovacs, 2012; McBratney et al., 2005). Candiago et al. (2015) have recently demonstrated how a UAV can collect multi-spectral data to obtain multiple vegetation indices for better assessing crop condition (e.g., canopy biomass, adsorbed radiation, chlorophyll content etc.). Similar approaches using sensing technology can enable and guide farmers to adjust and optimize the timing, amount and placement of nutrients, water, and herbicides when and where they are needed. However, while the use of new sensing technologies presents many new opportunities for assessing, modeling, and forecasting, it also introduces many new challenges for how such data and models are integrated, verified, and validated.

Agricultural crop irrigation scheduling, disaster response and water management during droughts or flooding extreme events, soil erosion and pollution monitoring making use of hydrological models, all require reliable predictions of daily and field-scale soil moisture. Phillips et al. (2014) discuss the feasibility and application of wireless sensor networks in measuring *soil moisture* for improved field, farm and regional-scale agroecosystem modeling and decision-support. *In-situ* sensor monitoring networks provide ground-truthing, or validation of soil moisture products derived from passive microwave satellite remote-sensing instruments (Champagne et al., 2014, 2010), with recent work exploring reactive monitoring, where the network responds to important events, such as precipitation, by increasing the monitoring frequency for the event duration (Cardell-Oliver et al., 2004). Considerable research efforts have been directed to measuring soil moisture with microwave remote sensing, as microwave bands X, C, and L are able to penetrate the soil surface to various depths. Soil moisture is measured with microwave remote sensing by modeling backscatter contributions of the soil surface and sub-surface. Inversion modeling of the scattering mechanisms allows for backscatter amplitude to be related to soil water content. Active microwave satellites generally have longer revisit times than passive microwave satellites but can acquire finer resolution data. Quantitative fine scale mapping of soil moisture using backscatter modeling techniques can be hampered due the presence of noise, inherent to radar data. Furthermore, active remote sensing requires ground truth data on surface roughness and crop type to calibrate and validate the methods of measuring soil moisture remotely over a particular area (Hosseini et al., 2015). The ability to only monitor at discrete time instances limits the applicability of remote sensing for agro-climatic monitoring. Remote sensing could potentially provide spatial information that is not captured by *in-situ* monitoring, but it lacks the temporal resolution that such localized monitoring networks can achieve.

In the future, wireless sensor networks could be extended into *mobile sensing platforms* that operate autonomously or in conjunction with several other units to collect critical ecosystem (e.g., soil, water, air) data for a particular site. Mobile robotic sensing platforms could have the desired sampling scheme pre-programmed before field deployment. The robotic units would communicate with each other in order to efficiently collect data from the desired sampling nodes. Envisioned components of a mobile soil moisture sensing platform would be a soil access tube, movement vehicle, radio transmitter and receiver, global positioning system and inertial measurement unit. Radio transmitters and receivers would be used for communication between the field operator and other mobile sampling units working on the same task. Global positioning system (GPS) and inertial measurement units would help keep a detailed record of each unit's movements and location that would be relayed to the operator and other units. The movement vehicle would move the platform between sampling locations. Sensors mounted on the robotic vehicle would collect observations via the access tubes, and the data would be sent wireless to an operator overseeing the field data collection (Phillips et al., 2014). Sensor networks can analyze and manipulate the data they collect, thereby undertaking *embedded processing* in performing error detection, data quality assurance and pre-scribed filtering procedures (such as identifies outliers) and altering experimental sampling design. With continued development, sensors that contain smart transducers containing actuators equipped with micro-controllers for sensing, computing, and communication, may provide embedded processing to enable sensor networks themselves to guide, optimize and adapt their network infrastructure in an intelligent way based on different sets of requirements and decision-making needs (Collins et al., 2006). Sensor-based sensing will not only revolutionize many scientific domains and industries (e.g., agriculture and the food industry and its product-chains) serviced by marine and terrestrial ecosystems, but also offers enormous potential benefits for measuring, assessing, and forecasting key ecosystem variables. They will also aid in human and environmental health within *urban* ecosystems in supporting an integrated, holistic understanding of environmental and social dynamics within urban spaces. Such sensing supports the development of "context-aware urban spaces" and adaptive frameworks for real-time monitoring and modeling of environment–human feedback loops (Sagl and Blaschke, 2014).

In addition to machine-intelligence, human intelligence and inter-agency cooperation, common data processing standards and long-term timely access to data are critical to support and enhance future development and expansion of integrated sensing and data sharing. The Open Geospatial Consortium (OGC) has developed a suite of publicly available standards for geospatial data, referred to as OpenGIS$^{(R)}$ standards[6]. The goal of OpenGIS standards is to enable interoperability of complex spatial information and services. The OpenGIS$^{(R)}$ Sensor Web Enablement (SWE) Common Data Model Encoding Standard defines data models for exchanging sensor related data. The representation, nature, structure and encoding of sensor related data can all be defined with the SWE Common Data Model. It consists of a set of eight standards[7], namely: Sensor Model Language (SensorML), Observations & Measurements (O&M), Transducer Model Language (TML), Sensor Observation Service (SOS), Sensor Planning Service (SPS), Sensor Alert Service (SAS), Web Notification Service (WNS), Sensor Web Registry, with an additional standard termed "SWE Common" under development[8]. SensorML provides a standard framework for describing sensors, systems and observation processing. These standards complement the core Web Feature Service (WFS), Web Map Service (WMS), Web Coverage Service (WCS) and Web Processing Service (WPS) standards that enable the sharing of geospatial data and information, alongside data assimilation and processing.

SWE, SOS and SensorML standards use XML (eXtensible Markup Language)[9] to define the set of rules for its encoding to ensure it is both human-readable and machine-readable, can be updated, reused and remixed in a standardized way, and enables human- or machine-learning. XMLencoding uses a "parent-child" class structure to encode information. In XML, classes are defined by a schema, and each parent class may have a valid number of child classes, with each child having its own child classes. Other international encoding standards are UML (Unified Modeling Language) that is supported by the distributed object programming, Object Management Group (OMG) and the Common Object Request Broker Architecture (CORBA), and the OpenGIS$^{(R)}$ Geography Markup Language (GML) that is a modeling language for geographic systems and format for online geographical transactions using XML-based grammar.

---

[6]OpenGIS, `http://www.opengeospatial.org/`
[7]SWE Common Data Model, `http://www.opengeospatial.org/standards/swes`
[8]SWE Common, `http://www.opengeospatial.org`
[9]World Wide Web Consortium (W3C), `www.w3.org/XML/`

4.3.3　Monitoring network design

Despite their wide applicability and use, there are still many complex statistical issues, problems, and challenges related to the sensor network operational deployment and design, coverage, and connectivity technological constraints, and experimental sampling in space and time to optimally identify, localize, detect, and track single/multiple signals and targets, sensor data assimilation and integration (Muller et al., 2013; Katenka, 2009). As the spatial and temporal heterogeneity of an ecosystem, environment, or landscape increases, so too does the density of sensors in a network required to resolve processes and patterns. A wide array of socio-political sensitivities and operational challenges may exist when more pervasive environmental monitoring is required. This is especially true, when monitoring urban areas (as compared to more rural ones). The "Common Scents" project aims to develop an over-arching infrastructure across sensor network applications toward the realization of a vision of a "digital skin" for the Earth and is anticipated, in the future, to have far-reaching impacts on urban monitoring systems through the deployment of ubiquitous and very fine-grained sensor networks, measuring air pollution, CO_2 concentration, air temperature, precipitation, relative humidity, noise, and other relevant ecosystem variables (Resch et al., 2015; Blaschke et al., 2011). The future transition from the use of non-renewable fossil fuels to renewable energies is expected to profoundly impact air quality and health. Such transitions require distributed high-resolution monitoring of multi-pollutant and GHG emissions, affecting human and environmental air quality and health, under current and future conditions. A reliable determination of key modifiable factors (e.g., urban commuting networks and usage patterns) associated with such impacts is also crucial for determining personal/individual exposures within the urban environment characterized by heat-islands and highly variable ambient levels of pollution. In the future, dynamic sensor networks, integrated with satellite remote sensing, UAVs and other technologies may monitor, measure, assess risk, and automate the complex decision-making of regulating air quality in urban ecosystems and mega-cities having populations typically over 10 million people (e.g., São Paulo, Mexico City, New Delhi, and Beijing) (de Fatima Andrade et al., 2015; Krzyzanowski et al., 2014; Janssen et al., 2012). Novel high-resolution air quality forecasting models integrate climate/atmospheric modeling, satellite remote-sensing and surface monitoring. Due to high variations in aerosol concentration, composition, and sources, high-resolution surface monitoring optimally requires the integration of fixed monitoring sites and mobile units, and wireless online multi-pollutant monitors that simultaneously measure air pollutants and GHGs (i.e., carbon monoxide, CO, carbon dioxide, CO_2, methane, CH_4, fine-particulate matter $PM_{2.5}$ (particles less than 2.5 micrometers in diameter), nitrogen dioxide, NO_2, ozone, O_3, sulfur dioxide, SO_2, and volatile organic compounds, and VOCs such as benzene and methylene chloride) (Resch et al., 2015). How to integrate different network measurement platforms (mobile/fixed stations, sensors, UAVs, satellite remote-sensing) remains an area of crucial importance—especially in tackling issues of high variability, microclimates, simultaneous/non-simultaneous responses, blocks of nonresponse, subsets of response variables at given nodes, collections of nodes or different networks, latent variables and the integration of different ecosystem monitoring networks.

Zidek et al. (2002) have explored the integration of networks of ambient monitoring stations set up to monitor environmental pollution fields but where each station measures a different subset of pollutants and their concentration within a geographical region. They develop a Bayesian framework for integrating the measurements of such station networks, yielding a spatial predictive distribution for both unmonitored sites and unmeasured concentrations at existing stations. A further issue with monitoring networks is the loss of spatial dependence between the response from an environmental process to its extremes across different sites or nodes—high pollution exposure levels between sites may occur, and have significant impacts on human and environmental health. Such potential risk also increase as the distance between monitoring sites increases, and such a potential deficiency raises concerns about the adequacy of air pollution monitoring networks whose primary role is the detection of noncompliance with air quality standards based on extremes designed to protect human health (Chang et al., 2007). Findings show that the field of extreme values can be accurately modeled as a multivariate Gaussian-Inverse Wishart (IW) hierarchical Bayesian distribution (i.e., conjugate prior for the covariance matrix of a multivariate normal distribution) to enable the design of monitoring networks that regulate air pollution compliance (Chang et al., 2007; Brown et al., 1994).

The natural conjugate prior distribution for covariance ($\mathbf{\Sigma}$) pdf following a *inverse Wishart* (IW) yields,

$$\mathbf{\Sigma} \sim IW(\mathbf{\Phi}, \nu) = W^{-1}(\mathbf{\Phi}, \nu) = \frac{|\mathbf{\Phi}|}{2^{\frac{\nu p}{2}} \Gamma_p(\frac{\nu}{2})} |\mathbf{\Sigma}|^{-\frac{\nu+p+1}{2}} e^{-\frac{1}{2} tr(\mathbf{\Phi}\mathbf{\Sigma}^{-1})}, \qquad (4.2)$$

where $\mathbf{\Sigma}$ and $\mathbf{\Phi}$ are $p \times p$ positive-definite matrices, $tr(\cdot)$ is the trace function and $\Gamma_p(\cdot)$ is the multivariate Gamma function. So $\mathbf{\Sigma}$ follows an IW pdf (i.e., $\mathbf{X} \sim W^{-1}(\mathbf{\Phi}, \nu)$) when its inverse $\mathbf{\Sigma}^{-1}$ has a Wishart distribution $W(\mathbf{\Phi}^{-1}, \nu)$.

The $\mathbf{\Phi}^{-1}$ is termed the scale-matrix, and ν degrees of freedom. The ν single hyper-parameter controls the precision of all the elements of Σ, and $\nu \geq r$ is required for the density to be proper (i.e., a density distribution that can be mathematically integrated). Erickson (1989) discusses inference and the use of proper and improper priors. The form of this pdf can be transformed to $IW(\mathbf{\Phi}, \delta)$ by introducing a shape parameter δ, where $\nu = \delta + p - 1$. So when $\delta > 2$, the distribution mean exists and is $E(\mathbf{\Sigma}) = \mathbf{\Phi}/(\delta - 2)$. Other, more flexible prior distributions for covariance have been proposed and applied. For example, an approach by Barnard et al. (2000) involves a strategy for separating variance (or standard deviations) from correlations. A truncated-Gaussian or lognormal distribution can be used as a prior on individual variances or standard deviations, and different families of IW priors can be used as priors for correlations. O'Malley and Zaslavsky (2008) have applied such a "separation strategy prior" in the analysis of US health care quality survey data that consists of large blocks of structured nonresponse and a widely varying response rate, such that there is a need to not only account for hospital/organizational unit or "domain-level" variation, but to account for patients/respondents or individual-level variation.

Sensor and station monitoring networks can be distinguished as centralized or decentralized systems, depending on whether a sensor network has a central processor (i.e., a centralized system) or a fusion center (i.e., a decentralized or distributed system). Sensor communication network topology (or alternatively, configurations, arrangements or architectures) affects not only sensor measurement bias, but also overall estimation and prediction performance. This is because a fusion center in a distributed network only has partial, not complete, information as communicated by sensors, even though losses can be minimized by optimally processing data/information at sensor nodes (Viswanathan and Varshney, 1997). Given N sensors deployed at fixed locations, distributed across a two-dimensional monitoring region Ω, and a single target located at $v \in \Omega$ that emits a signal detected at location i, $S_i = S_i(v)$, $(i = 1, ..., N)$, whereby signal strength is assumed to decrease monotonically with increasing separation distance from the target with normally-distributed stochastic/random noise. In turn, $a_i = S_i + \epsilon_i$ is the signal strength measured by the ith sensor.

Let the hypothesis test have a null, H_o: no target is present ($S_i = 0$ for all i) and an alternative H_a, that there is a target present ($S_i > 0$ for some i). In the context of sensor networks, the detection probability (TN) is the conditional probability that the sensors correctly report the presence of the target. The false alarm (FP) and detection probabilities (TN) may not be known and/or equal for all sensors. Given a target in $\mathbb{R}^2$, the objective of the sensor network is to maximize the probability of detection, while controlling the network of all false alarm probability (FP), as the conditional probability that the sensors detect the target given that there is no target in the monitored region, is then $P_i = I(a_i \geq \tau_i)$, for a specified threshold, τ_i, and indicator function, $I(\cdot)$ (Katenka, 2009). Further, let an uncertain quantity, θ be a *state of nature* affecting a decision process, whereby Θ denotes the set of all possible states of nature. Such a state, for example, may be associated with an experimental measurement trend and/or an observable event or a set of consecutive events. Similarly, let a decision *action*, a_i that is associated with a set of all possible actions under consideration be denoted $\mathbf{A}$. When an action (a_1) that is taken, and θ_1 is a true state of nature, then a loss, $L(\theta_1, a_1)$ is incurred. A loss function therefore specifies how costly every possible action is.

An inter-comparison of fundamental aspects and statistical performance metrics for distributed detection and multiple sensor network topologies, namely: parallel (with and without a fusion center), serial and tree-based, is shown in Figure 4.7. More complex network designs are possible, as these four basic topologies assume no feedbacks from a fusion center or from individual sensors to other sensors, and no ability of sensors to "rethink" or adjust their decisions after relaying them. The ith sensor is represented as s_i associated with the observational vector, θ_i, and the decision or action vector, a_i. *Parallel* topologies assume sensors do not communicate with each other, and there is no feedback from a *fusion center* to any sensor. In contrast, *serial* topologies assume that sensor observations are conditionally independent, and that sensors pass quantized information to each other. In this case, a first sensor uses only its observation to derive its quantized data for another sensor through the network to the last sensor, making a decision as to which of two possible

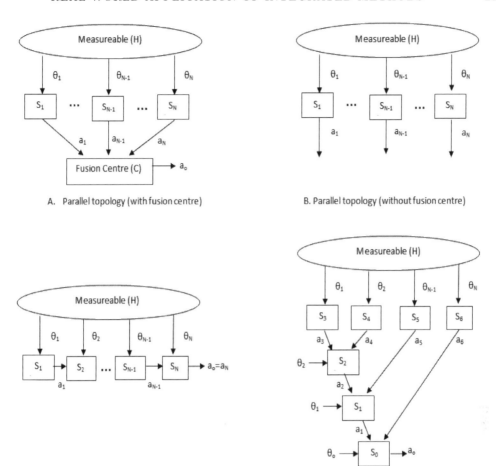

FIGURE 4.7: Distributed detection and multiple sensor network topologies: parallel (with and without a fusion center, C), serial, and tree-based. The ith sensor is represented as s_i with an observational vector, θ_i, and decision or action vector, a_i.

hypotheses the observations of the sensors correspond to. *Serial* topologies suffer from an accumulation of processing and communication delays, which can only be overcome by modifying the sensor communication network structure. Nonetheless, sensor failures, biases, anomalies, and outliers can degrade the performance of parallel topologies. Barbarossa et al. (2013) and Viswanathan and Varshney (1997) further explore statistical issues involved in the design and application of more dynamic sensor networks. In particular, Barbarossa et al. (2013) showcase the importance of solving the issues to ensure reliable future transport, communication, and energy *smart grids*.

The important assumption in parallel networks is that the *conditional independence* of sensor observations implies that the joint probability of observations satisfies,

$$p(\theta_1, \theta_2, ..., \theta_N | H_l) = \prod_{i=1}^{N} p(\theta_i | H_l), l = (0, 1). \tag{4.3}$$

Based on this assumption, given an *optimal* value for the threshold, τ_o, and stochastic noise, ϵ_o, a minimal false alarm (FP) probability, $P_o = \alpha$, can be achieved, such that the indicator function, I, is given by,

$$I(\mathbf{A}) = \frac{p(\mathbf{A}|H_a)}{P(\mathbf{A}|H_o)} = \prod_{i=1}^{N} \frac{p(a_i|H_a)}{P(a_i|H_o)} = \begin{cases} > \tau_o & \text{select } H_a, \text{ or set } a_o = 1 \\ = \tau_o & \text{randomly select } H_a \text{ with probability } \epsilon \\ < \tau_o & \text{select } H_o, \text{ or set } a_o = 0 \end{cases}$$

$$\tag{4.4}$$

The above indicator function for a sensor network having an optimal fusion center is the likelihood "odds ratio" (i.e., ratio of likelihoods) and forms the likelihood ratio test (LRT) for binary decisions or two-category classification. This indicator function considers only likelihood probabilities and assumes that prior probabilities are uniform. A general indicator for signal or information loss is risk (i.e., recall from Chapter 1 that an expected loss is a "risk").

Decision-making under uncertainty relies on Bayesian decision theory as a fundamental statistical approach to the problem of pattern classification (Berger, 2013; Duda et al., 2001). The *Bayes factor*, K, involving decision-making actions under a hypothesis test is given by,

$$K = \frac{p(\mathbf{A}|H_a)}{p(\mathbf{A}|H_o)} = \frac{\int p(\mathbf{x}|\theta_1, M_1)p(\theta_1|M_1)d\theta_1}{\int p(\mathbf{x}|\theta_2, M_2)p(\theta_2|M_2)d\theta_2}. \tag{4.5}$$

The above equation represented in the context of decision-making actions associated with a given hypothesis or a "hypothesis space" can be translated to "model space" within the context of deciding between competing model representations of nature associated with a given set of evidence, as previously presented in Chapter 3. In this way, evidence/data or feature vector $\mathbf{x}$ would be substituted for the vector $\mathbf{A}$, and competing models M_1 and M_2 having different parameter sets, θ_1 and θ_2 would be substituted for H_a and H_o. Unlike a likelihood-ratio test (LRT), a statistical classification, discrimination, or decision-rule test that is based on the Bayes factor integrates over all parameters in each model, so does not depend on any single set of parameters, thereby also avoiding modeling more noise than signal (i.e., overfitting).

Zhang et al. (2012) demonstrate how integrating spatial and temporal aspects of outlier detection improves the reliability of sensor network data collected from sensor nodes with varying data quality, limited resources and deployment environments that vary. They compare "temporal and spatial real-data-based outlier detection, TSOD" with the detection rate (DR) and false positive rate (FPR) as accuracy metrics to a set of existing, online, and distributed techniques for statistical outlier detection. In their study, they distinguish between error and event-type outliers. Outliers that occur in a consecutive time sequence are outlier events, while isolated outliers detected by a node would be labeled as outlier errors. An effective outlier

detection technique should achieve a high DR and a low FPR. They further emphasize the importance and critical need for a near-real-time labeling technique that can
accurately distinguish between a wide range of different types of outliers in sensor
network data.

Jindal and Psounis (2006) have applied Markov Random Fields (MRF) (i.e., for
which the multivariate Gaussian model is a special case) to capture spatial correlation in sensor network data of any degree, irrespective of the granularity, density,
number of source nodes, or topology (Jindal and Psounis, 2006). A MRF is a random field superimposed on a graph or network structure, or more specifically an
undirected graphical model in which each node corresponds to a random variable
or a collection of random variables, and its edges represent conditional probability
dependencies. The Isling model of ferromagnetism in statistical mechanics/physics
is a classic MRF model (Cipra, 1987). The great attraction of MRF models is that
they leverage the so-called "knock-on effect", whereby explicit short-range conditional
dependencies give rise to extended, long-range correlations, providing solutions to
many problems in vision science, including image reconstruction, image segmentation, 3D vision, and object labeling (Blake and Kohli, 2011).

For temporal data, statistical dependence is often introduced by fitting a Markov
model, that is, a model with the property that the present, conditioned on the past,
depends only on the immediate past such that a Markov process is a memoryless
process. Or stated another way—given a present state, the future and past states are
independent. Let I be a countable set (i.e., state-space) and every $i \in I$ a identifiable
state; then X_n $(n \geq 0)$ is a *Markov chain* (denoted $Markov(\lambda, P)$), with an initial
distribution λ and a transition matrix, P, if, $\forall n \geq 0$ and $i_0, i_1, ..., i_{n+1}, n \in I$. The
index n, for example, can denote the sites of a lattice, or discrete points in time

$$P(X_0 = i_0) = \lambda_{i_0} \quad (4.6)$$

$$P(X_{n+1} = i_{n+1} | X_0 = i_0, ..., X_n = i_n) = P(X_{n+1} = i_{n+1} | X_n = i_n) = p_{i_n i_{n+1}}. \quad (4.7)$$

The initial distribution satisfies: $0 \leq \lambda_i \leq 1$, $\forall i$ and $\sum_{i \in I} \lambda_i = 1$. The transition
matrix $P = (p_{ij} : i, j \in I)$ with $p_{ij} > 0, \forall i, j$.

A spatial analog for the temporal Markov property can be obtained (Cressie and
Lele, 1992). For a stochastic process $\mathbf{Z} \equiv (Z(s_1), Z(s_2), ..., Z(s_n))'$ on the set of
spatial locations or sites, $(\mathbf{s_i} : i = 1, ..., n)$ with the pdf $P(Z(\mathbf{s_1}), Z(\mathbf{s_2}), ..., Z(\mathbf{s_n}))$,
and conditional probabilities, $P(Z(\mathbf{s_i}) | (Z(\mathbf{s_j}) : j \neq i)$, and $i = (1, 2, ..., n)$. The
neighborhood N_i of the ith site is the set of all other sites s_j $(j \neq i)$, whereby

$$P(Z(\mathbf{s_i}) | (Z(\mathbf{s_k}) : k \neq i)) = P(Z(\mathbf{s_i}) | (Z(\mathbf{s_j}) : j \in N_i)), i = (1, ..., n). \quad (4.8)$$

This spatial Markov property implies that the ith site, conditioned on all other sites,
depends only on its neighboring values $(Z(\mathbf{s_j}), j \in N_i)$. An MRF is therefore defined
by the set of neighbors for each site, and conditional probabilities between each site
and its neighboring sites.

A fundamental theorem of random fields is the *Hammersley-Clifford Theorem
(HC Theorem)* that specifies the necessary and sufficient conditions for *positive*
probability distributions that can be represented as MRFs (Cressie and Wilke, 2011;

Cressie and Lele, 1992; Besag, 1974; Hammersley and Clifford, 1971). Given a neighborhood N_i, a subset of members c forms a *clique* if each member within this set is a neighbor of each of the other members. The HC Theorem states that for $P(\mathbf{X}) = P(X_1, X_2, ..., X_n) > 0$ is a joint probability distribution of a MRF iff (i.e., if and only if)

$$P(\mathbf{X}) = \frac{1}{\kappa} exp \left(\prod_{c \in \mathbf{C}} \phi_c(\mathbf{X_c}) \right), \tag{4.9}$$

where $\mathbf{C}$ is a collection of cliques and the value $\phi_c(\mathbf{X})$ only depends on the members that belong to a given clique c (i.e., a potential function). Here, $\mathbf{X}$ can comprise spatial locations $(Z(\mathbf{s_1}), Z(\mathbf{s_2}), ..., Z(\mathbf{s_n}))$. Also, κ is a normalization constant or "partition" function. If $\mathbf{W}$ is the set of all possible configurations of the values of the vector of state variables $\mathbf{X}$, then the partition function, κ is,

$$\kappa = \prod_{\mathbf{X} \in \mathbf{W}} \sum_{\mathbf{X} \in \mathbf{c}} \phi_c(\mathbf{X_c}). \tag{4.10}$$

The *exponential* functional form (e.g., multi-parameter beta, gamma and log-Gaussian exponential-type conditional distributions) is termed a neighborhood *Gibbs measure*, so that the HC Theorem says that all MRFs can be represented or decomposed into (normalized) neighborhood Gibbs probability measures or potential functions. Cressie and Lele (1992) highlight how obtaining a normalizing constant for the conditional distribution can be difficult and have developed a flexible class of "automodels" for MRFs (i.e., comprising mixtures of Gaussian conditional probability distributions) based on sufficient conditions for the conditional probability specifications that yield MRFs.

$$log \left(p(Z(\mathbf{s_i})|Z(\mathbf{N_i})) \right) = G(Z(\mathbf{s_i})) + \beta \sum_{j \in N_i} Z(\mathbf{s_i})Z(\mathbf{s_j}) - \kappa_G \left(\beta, \sum_{j \in \mathbf{N_i}} Z(\mathbf{s_i}) \right) \tag{4.11}$$

The first term in this expansion captures dominant features of the random variation in $Z(\mathbf{s_i})$, while the second term specifies its spatial dependence within a neighborhood $\mathbf{N_i}$ where the parameter β measures the strength and direction of the spatial dependence, and the third $exp(-\kappa(\cdot))$ term is the normalization constant. For this model, the conditional mean, $E(Z(\mathbf{s_i})|Z(\mathbf{N_i}))$ is given by,

$$E(Z(\mathbf{s_i})|Z(\mathbf{N_i})) = \frac{\sum_{i=1}^{k} \pi_l(\mu_l + \sigma^2 \beta y_i) M_l(\beta y_i)}{\sum_{l=1}^{k} \pi_l M_l(\beta y_i)} \tag{4.12}$$

for the moment generating function, $M_l(x) \equiv \exp(\mu_l x + (\sigma x)^2)$, $y_i \equiv \sum_{j \in N_i} Z(s_j)$, $(l_1, l_2, .., l_n)$ is a collection of numbers sampled with replacement from $(1, 2, .., k)$ and $\sum_{l=1}^{k} \pi_l = 1$, $\pi_l > 0$. For every k, there exists a MRF with this conditional mean for $Z(\mathbf{s_i})$ given its neighborhood $Z(\mathbf{N_i})$. Fixing k, optimal values for the parameters σ and $(\mu_l, \pi_l) : l = 1, ..., k$ can be obtained.

4.3.4 Spatio-temporal interpolation

The inter-comparison and validation of alternative ways of representing dominant random variation and neighborhood spatial dependence within a network of monitoring sites has been investigated in terms of their accuracy and reliability for spatially interpolating daily precipitation and temperature across Canada. This has involved generating a sequence of consistent spatially-interpolated grids for each day of the historical climate network record (Newlands et al., 2011). Validated, interpolation surfaces can be used to generate a 'computing grid' at a given resolution for geospatial or spatially-explicit ecosystem modeling, assessment, and forecasting (Wang et al., 2005). An interpolation problem involves approximating an unknown function by an interpolation function whose form is postulated in advance either explicitly (e.g., second-order polynomial), or implicitly (e.g. under a condition of minimum curvature). Parameters of the interpolation function may be optimized under deterministic (i.e., exact fit at points) or stochastic criterion (i.e., least-squares). Unlike the classic interpolation problem, classic kriging starts with a statistical model rather than postulating an interpolation function. Kriging represents a family of statistical interpolation techniques in which correlation or covariance functions are specified to allocate weights to minimize variance and bias in interpolated estimates. Thin-plate splines are polynomial functions that fit a smooth surface through the data points with minimum curvature, and are a generalization of a multivariate linear regression model where a nonparametric function is specified. Recent reviews of the strengths and weakness of a select set of interpolation methods conclude that the performance of these methods varies according to the relative influence of key forcing factors at different spatial and temporal scales (Daly, 2006; Lobell et al., 2006; Jolly et al., 2005). Functional and statistical approaches for spatio-temporal interpolation include universal kriging and co-kriging (Chilès and Delfiner, 2012; Host et al., 1995), spline-fitting (Hutchinson, 1998a,b, 1995), distribution-based (Thornton et al., 2000), and other simpler methods (e.g. nearest neighbor assignment, NN, inverse-distance weighting, IDW), hybrid or integrated approaches (Hasenauer et al., 2003; Ninyerloa et al., 2000; Shen et al., 2005), including more complex expert rule-based methods like the PRISM (Parameter-elevation regressions on independent slopes model) methodology (Daly et al., 2008; Daly, 2000; Johnson et al., 2000).

Due to an array of influences, climate distributions can be highly skewed, with irregular spacing between network sites/locations, and irregular sampling in time (i.e., data sparse). Also, determining the neighborhoods within such networks requires accurate estimation of the spatial correlation range over typically heterogeneous or inhomogeneous areas and with significant non-stationarity. This is a challenge, especially for a variable like precipitation that is highly stochastic. Findings from the inter-comparison of different interpolation models indicate they are inaccurate in estimating daily mean precipitation especially during the summer, when there is high variability due to convective control of local rainfall events. This inference is supported by previous validation work (McKenney et al., 2006). Daily precipitation, accounting for dry and wet episodes, in warmer months exhibits a mixed exponential distribution, while in colder months, is Gamma-distributed (Wan et al., 2005). Also, an empirically average spatial correlation function derived from information over all of North America and approximated at the monthly scale that varies from 250–1050 km, with correlation extending in winter compared to summer months (Groisman and Easterling, 1994).

The use of daily climate data has increased considerably over the past two decades due to the rapid development of information technology and the need to better assess impacts and risks from extreme weather and accelerating climate change. Daily station data is now regularly used as an input to biophysical and biogeochemical models for the study of climate, agriculture, and forestry. However, many questions still remain on the level of uncertainty in using such data, especially for predictions made by spatial interpolation models. Local orography/elevation (e.g., temperature lapse rate), slope and aspect (topography), the effect of large water bodies (e.g., ocean and lake-effects), vegetation/land cover, and wind conditions all influence the accuracy and reliability of monitoring data that is used in urban, forestry and agricultural ecosystem assessment and forecasting. As highlighted by Daly, spatial climate patterns are most affected by terrain and water bodies *at a spatial scale below 10 km*, primarily through the direct effects of elevation, terrain-induced climate transitions, cold air drainage and inversions, and coastal effects (Daly, 2006). The 10 km spatial scale marks an identified transition whereby terrain and water-bodies dominate climate spatial patterning (Newlands et al., 2011).

Long-term daily climate meteorological data across Canada was historically (1891–2004) gathered by a network consisting of 7514 stations. When station density or coverage and data quality aspects are considered, this dataset reduces to 6616 stations. When further filtered for stations situated on or outside national boundaries, duplicate records, and station locations that have no measured elevation, there are a total of 6600 stations in this monitoring network (i.e., varying in time between 1200 and 2800 stations). Fewer than 10% of the stations are located at elevations greater than 1500 meters. Systematic measurement uncertainties include station positioning error, changes in the definition of climate day (i.e., there was a nationwide change in observing time in July 1961) and other incremental adjustments and improvements that have been made to the monitoring equipment (Hopkinson and McKenney, 2011; Vincent et al., 2009, 2002; Vincent and Gullett, 1999; Mekis and Hogg, 1998; Vincent, 1998). A homogenized dataset of Reference Climate Stations (RCS) has been generated consisting of 368 high-quality stations for Canada (Milewska and Hogg, 2001). For model inter-comparison and validation, a dataset of validation reference stations consisting of 150 stations is obtained when record lengths of 27 years and 90% temporal coverage are specified (Newlands et al., 2011). Also, monthly temperature datasets for trend analysis have been generated (n =338 stations) (Vincent et al., 2012)[10].

An approach that does not assume a neighborhood is first presented—called smoothing splines (McKenney et al., 2006; Hutchinson, 1995; Wahba, 1990; Craven and Wahba, 1979). Boer et al. (2001) compare three forms of thin-plate smoothing splines (e.g., bivariate, trivarate and partial thin-plate splines). The ANUSPLIN

---

[10]EC, `www.ec.gc.ca/dccha-ahccd/`

model and software implements this method[11]. It is a statistical modeling approach that does not require specification of spatial correlation lengths of variables and assumes a global rather than a local neighborhood. A smoothing spline function, f, has $m - 1$ continuous derivatives (i.e., generally $m = 2$ and second-order derivatives are used to specify its boundary conditions). For n monitoring locations, $i = (1, 2, ..., n)$, and uncorrelated, normally-distributed error ϵ with zero-mean and unknown variance σ^2 (i.e., spatial covariance is $\mathbf{V}\sigma^2$, for $\mathbf{V}$ as a known $n \times n$ matrix that is positive-definite),

$$Z(\mathbf{s_i}) = f(\mathbf{s_i}) + \epsilon_i \tag{4.13}$$

$$\frac{1}{n}\sum_{i=1}^{n}(y_i - f(\mathbf{s_i}))^2 + \rho\int_a^b f^m(s)^2 ds, \rho > 0. \tag{4.14}$$

where ρ is called the "smoothing parameter" and is the inverse of the SNR ratio.. This statistical method permits a degree of flexibility in specifying the functional form of the smoothing spline and for integrating additional dependencies and sub-models of covariates, and is derived from co-kriging where it is assumed that the mean of the dependent variable is varying and unknown, and the error covariance is also varying, independent (i.e., uncorrelated) and unknown. Spline smoothing typically assumes that random errors are independent. Thus, there is no temporal or spatial correlated random error, and correlation in data does not affect the SNR ratio or its inverse, the smoothing parameter ρ. A spatially-correlated error can be estimated, however, by specifying a non-diagonal covariance matrix, investigated further by Wang (1998). Assuming a second-order polynomial spline function ($p = 2$) and three spatial dimensions ($\lambda = 3$) (i.e., latitude, longitude, elevation), the corresponding spatial covariance function, $C(r)$, for a separation distance from a point i is r and is co-kriging under isotropic assumptions (i.e., the space is homogeneous) (Chilès and Delfiner, 2012),

$$\frac{C(r)}{C(0)} = \frac{1}{\lambda}|r|^p log(r) \tag{4.15}$$

Because daily climate fields are highly heterogeneous with data gaps they do not strictly obey the above power-law decay, but their spatial covariance can be benchmarked against such homogeneous covariance. Often, to reduce the effect of skewness in daily precipitation data, a square-root transformation (i.e., pre-conditioning) needs to be applied that introduces small positive correlations in residual error and positive bias, but that substantially reduces interpolation error by up to 10%, even though a degree of systematic bias is introduced. Generally, a logistic model is assumed for precipitation occurrence. This method can suffer from over-smoothing in areas where there is high curvature in topography/orography, thereby missing localized pronounced extreme values, while under-smoothing in flatter areas. For example, zero values of precipitation occur when there are more than three subsequent days of evaporation loss, and values close to zero (i.e., 0.2 mm) are generally

---

[11]ANUSPLIN, `http://fennerschool.anu.edu.au/research/products/anusplin-vrsn-44`

assigned to be zero. Values greater than 0.2 mm take on positive interpolation values. Trace amounts can vary in the range of 0.1 to 0.7 mm, depending on the station latitude and longitude and temperature.

A hybrid interpolation method, integrating inverse-distance weighting (IDW) and nearest-station assignment (i.e., *natural neighbor* or the Thiessen polygon method (i.e., not to be confused with the *nearest-neighbor*, NN method) has been developed and validated (Shen et al., 2005). It interpolates temperature differently than precipitation, to better account for precipitation variability in space and time. For station i, where $(i = 1, 2, ..., M_j)$ of a total number of stations, M_j in relation to a grid point j, the estimated temperature at the interpolated grid point, j, follows an inverse-distance weighting function for $i \neq j$ given by,

$$E(T_j) = \begin{cases} T_i, & i = j \\ \left(\sum_{i=1}^{n} \frac{1}{d_{ij}} \right)^{-1} \sum_{i=1}^{M_j} \frac{T_i}{d_{ij}}, i \neq j \end{cases} \tag{4.16}$$

where d_{ij} is the station-to-grid point distance and T_i is the observed temperature at station i. M_j is selected to comprise the first n^{th} nearest stations that lie with temperature and precipitation correlation length scales. Typically M_j varies up to 8 nearest stations for the temperature spatial correlation length, $d_{ij,T} \leq 200$ km and precipitation spatial correlation length, $d_{ij,P} \leq 60$ km, respectively. Because the total number of stations is variable and determined within search radii, this model inherently adjusts to station density. When a station and grid point coincide $(i = j)$ the interpolated value is the observed value. When no stations are situated within the respective climate variable correlation length scale to a grid point (i.e., $M_j = 0$) then the nearest-station assignment is applied whereby the interpolated value is assigned the value of the first nearest station, one of the hybrid features of this model. Also, as inverse-distance weighting overestimates the number of precipitation days and underestimates daily precipitation amount and is not able to represent observed temporal and spatial variance in precipitation. Instead, an integration of inverse-distance and nearest-station methods are used to estimate daily precipitation amount and precipitation frequency. Monthly total precipitation for an area polygon is determined by inverse-distance, and precipitation for a polygon is assigned as the first nearest station to its centroid to estimate daily precipitation amount for grid points j contained within a defined polygon $(P_{t,centroid})$ and daily precipitation frequency $(P_{t,polygon})$,

$$E(P_{m,polygon}) = \sum_{t} \left(\left(\sum_{i=1}^{N} \frac{1}{d_{ij}} \right)^{-1} \sum_{i=1}^{M_j} \frac{P_i}{d_{ij}} \right), i \neq j \tag{4.17}$$

$$\left(\frac{P_t}{P_m} \right)_{centroid} = \left(\frac{P_t}{P_m} \right)_{polygon} \tag{4.18}$$

Precipitation occurrence for this model is determined by a binomial interpolation function weighted by the observed occurrence of surrounding stations and the critical precipitation amount (i.e., trace amount). For this model, isotropic spatial covariance is represented as the contribution of two terms—the first from inverse-distance decay and the second due to nearest-station assignment as,

$$\frac{C(r)}{C(0)} = e^{-\left(\frac{r}{a}\right)^2} - \mu e^{-2\mu r}. \tag{4.19}$$

The model has a very different spatial variance from the power-law decay isotropic field described by Equation 4.15, especially as the station-to-grid point separation distance r increases.

An alternative method that uses a weighted-truncated Gaussian filter has also been employed to generate high-resolution (i.e., 1 km scale) surfaces of daily meteorological variables over large regions of complex terrain (Thornton et al., 2000)[12]. It has also been validated at the 10 km scale and inter-compared to other interpolation methods (Newlands et al., 2011). For a grid point j and a total of M_j stations with station-to-grid distances $d_{ij} \leq d_{ij,T}$, temperature is estimated by exponential decay weighting of the observed temperature at a station location i regressed by observed station-to-grid point elevation h_i, h_j differences,

$$E(T_j) = \left(\frac{\sum\limits_{i=1}^{M_j} e^{-(d_{ij}/d_{ij,T})}(T_i + \beta_0 + \beta_1(h_j - h_i))}{\sum\limits_{i=1}^{M_j} e^{-(d_{ij}/d_{ij,T})}} \right) \tag{4.20}$$

where β_0, β_1 are temperature-elevation regression coefficients. Similarly, daily precipitation, conditional on precipitation occurrence (PO_i) at station i and grid j points, is estimated also according to a weighted regression (i.e., precipitation-elevation) as,

$$E(P_j) = \left(\frac{\sum\limits_{i=1}^{M_j} (e^{-(d_{ij}/d_{ij,P})} - e^{-\alpha})\left(\frac{1+f}{1-f}\right)PO_j}{\sum\limits_{i=1}^{M_j} (e^{-(d_{ij}/d_{ij,P})} - e^{-\alpha})PO_i} \right) \tag{4.21}$$

where $f = \beta_0 + \beta_1(h_j - h_i)$, $|f| < f_{max} < 1$ is termed the "roughness penalty" and f_{max} is the maximum value for precipitation regression extrapolations. As with the hybrid interpolation method, precipitation occurrence for this model is determined by a binomial interpolation function weighted by the observed occurrence of surrounding stations and trace amounts. The isotropic spatial covariance for this model contains an additional scaling or shape parameter, α, compared to the hybrid interpolation model. This modifies the spatial dependence of the climate variables with a distance-independent contribution, whereas the scaling parameter, μ, in the hybrid method modifies only the magnitude of spatial variance.

$$\frac{C(r)}{C(0)} = e^{-\left(\frac{r}{R_p}\right)^2 \alpha} - e^{-\alpha} \tag{4.22}$$

---

[12]DAYMET, http://daymet.ornl.gov/

This model interpolates daily values using data from stations located within the neighborhood or "truncation radius," R_p, which depends on the density of stations around an interpolation location and on the shape parameter. To parameterize the model, an iteration procedure starts with an initial truncation radius, a given number of iteration steps, and a target number of stations required for interpolation for temperature and precipitation. Thus, if station density is high, the truncation radius is low, and vice versa. Based on changes in station density, the truncation radius is varied from the initial value to ensure the targeted number of stations for interpolation, so as to minimize interpolation uncertainty.

4.3.5   Modeling extremes

Findings from the inter-comparison of the smoothing spline, the hybrid inverse-distance weighting with a natural neighbor and a weighted, truncated Gaussian model lends support for decomposing or partitioning climate variability into neighborhoods or zones having separate potential functions to better capture localized, dominant trends or drivers as well as spatially-dependent changes in the correlation lengths of temperature and precipitation, as well as the anisotropy or directional bias in these variables (Newlands et al., 2011). One notable region with strong localized influences on climate would be the mountainous region of British Columbia (Neilsen et al., 2013; Duke et al., 2009; Porcelli, 2008). Representative (n=13) climate zones within six major watersheds are shown in Figure 4.8. High-elevation data from the stations shown has been integrated and statistically verified for quality and consistency from Environment Canada (EC) ($n = 208$), the BC Ministry of Forests and Range (BCFR) ($n = 4$), BC Ministry of Environment ($n = 34$), and the BC Ministry of Transportation and Highways (MOTH) ($n = 11$) (Porcelli, 2008). In addition to reliably capturing temperature changes with elevation or lapse-rate inversions, precipitation extremes, there is significant heteroscedasticity (i.e., unequal variance), evident as a non-uniform pattern of model residuals, especially with those that assume stationarity, when interpolating climate at high-resolution across mountainous zones (Diaz and Bradley, 1997). Some stations also have a stronger influence (i.e., high-leverage) on interpolation accuracy. In addition to deterministic trends, spatially-correlated error or stochastic variation could be included, as is represented in the joint probability functional underlying MRF models (Equation 4.11). Such a decomposition via subnetworks or patterning of climate neighborhoods may also capture better the distribution of extreme values in climate variables. In particular, when separating the climate signal from noise, a wavelet or coherence spectral analysis of high-elevation stations across British Columbia's mountainous zones reveals that its climate variability exhibits substantial non-stationarity, with changing spatial and temporal correlation structure for minimum and maximum temperature and precipitation. Such correlation is further confounded by noise that increases at shorter time periods. Interpolation at the monthly (rather than daily) scale reduces such noise and increases temperature and precipitation cross-correlation.

The Generalized Extreme Value (GEV) Distribution is a flexible family of models (i.e., it combines Gumbel, Fréchet, and Weibull *maximal* extreme value distributions)

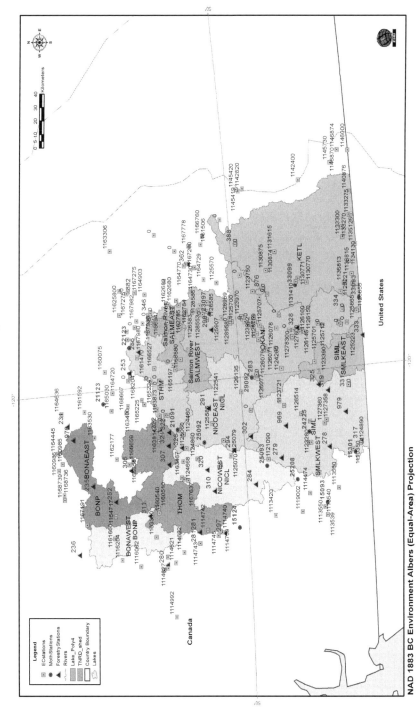

FIGURE 4.8: Distribution of high-elevation stations integrated across the climate zones of British Columbia's mountainous region (i.e., Okanagan, Similkameen East/West, Nicola East/West, Thompson, South Thompson, Bonaparte East/West, Salmon River East/West, Kettle Valley, associated with six major watersheds).

having the pdf (Stephenson and Gilleland, 2006; Beirlant et al., 2004; Coles et al., 2003; Coles, 2001),

$$GEV(X; \mu, \sigma, \xi) = \begin{cases} \frac{1}{\sigma}(1 + \xi Z)^{-1-1/\xi} e^{-(1+\xi Z)^{-1/\xi}}, \xi \neq 0 \\ \frac{1}{\sigma} e^{-Z} - e^{-(e^{-Z})}, \xi = 0 \end{cases} \quad (4.23)$$

where $Z = (X - \mu)/\sigma$, and ξ, $\sigma > 0$, μ are the shape, scale and location parameters, respectively. For $\xi = 0$, $-\infty < x < +\infty$, and for $\xi \neq 0$, $\left(1 + \xi \frac{(x-\mu)}{\sigma}\right) > 0$. The Gumbel ($\xi = 0$), Pareto or Fréchet ($\xi > 0$) and reversed Weibull ($\xi < 0$) extreme (i..e., maximum) value distributions are obtained from this general pdf. The return level is the block maximum value, or the level that is exceeded on average every T years (i.e., m^{th} period of time), is then,

$$P(y > z_m) = 1 - GEV(y \leq z_m) = \frac{1}{m} \quad (4.24)$$

$$z_m = \mu - \sigma log\left(-log\left(1 - \frac{1}{m}\right)\right) \quad (4.25)$$

While block maximum values converge to a GEV pdf (i.e., $max(X_1, X_2, ..., X_n)$) for $n \to \infty$, the conditional values $p(X_i - u | X_i > u)$ termed the "peak over threshold" (POT) of exceedences over a specified threshold u, converge to the Generalized Pareto Distribution (GPD),

$$GPD(X; \mu, \sigma, \xi) = \frac{1}{\sigma}(1 + \xi Z)^{-(\frac{1}{\xi}+1)} \quad (4.26)$$

where σ is a scale and ξ is a shape parameter, respectively. Depending on the choice of shape parameter, ξ, the GPD distribution also results in a Gumbel ($\xi = 0$), Pareto or Fréchet ($\xi > 0$) and Weibull ($\xi < 0$).

Findings from a GPD-based analysis of precipitation extremes across the BC mountainous zone reveals large variation in the scale and shape parameter between the climate zones. At lower elevations within the semi-arid Okanagan basin, daily precipitation extreme events of 60 mm can be expected, based on the integrated historical data, every 100 years. Across the larger region, 100 yr return levels range approximately between 45–60 mm/day, and 10 year return levels are 25 mm/day. This compares to annual maximum values of extreme daily rainfall/precipitation within the Alpine Region of Switzerland of 43–53 mm day (Frei and Schar, 2000). These estimates within mountainous zones compare to 10-yr and 100-yr return levels of annual maximum daily rainfall across the central coast of Venezuela, for example, that are roughly 100 mm and 130 mm from an extremes analysis (Coles et al., 2003). In general, multiple extreme precipitation events that occur over successive days often are associated with droughts and floods, and this needs to be given more focus in the analysis of extremes. Hunter and Meentemeyer (2005) have tested a multi-scale approach, involving the integration of longer-term (30-yr) averaged PRISM-generated climate surfaces (i.e., that account for adiabatic lapse rates, orographic effects, coastal proximity, and other environmental factors), with daily interpolated surfaces at the 2 km scale, to predict daily temperature and precipitation extremes (Hunter and Meentemeyer, 2005). They find that prediction errors show a strong seasonal dependence, alongside substantial reductions in interpolation error.

For the smoothing spline, hybrid, and truncated-Gaussian models, interpolation errors for temperature and precipitation at the *daily and 10 km* scale ranged between 1.5–3.5°C and 3–5 mm root-mean-squared error (RMSE) with bias ranging between ±0.4°C, ±0.3 mm, with the largest errors for winter temperature (over-estimation) and summer precipitation (under-estimation) (Newlands et al., 2011). These daily error ranges are comparable to interpolation error ranges at the *daily and 500m* scale across the Okanagan Basin of British Columbia of 1.3–1.8°C (T_{max}) and 1°C (T_{min}) with 10–18% monthly and 6.2% (27 mm) annual precipitation (mean-absolute error, MAE) (Neilsen et al., 2013). These errors compare also to *monthly and 10 km* scale error estimates of 0.5–1.5°C and 15–30 mm across Canada using smoothing splines (McKenney et al., 2006). The "climatological day" or "climate day", before 1961, was defined as ending at 1200 UTC for maximum temperature and 0000 UTC for minimum temperature, but it was redefined for minimum temperature to end at 0600 UTC (coordinated universal time) at all synoptic stations (i.e., stations situated at airports) across Canada on July 1, 1961. This redefinition introduced a systematic bias in the annual and seasonal means of daily minimum temperature across Canada (Vincent et al., 2009). This, in turn, required the readjustment in interpolated or gridded temperature and precipitation data (Hopkinson and McKenney, 2011; Newlands et al., 2011). While 77–85% of precipitation occurrence is explained by the daily interpolation models, strong regional-scale differences in precipitation frequency (i.e., number of days with significant precipitation) in summer months (May–August) between these models is evident with an estimated 10% error trade-off within transition periods (i.e., Spring and Fall) in the ability of these models to track temporal (i.e., seasonal variability) trends versus vertical/elevation and horizontal/topography spatial trends (Figure 4.9).

Integrating the GEV or GPD distributions (stationary or non-stationary assumptions) when modeling and interpolating climate variables is crucial to capture extreme behavior, in addition to capturing dominant/mean, spatial variance and temporal autocorrelation effects. Bias is attributed in anthropogenic contributions to climate change due to an observed widespread intensification of precipitation extremes globally, but especially within the Northern Hemisphere (Zhang et al., 2013; Alexander et al., 2006). Given the human (e.g., GHG, urban heat-island) contributions to climate extremes, accurate partitioning of climate monitoring data is likely to be of paramount importance in the future, to ensure reliable and more robust predictive ecosystem modeling and interpretations. This is especially important as many ecosystem modeling studies ignore the uncertainty that is present in the underlying interpolated long-term climate surfaces because comprehensive estimates are not widely available. Recently, Wilson and Silander (2014) demonstrate how Bayesian kriging interpolation is capable of propagating interpolation uncertainties to estimate ecologically relevant climate metrics. They emphasize the importance of

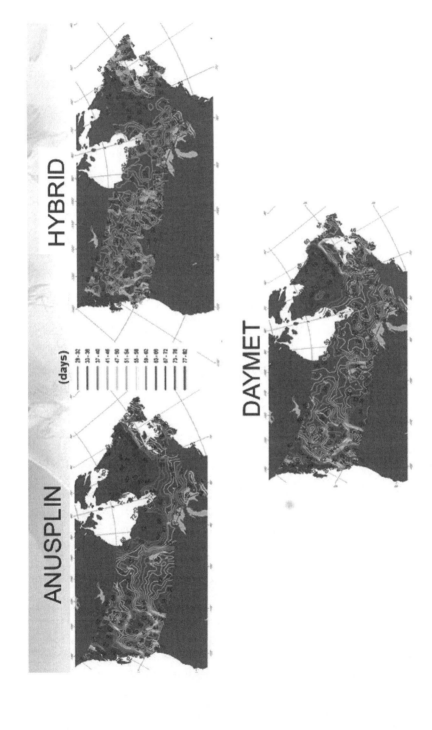

FIGURE 4.9: Regional variation in precipitation frequency during summer months (May–August), coinciding with the growing season across many regions of Canada between the interpolation models (smoothing spline, hybrid, and truncated-Gaussian).

representing, integrating and propagating all uncertainties to represent the overall confidence in forecasts for ecosystem decision-making.

Because many geostatistical interpolation methods (e.g., kriging) are sensitive to outliers and extreme observations they are strongly influenced by the marginal distribution of underlying stochastic/random field variation. This makes them unreliable when applied to extreme value or multi-modal pdfs. Kazianka and Pilz (2010) have applied multivariate *copulas* (i.e., multivariate joint pdfs or cumulative distribution functions (cdfs) with uniform marginal probability distributions of each variable). *Sklar's Theorem* states that any multivariate joint distribution can be written in terms of univariate marginal distribution functions and a "copula" that represents the dependence of the random variables. This approach enables modeling of the dependence structure (e.g., spatially-dependent covariates) separately, especially when extremes of two or more random variables depend on each other (i.e., simultaneous extremes). It also does not necessarily rely on the Gaussian or normality assumption that often does not hold for many ecosystem/environmental processes. In summary, real-world applications of sensor and station monitoring network data demonstrate the importance of context regarding how variables are measured, and causality assumptions. They also showcase how variability and spatial dependence can vary substantially as dominant influences trade-off as they evolve dynamically. This suggests the need to embed conditional probability into network models (e.g., Bayesian Belief or Deep Belief Networks).

Cannon (2009) has developed a flexible, *nonlinear* modeling framework using the GEV distribution and artificial neural networks (ANNs) to model non-stationarity in extreme values. Correa et al. (2009) highlight how ANNs, unlike other empirical and statistical methods, can generate unreliable predictions, especially when making real-time forecasts in real-world operational deployment or applications. In general, there are two main types of ANNs, namely: feed-forward (acyclic) (FNNs) and recurrent (cyclic) (RNNs) types. Kotsiantis et al. (2006) review ANNs in the context of other supervised, machine learning models, techniques and algorithms—and substantial gains in prediction achieved by integrating different learning classifiers. A standard neural network (NN) consists of many simple, connected processors called neurons, each producing a sequence of real-valued activations. Input neurons get activated through sensors perceiving the environment, other neurons get activated through weighted connections from previously active neurons. Neurons may influence their environment by triggering actions. ANNs have had difficulty forecasting beyond the historical reference, training or calibration range and must be carefully re-calibrated accounting for data quality, unanticipated events and trend patterns. Opportunities and limitations of the ANNs in ecological modeling are critically discussed by Schultz et al. (2000), including their requirement for large amounts of data for reliable training and validation. Where sufficient and representative data is not available, statistical bootstrapping or the use of genetic algorithms to optimize input data can help improve their prediction. Nonetheless, rapid progress and improvements are being made in ANNs involving *Deep Learning (DL)* for accurately learning and assigning (i.e., reinforcing) network weights (i.e., developing accurate *credit assignment paths* of causal links between actions and effects) across many computational stages, where each stage nonlinearly transforms the integrated or aggregate activation of the network (Schmidhuber, 2014). Depending on the real-world problem the interconnection of neurons (i.e., network design), long causal chains may be necessary to achieve desired levels of accurate prediction/forecasting. Huang et al. (2015) provide a comprehensive review of recent progress on Extreme Learning Machines (ELMs) for calibrating or training FNNs (feedforward neural networks) having a single hidden layer of neurons. Deo and Sahin (2015) have recently applied an Extreme Learning Machines (ELM) model to predict drought duration and severity, offering considerable improvements beyond standard ANNs. Learning strategies for adaptation within network models whose topologies and states coevolve are being explored in real-world CAS (Chandrasekar et al., 2014; Sayama et al., 2013).

Cumulative effects that arise from longer-term temporal autocorrelation (e.g., inter-annual) involving multiple, non-local climate teleconnections (e.g., ENSO) can add additional uncertainty, as well as helping to extend the forecasting window and improve early-warning forecasting, depending on the forecasting methodology employed. Climate teleconnections involve periodicity or oscillations in the variability in sea-level pressure (SLP), sea-surface temperature (SST) and/or normalized geopotential height (GPH) (i.e., anomaly departures) impact regional climate and drive extreme event activity (e.g., droughts, floods, heat-waves, cold spells). Recently, their spatial impacts or "tele-footprints" have been mapped using functional data analysis (FDA) to guide improvements in monitoring model-based inferences by accounting for multi-scale temporal (i.e., daily, monthly, annual, interannual) and spatial (i.e., station, subregion, region, national) variation and uncertainty (Newlands and Stephens, 2015; Bonner et al., 2014; Shabbar and Skinner, 2004).

4.4 Integrating Causality

A Bayesian (belief) network (BN) (also called belief networks or causal networks) and MRFs are probabilistic graphical models. A BN is similar to a Markov random field (MRF) in its representation of dependencies, but while MRFs are *un-directed* networks and can be *acyclic* and/or *cyclic*, BNs are *directed* networks and strictly *acyclic*. BNs, by considering directionality, can consider more complex representations of conditional independence (i.e., causality) and use the Bayes rule for inference and can utilize both frequentist and Bayesian methods for estimating parameters of its conditional probability distributions. A BN is termed a *directed acyclic graph (DAG)* that contains a probabilistic description of relationships among variables in a given system.

Given a DAG and a joint distribution P over a set $\mathbf{X} = (X_1, X_2, ..., X_n)$ of discrete random variables, we say that G represents P if there is a one-to-one correspondence between the variables in $\mathbf{X}$ and the nodes of G, such that P admits a recursive product decomposition (Pearl, 2004),

$$P(X_1, X_2, ..., X_n) = \prod_i P(X_i | pa_i) \qquad (4.27)$$

where pa_i are the direct predecessors (i.e., parents) of $X_i \in G$. For the parent set, pa_i of nodes, each variable X_i is conditionally independent from predecessors, depending only on the immediately preceding state (i.e., satisfying the Markov temporal property), whereby a child-parent family in a DAG, G, represents a function, $X_i = f_i(pa_i, \epsilon_i)$. The graphical structure of a BN explicitly represents cause-and-effect assumptions that allow a complex causal chain linking actions to outcomes to be factored into an articulated series of conditional relationships. Underlying the graphical representation of the model as a network with flows/arcs, is a hierarchical model of mathematical/statistical equations that relate input, intermediate and output parameters and variables. The conditional independence implied by the absence of a connecting arrow simplifies an overwhelming, complex modeling process by allowing separate submodels, which could be, in theory, integrated at a later stage. Each interrelationship can then be independently quantified using a submodel suitable for the type and scale of information available. For example, submodels can be developed from any combination of process knowledge (data), statistical interpretation (regression analysis) and/or expert judgment.

Unlike artificial neural network models (ANNs) that have the disadvantage of not having symbolic reasoning and semantic representation, BNs enable exploring and inferring real-world causality assumptions to obtain a greater integrated understanding of the structure and function of ecosystems (Correa et al., 2009). They can be readily applied to integrated policy development and analysis, because they can explore *counter-factual reasoning* involved in real-world planning and rapid appraisal. This reasoning is needed to develop a robust action, remedial, or retrofit plan, especially when abrupt and/or large deviations (e.g., extreme events) occur between what is anticipated or expected and real conditions (measured or observed) and what caused the deviation needs to be identified (Pearl, 2004). They have been applied in a diverse range of real-world problems spanning many scales—from natural resource management (Barton et al. (2012); McCann et al. (2006)), landscape and urban planning (McCloskey et al. (2011)), multi-objective remediation involving non-point source pollution (Dorner et al. (2007)), the prediction of regional energy feedstock distributions (Newlands (2010)) and the modeling of biological processes in cells involving gene regulations, signal transductions and interactions between proteins. In genetic studies, ANNs have been used to predict dominant protein-protein interactions in genomic data (Jansen et al. (2003)) and gene regulatory networks from DNA microarray data (Huang et al., 2007). Nariai et al. (2005) integrate BNs and MRFs in an *integrated* method for estimating gene-regulatory networks and protein-protein interaction networks *simultaneously* according to the reliability of each biological information source, assigning a penalty to coexisting directed and undirected edges between genes in a BN network. BNs offer many benefits, including:

- Enabling diagnostic-reasoning with interdisciplinary science, industry and policy stakeholders

- Providing a flexible, scalable modeling framework
- Making conditional dependencies explicit and dynamic (contemporaneous vs. non-contemporaneous)
- Gauging model structural uncertainty, in addition to parameter uncertainty
- Automating the learning process making generating new predictions more efficient when new data or knowledge becomes available

Key challenges in applying BNs in the environmental sciences are: 1) their limited ability to handle continuous data, whereby data generally needs to be discretized, thereby impacting the resulting model due to its dependence on the number of intervals and division thresholds, 2) collecting and structuring expert knowledge on a system is difficult, especially where data is sparse or where there is a lack of consensus on interpretation, 3) representing system knowledge in terms of probability distributions and inter-dependencies, given a lack of training data, and 4) little support for feedback loops, whereby dynamic BNs are discontinuously staged in time as a series of separate networks (Uusitalo, 2007). Johnson et al. (2011) provide best-practice examples for integrating BNs and Geographical Information Systems (GIS) that offers great benefits in dealing with multi-scale input data and model outputs, and enabling a platform for visualization that helps to reduce cognitive biases in expert judgment and at the same time, facilitates posing more complex and integrated hypotheses.

BN methods assess model structural uncertainty by searching, proposing, and testing candidate models/BN networks with a search heuristic or score (i.e., commonly rules that govern reversal, removal, addition of arcs and/or nodes). Structural learning is crucial in assessing a model's representational power, appropriate level of complexity (number of nodes and arcs), and generating useful/insightful inferences in various applications. A scoring function is then applied to assess, for any two candidate networks, the model structure with the highest likelihood, given the observation (training) data. Several different search-and-score algorithms exist that have different extents of search supervision and conditional probability (i.e., node and arc correlation constraints): Greedy-Equivalence (Chickering, 2002), Gradient/Hill-climbing (Friedman et al., 1997), k-NN Nearest-Neighbor (Pernkopf, 2005), Naive-Bayes (NB), Tree-Structured Naive-Bayes (TAN), General Bayes (GB) (Madden, 2009), TAN with abstraction hierarchies (Desjardins et al., 2008), and global optimization approaches (i.e., genetic algorithm (GA), simulated annealing (SA), particle swarm optimization (PSO)) (Oliveira et al., 2015; Panigrahi et al., 2011; Kang et al., 2008; Tripathi et al., 2007; Van den Bergh and Engelbrecht, 2006).

Just like ecological sensor networks, and human social groups, animals organize themselves to communicate and share information to minimize risks and trade off different survival actions having different costs within an uncertain environment. Collective movements are driven by shared decision making in a wide range of animal groups, where false alarms occur as groups with different movement and sensory abilities adapt their collective response to changing hazard probabilities and levels of uncertainty (Rosenthal et al., 2015; Parrish and Hamner, 1997). For example, highly migratory ocean predators, such as Atlantic bluefin tuna (*Thunnus thynnus*

L.) form distinct schooling structures as individuals take actions to dynamically adjust the school size, shape, and structure linked with a collective trade-off between group compactness and elongation, conferring hydrodynamical/energetic and visual advantages to its member individuals (Newlands and Porcelli, 2008). Shared decision making within human societies and social/cultural groups has added complexity (compared to animal ones), because human language allows the expression of much more complex decision contents (Conradt and List, 2009). Linguistics, in addition to differences in the level of knowledge development, type of information sharing mechanisms, rationality/irrationality, and incentive and reward pay-off structure. As Conradt and List (2009) explain, "One of the most significant differences between human and non-human group decisions lies in the role that language can, or cannot, play in such decisions. While humans and non-humans share the capacity both to communicate prior to making a decision and to decide, by voting or acting, to bring about a particular outcome, the nature of the communication and decision in the two cases is very different. In the non-human case, communication takes the form of the exchange of relatively simple signals and the subsequent decision consists of the support for one particular option or in the choice of a concrete behavioral strategy. In the human case, by contrast, the expressive resources of language can make the communication stage and the decision stage much more complex." This implies that potential states and behavior (e.g., decision-making actions), and losses must be integrated.

4.5   Statistical Attributes of Integrated Risk

In Equation 1.4 (Chapter 1), a general equation for *integrated risk* that comprised a superposition of hazard and cost probability distribution functions (pdfs) over all possible (i.e., considered) potential events and associated costs or consequences of such events was presented and discussed. With the added context on environmental losses/costs provided in Chapter 2, and the background on Bayesian theory provided in Chapter 3, we now detail further this *integrated risk* function in terms of different statistical attributes of decision-making processes, and the extent that such decisions consider prior information and uncertainty.

In general, the *conditional expected loss* or *risk*, associated with taking an action a_i, and loss or cost function, $L(a_i|\theta_j)$, when the hazard probability that the *true* state of nature is $p(\theta_j|\mathbf{x})$ of a possible set of c possible states, based on evidence or a feature vector, $\mathbf{x}$ is given by,

$$R(a_i|\mathbf{x}) = \sum_{j=1}^{c} L(a_i|\theta_j)p(\theta_j|\mathbf{x}) \qquad (4.28)$$

This implies that an action can be selected that minimizes the conditional risk based on evidence, such that an action vector can be formed with actions associated with a given evidence feature vector $\mathbf{x}$, yielding the equation for overall or *integrated* risk,

$$R = \int_i R(a_i|\mathbf{x})p(\mathbf{x})d\mathbf{x} = \int_i L(a_i|\theta_j)d\{p(\theta_j|\mathbf{x})p(\mathbf{x})\}. \tag{4.29}$$

To minimize the integrated risk, it is necessary to compute the conditional risk of every possible action and its associated costs and hazard probability, and then to select the action a_i for which $R(a_i|\mathbf{x})$ is a minimum. This minimum or optimal risk, $R(\hat{a})$ is called the *Bayes risk* and is the best performance that can be *realistically* achieved. When the average or expected loss is minimal, the optimal action or decision rule is termed the *Bayes rule* and satisfies, $R(\hat{a}) \geq R(a)$, $\forall a \in \mathbf{A}$ such that,

$$\hat{a} = \arg\min_{a_i \in \mathbf{A}} R(a_i|\mathbf{x}) = \arg\min_{a_i} \sum_{j=1}^{c} L(a_i|\theta_j)p(\theta_j|\mathbf{x}). \tag{4.30}$$

Empirical risk (R_e) can be defined, involving a prediction function, f as,

$$R_e = \frac{1}{N}\sum_{i=1}^{N} L(a_i|f(x_i, \theta_j)), \tag{4.31}$$

and is constructed using a training dataset. This involves the induction principle of empirical risk or structural risk minimization (ERM), that assumes that the prediction function, $f(x, \hat{\theta}_j)$ that minimizes the empirical risk, R_e, results in a risk, R that is close to its minimum. In the above equations for risk, a_i and θ_j in a decision-making framework or context can be translated into the statistical inference context, as θ, a model parameter and $\hat{\theta}$, its best possible estimator, respectively. The *VC bound*, as the upper bound of integrated risk is related to the empirical risk and can be used as a risk performance measure. With probability $1 - \eta$, $(0 \leq \eta \leq 1)$ and ν defined as the *VC dimension* of a learning machine,

$$R \leq R_e + \sqrt{\frac{\nu(log(2N/\nu) + 1) - log(\eta/4)}{N}}. \tag{4.32}$$

The VC dimension is a measure of the capacity of a statistical classification algorithm as the difference between integrated (Equation 4.29) and empirical (Equation 4.31) risk (Huang et al., 2015; Vapnik, 2000).

Many different loss functions are possible. They are generally assumed to be non-negative functions, whereby $L(a_i, \theta_j) > 0, \forall a_i$ and θ_j. A zero-one or a binary loss function is given by,

$$L(\theta|\hat{\theta}) = 1 - \delta_{ij} = \begin{cases} 1 & \text{if } i \neq j \\ 0 & \text{if } i = j \end{cases}. \tag{4.33}$$

Zero-one loss corresponds to the minimum *error rate* where all errors are equally costly and the risk is the *expected average* probability of a type I or false alarm error (refer to Figure 4.3 representing the case of two class probability distributions). The risk or expected loss is the *maximum* of the posterior probability or a *max a posteriori* MAP estimate that penalizes all errors by the same cost. Furthermore, if error is penalized by the same cost and prior probabilities are assumed to be uniform, then the maximum likelihood (ML) estimate is optimal (Viswanathan and Varshney, 1997). In addition, absolute and L^p loss functions are given by, $L(\theta|\hat{\theta}) = |\theta - \hat{\theta}|$ and $L^p(\theta|\hat{\theta}) = |\theta - \hat{\theta}|^p$, respectively. In the case of zero-one and absolute loss, the Bayes estimator is the median and mode of the posterior distribution, $p(a_i|\theta_j(\mathbf{x}))$, respectively. A square loss function has $L(a_i|\theta_j) = (a_i - \theta_j)^2$. The risk for this loss function is the *mean squared error (MSE)*, which decomposes into a variance, $\sigma^2(\cdot)$ and a bias, $\lambda^2(\cdot)$ contribution (i.e., bias-variance trade-off).

$$MSE = E_{a_i}[(a_i - \theta_j(\mathbf{x}))] = \sigma^2_{ij,a_i}[\theta_j(\mathbf{x})] + \lambda^2_{ij}[\theta_j(\mathbf{x})] \quad (4.34)$$

This implies that in the case of square loss, the minimal expected loss is the *covariance* of the posterior distribution.

To perform well over a range of prior probabilities such that the worst overall risk for any value of the priors is as small as possible, one can minimize the maximum possible overall risk (so-called *minmax* or *minimax* risk). Minimax is a strategy (i.e., sequence of actions) of always minimizing the maximum possible loss which can result from a choice that a player makes, as in the case in a finite two-player, zero-sum game, whereby a person can win only if the other player loses and no cooperation is possible. In such two-person, zero-sum games having a finite number of strategies, there exists a value, λ and a *mixed* strategy for each player, whereby the best payoff possible for one player is $+\lambda$, while the best payoff possible for the second player is $-\lambda$ (i.e., zero-sum). This optimal solution is termed the *Nash equilibrium*, whereby each player receives a payoff that is equal to both their maximin and minimax values (Nash, 1951; Von Neumann, 1928). *Alpha-beta pruning* dramatically increases the computational efficiency in obtaining minimax optimal solutions of games by decreasing the number of nodes that are evaluated within the search trees by stopping to evaluate moves (i.e., pruning of potential branches) when such moves are worse than previously evaluated ones. This algorithm belongs to the "branch-and-bound" class of optimization search methods (Russell and Norvig, 2010; Pearl, 1982). The *minimax risk* is given by,

$$R = \inf_{\hat{\theta}} \sup_{\theta} R(\theta, \hat{\theta}), \quad (4.35)$$

where the *inf* or *infimum* is over all estimators, and an estimator $\hat{\theta}$ is a *minimax estimator* if,

$$\sup_{\theta} R(\theta, \hat{\theta}) = \inf_{\hat{\theta}} \sup_{\theta} R(\theta, \hat{\theta}). \quad (4.36)$$

In the above equations, *sup* is the *least upper-bound*, and *inf*, the *greatest lower-bound*, and it generalizes the definition of *minimax* risk for functions with no maximum or minimum. For the case where R has a global minimum and maximum, $inf \rightarrow min$ and $sup \rightarrow max$.

Different strategies can be associated with different risk measures (i.e., the *average or expected risk* (realist), *maximax* (optimist or risk seeker), *maximin* (pessimist or risk adverter) and *minimax regret* (opportunist or risk neutral). *Value-at-risk* (VaR) and *conditional-value-at-risk* (CVaR) from economic portfolio theory are downside (maximin-type) risk measures based on distribution quantiles. VaR is defined as a threshold value associated with a specified confidence level of outcomes. Given a cumulative distribution function (cdf) of $\mathbf{X}$, $F(\mathbf{X}) = P(\mathbf{X} \leq \mathbf{x})$, the VaR with confidence level, $\alpha \in [0,1]$ (Heckmann et al., 2015), is,

$$VaR_\alpha(\mathbf{X}) = min_\alpha(\mathbf{Z}|F_\mathbf{X} \geq \alpha). \qquad (4.37)$$

$CVaR_\alpha(\mathbf{X})$ is the conditional expectation of $\mathbf{X}$ subject to $\mathbf{X} \geq VaR_\alpha(\mathbf{X})$, or the *average loss* beyond a specified confidence level (i.e., lower α-percentile). While a great deal of research has been undertaken involving two-player games, many games and most real-world domains involve far more than two parties, whether in cooperation or in competition minmax risk in multiplayer games, often with a tree network topology (Sturtevant, 2003). It is also generally assumed that actions coincide with posterior probability distributions for states of nature θ_j and that actions coincide with the time of decision-making. This is a very strong assumption, and realistically there are significant time-delays between the integration of scientific evidence and determination of state probabilities, to when decision-making processes occur, to when actions take place. Hickey et al. (2009), for example, have applied loss functions in assessing threshold toxicity factors for different species in an ecosystem based on the Bayes risk factor, as a way to statistically extrapolate from specific concentration data from a small sample of species to values of hazardous concentration to a toxic substance p denoted p% (termed HC_p) of a community assemblage of biological species. In their study, they assess loss functions which punish over- and under-estimation that can be asymmetric and non-linear to better represent the reality of risk assessment scenarios. Norstram (1996) has defined an asymmetric loss function called a *precautionary loss* function (PLF) applicable when underestimation is anticipated to have serious potential negative impacts. Hasan et al. (2013) has also more recently explored the application of the modified squared error loss functions (MSELF) and PLF, finding that they are better choices for optimal decision making in mixture experiments, where response variables are logistic or Poisson-distributed.

4.6   Stochastic Sustainability

Recall from Chapter 1, Section 1.3, that sustainability assessment frameworks (SAFs) integrate *metrics* as "standard of measurement" (i.e., ecological metrics, economic metrics, and sociological metrics) and *indicators* that measure the state of a metric for tracking progress towards sustainability objectives (Ahi and Searcy, 2015, 2013). Indicators of the state of a system are *content* indicators, while those that track the behavior of a system, are *performance* indicators (Sikdar, 2003). While sustainability planning, identifying needs, potential risks, impacts, and actions may possess a degree of uncertainty in terms of the state of Nature and future possible states, measurable loss functions, causality between variables, scientific and technological capacity and learning rate and more—such a decision-making process is far from fully stochastic. Nonetheless, the integration of stochasticity and uncertainty *explicitly* within SAFs ensures that they are reliable under real-world, CAS dynamics.

Hierarchical frameworks construct 2-D and 3-D metrics and indicators based on trade-offs and intersections between 1-D metrics and indicators (i.e., simultaneously measuring more than one aspect of an ecosystem). Here, the goal is to have a reduced set of sustainability metrics that are independent of each other as much as possible. Metrics and indicators are optimally designed based on the general criteria: 1) simple (not requiring large amounts of time and manpower to develop), 2) useful (management decision-making and relevant to business objectives), 3) understandable (to a wide variety of audiences, stakeholders, with statistical and quantitative knowledge expertise), cost-effective (in terms of data collection, verification and validation), reproducible (integrating decision rules that produce consistent outputs that can be inter-compared across different assessment and forecasting methodologies), robust (causally linked to improved sustainability), scalable or stackable so that they can be applied across supply-chains, not dependent solely on proprietary information (an open, transparent way to enable independent verification and validation) (Schwarz et al., 2002). Heckmann et al. (2015) defines *supply chain risk* as, "Supply chain risk is the potential loss for a supply chain in terms of its target values of efficiency and effectiveness evoked by uncertain developments of supply chain characteristics whose changes were caused by the occurrence of triggering-events."

An example of an Environmental Sustainability Index (ESI) (ESI Score) that encompasses environmental health and ecosystem vitality, sustainability objectives, has been developed by Yale University's Center for Environmental Law and Policy, and Columbia University's Center for International Earth Science Information Network in collaboration with the European Commission's Joint Research Centre (JRC) and the World Economic Forum since 2005. Originally, in 2005, the ESI index spanned 5 thematic categories equally weighted across 21 indicators, and 76 variables, but this has been broadened to include 9 categories, integrating 20 indicators (Hsu, 2016; Esty et al., 2005). ESI categories are extensive, but not necessarily comprehensive. For example, within the *Agriculture* category, there are only two metrics, namely agricultural subsidies and pesticide regulation, whereas, similar to other context and content gaps related to industrial sectors, there are many more key metrics that dictate the state and behavior of agricultural production systems (refer to Figure 1.16 in Chapter 1) (Kouadio and Newlands, 2015; Newlands et al., 2013). Also, typically sustainability frameworks do not include risk measures as metrics or indicators. Nonetheless, indicators measure both system current state and changes in state or process improvements via reduction gradients across its main components: 1) Air Quality, 2) Water and Sanitation, 3) Water Resources, 4) Agriculture, 5) Forests, 6) Fisheries, 7) Biodiversity and Habitat, 8) Climate and Energy, and 9) Health Impacts (see Figure 4.10). Following data pre-conditioning with logarithm, square-root or power-function transformations, a Bayesian approach is used involving MCMC simulation to impute or fill gaps in data, assuming multivariate normality and samples from the posterior distribution of missing data given the observed data. Imputation affects the sensitivity (or relative robustness) of the

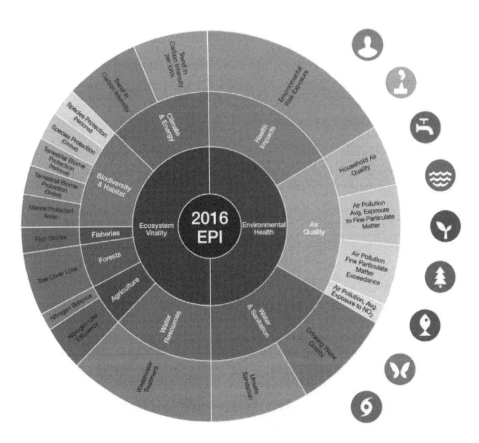

FIGURE 4.10: The 2016 EPI framework includes 9 issues and more than 20 indicators. Access to Electricity is not included, as it is not used to calculate country scores (Hsu, 2016).

ESI index. The ESI index weights indicators and variables equally, and variables are aggregated on the same scale (i.e., normalized) in minimizing distortions for variables that have large values or variances. This is a great simplification—findings from network models and Integrated Assessment Models (IAMs) could help to inform how to specify and learn weighting values alongside justifications for causal linkages. Jain (2005) suggest superimposing a preferential structure in forming utility or value trade-offs in ecosystem services and associated short, medium, and longer-term policy choices as context-specific as possible—such as to a specific industrial sector and region within a country. At the country level, it is expected that indicators have different weights, but a generalizable (i.e., globally applicable) differential set of weights has not been devised yet to enable an unbiased international inter-comparison. Quality assessment of the ESI variables can be used to identify, for example, urban air quality and terrestrial water quality measures, despite the need for substantial improvements in their spatial coverage. As well, quality assessment indicates there is a current need for more consistent times-series of biodiversity indices, land salinization and acidification. Further uncertainty and sensitivity analysis, whereby key assumptions of the ESI index are relaxed were also explored, such as: (1) variability in the imputation of missing data (2) equal versus expert weighting of indicators (3) aggregation at the indicator versus the component level, and (4) linear versus non-compensatory aggregation schemes. Improvements in this index aim to address aspects of poor quality and coverage of available data, data skewness and heteroscedasticity and inconsistent methodologies. *Heteroscedasticity* is defined for a finite sequence of random variables, Y_t and X_t, associated with the conditional expectation, $E(Y_t|X_t)$; the sequence Y_t is termed heteroscedastic if the conditional variance, $\sigma^2(Y_t|X_t)$ changes with t. In the latest, 2014 EPI index, metric and indicator weights have been introduced, along with a "proximity-to-target" methodology for assessing how close a particular country is to an identified policy target, such as a high-performance benchmark aligned with international or national policy goals or established scientific thresholds.

A composite or integrated environment, health and sustainability or EHS Index, comprising 65 environmental health and 10 socio-demographic metrics, has also been constructed by the EPA, in collaboration with the Partnership for Sustainable Communities with the U.S. Department of Housing and Urban Development (HUD) and the U.S. Department of Transportation (DOT) and used to gauge and compare the overall performance ranking of 50 US major cities (Gallagher et al., 2013). Findings reveal strong associations between public health metrics and environmental factors, including transportation systems, land-use, recreational spaces (e.g., parks), housing, and energy production. Many of the health metrics integrated are recommended by the WHO European Healthy Cities Network (WHO-EHCN). GIS mapping was also employed to measure spatial patterns and the relative distances to environmental health hazard sources, and urban neighborhoods with varying levels of health inequity. Urban ecosystems (e.g., cities) with the highest gains in population had a more inadequate water supply, congestion, and poorer air quality.

Findings from the assessment of index-based sustainability approaches (such as the EHS Index across urban areas) highlight the importance of utilizing causal-based weights, determined from integrated data or models. There is also a need for preferential indicators, specific to different geospatial aspects and contexts, in order to ensure that sustainability indices (and their frameworks) can be inter-compared and cross-validated (e.g., using interviews, observational surveys, case-based analyse, and experimental studies). Such adaptive verification and validation seeks to enhance their reliability, and in turn, their wider adoption, extent, and reporting quality. As Hahn and Kühnen (2013) identify in a review of many frameworks and research studies, quality assessment and issues in sustainability reporting are often neglected, and while research studies and indices strive to be integrative, how they address stakeholders' informational needs and can be coupled to informing policy decisions and empowering sustainability actions (e.g., site assessment and remediation) remains a challenge (Martins et al., 2007).

Inconsistency is a real issue if sustainability frameworks do not include metrics and indicators that are major drivers. Such variables must be integrated to assess and forecast component states and behaviors within IAM as well as life-cycle assessment and forecasting models. Sustainable frameworks therefore need to be developed in congruence with IAM, Life Cycle Assessment (LCA) and forecasting methodologies and their findings. This would build capacity for more informed and coordinated action, whether at the global, national, regional, community, ecosystem or site level. *Environmental Site Assessment (ESA)* and *remediation* involve the removal of pollution or contaminants from environmental media such as soil, groundwater, sediment, or surface water. This would mean that once requested by the government or a land remediation authority, immediate action should be taken as this can impact negatively on human health and the environment. *Remediation* is generally subject to an array of regulatory requirements, and also can be based on assessments of human health and ecological risks where no legislated standards exist or where standards are advisory. Clarvis et al. (2015) emphasize the current need for greater integration of environmental risks into investment decisions and outline an analytical methodology called the E-RISC (Environmental Risk in Sovereign Credit) framework to identify and quantify economic risks that a country is exposed to because of its natural resource consumption. This framework accounts for the role credit rating agencies (CRAs) have on sovereign credit ratings and their interest rates, borrowing costs, debt sustainability, and the ability to finance and invest in new sustainability action initiatives involving emerging environmental risks. It integrates 20 qualitative and quantitative indicators across four dimensions (resource balance, trade-related, degradation-related risk, and financial resilience). *Resource balance* measures the ratio of a nation's ecological footprint to its biocapacity, *trade-related risk* evaluates its exposure to natural resource price volatility and supply disruption, *degradation-related risk* assesses exposure to declining productivity in ecological assets due to resource over-harvesting, and *financial resilience* appraises the ability of a nation to respond to adverse macroeconomic shocks or extremes. Clarvis et al. (2015) apply the E-RISC framework to Brazil, France, India, Japan, and Turkey and demonstrate the wide variety of resource profiles that countries possess. In many countries, agricultural production is increasing in its risk level, as despite large economic outputs, large populations of people are increasingly dependent on it.

As incremental actions, adaptations, and progress toward sustainability goals take place at different scales, one way to reflect such changes is to have sustainability frameworks be informed more directly by findings from process-rich IAMs, context-rich LCA and geospatial-rich or regional-specific, ecological footprint assessments. LCA is an approach that has become increasingly adopted to identify, quantify, and evaluate the total potentials of environmental impacts on industrial production processes or products—from the procurement of raw materials (the cradle), to the production and utilization (the gates) and their final storage (the grave), as well as for determining ways to repair damage to the environment (Skowronska and Filipek, 2013). The LCA can be defined as a process of gathering and evaluating input and output data and assessing a product's potential effects on the environment during its life-cycle. According to the guidelines of ISO 14040, LCA consists of four main phases:

- Determining the goal and scope of the study (choosing the functional unit and system boundaries)
- Analysis of an inventory of inputs and outputs (analysis of the technological process, balance of flows of raw materials, energy, and auxiliary materials, waste balance, and identification of their potential sources)
- Assessment of the environmental impact of the life cycle (transforming the data collected into impact category or damage category indicators)
- Interpretation (conclusions and verification of results)

In addition to LCA, another methodology for estimating human demand compared to ecosystem's carrying capacity is *Ecological Footprint* accounting. Rather than speculating about future possibilities and the limitations imposed by carrying capacity constraints, Ecological Footprint accounting provides empirical, non-speculative assessments of the past. Ecological Footprints represent how much biologically productive area is required to produce the resources required by the human population and to absorb humanity's carbon dioxide emissions. Approximately 90% of all leading Ecological Footprint practitioners worldwide have joined the Global Footprint Network (GFN) and have agreed to adhere to these standards and to use a common set of data [13]. The Ecological Footprint Standards (2009) are the current operational standards that we use with all of our partners and businesses. The 2009 Standards build on the first set of internationally recognized Ecological Footprint Standards, released in 2006, and include key updates—such as, for the first time, providing standards and guidelines for product and organizational Footprint assessments. The Ecological Footprint is often used to calculate global ecological overshoot, which occurs when humanity's demand on the biosphere exceeds the available biological capacity of the planet. By definition, overshoot leads to a depletion of the planet's life supporting biological capital and/or to an accumulation of CO_2 emissions.

Because sustainability extends beyond the boundaries of any one organization,

---

[13]GFN, www.footprintnetwork.org/en/index.php/GFN/

assessing the performance of supply-chains is more complex as they integrate a wide array of different organizations and real-world components, steps, and processes, and producers, suppliers, distributors, retailers to customers. A recent comprehensive review of green supply chain management (GSCM) and sustainable supply chain management (SSCM) by Ahi and Searcy (2015) reveals a total of 2555 unique metrics (on the basis of scientific literature up to 2012) and a lack of consensus on how to measure sustainability performance across supply-chains and propose a well-informed, conceptual framework for GSCM and SSCM sustainability assessment now available for further application and broader validation. This is compounded by a lack of agreement on the definitions of GSCM and SSCM whereby there are a total of 22 definitions for GSCM and 12 definitions for SSCM being used and applied (Ahi and Searcy, 2013). While a diversity of perspectives in terms of definitions is useful, at the same time, inconsistent definitions may potentially confuse and limit consensus-building. Inconsistent definitions can also potentially prevent the integration of different (or "hierarchical") metrics with the aim to improve the reliability of sustainability assessments of supply-chains and to monitor incremental progress (Sikdar, 2003; Graedel and Allenby, 2002). Metrics must be measurable (Rossi, 2007).

While there exists some consistency in the use of air quality, GHGs, energy use, and energy consumption, over one-third of identified metrics were found to be cross-cutting (i.e., integrating or linking multiple characteristics at once). Moreover, no metrics were context-based that benchmark what sustainable impacts over time should be or profile what they should look like, and there is a need for metrics that address the entire spectrum of supply-chain considerations. Similarly, a study of sustainability reporting by Canadian corporations indicates the use of a total of 585 different indicators in 94 reports from 2008, and a strong dependency on indicator selection within specific industrial sectors (Roca and Searcy, 2012). A total of 31 of the 94 reports included indicators linked to those specified by the Global Reporting Initiative (GRI) indicators[14]. Findings from this study point to the need for benchmarking and the need for corporations, industry associations, and levels of government to validate and refine sector-specific, supply-chain indicators, alongside other cross-cutting ones. (Waheed et al., 2009) discuss "Driving Force-Pressure-State-Exposure-Effect-Action (DPSEEA)" linkage-based sustainability assessment frameworks within the context of other current methodologies (e.g., objective-, impact-, influence-, process- and LCA-based approaches). Linkage approaches are developed from, "causal indicators that present a complete range of metrics to identify and measure a cause that create particular conditions affecting sustainability, the impacts of these causes, and the corrective actions that can be taken to address them." They propose integrated assessment to evaluate specific monitoring programs where goals and objectives are clearly defined. Niemeijer and de Groot (2008) also emphasize the need for causal-based network models for linking sustainability indicators. This admits that real-world (e.g., industrial) process-

---

[14]GRI, `www.globalreporting.org/`

changes of sufficient succession length and amplitude, via underlying causal linkages, can induce beneficial structural changes that were not previously evident.

Given a *deterministic* sustainability policy *objective* $\boldsymbol{\pi} = \pi(\beta, d, \nabla\lambda)$ having a social preference strength, β_i, anticipated reward d_i, and social adoption or social learning rate $(\nabla\lambda)$ across a time-horizon, $\nabla t = [t_i, t_f]$, under *stochastic* variability, $\boldsymbol{\xi} = \boldsymbol{\xi}(\mathbf{X}, \mathbf{Z})$, $\boldsymbol{\xi} \in \Omega$ and m potential outcomes or SES system realizations, Ω, the optimal policy scenario is given by,

$$I^* = \min_{\boldsymbol{\xi}} E\left[I_o(\boldsymbol{\pi}, \boldsymbol{\xi})\right] = \int_\Omega I_o(\boldsymbol{\pi}, \boldsymbol{\xi}) p(\boldsymbol{\xi}) d\boldsymbol{\xi} \qquad (4.38)$$

$$\text{subject to } E\left[I_i(\boldsymbol{\pi}, \boldsymbol{\xi})\right] \leq 0, i = (1, 2, ..., m) \qquad (4.39)$$

I_i is the objective function, and I_i are feasible constraints. At the core of such a stochastic sustainability index, is its reliance on a SES system scaling relation (i.e., decision or design vector), $\boldsymbol{\xi} = \boldsymbol{\xi}(\mathbf{X}, \mathbf{Z})$ (Ahi and Searcy, 2014; Krokhmal et al., 2011; Nakau, 2004) This is a link function involving a separable/non-separable matrix of the key statistical elements (i.e., variables) of ecosystem CAS pattern and process, $\mathbf{X}$ (i.e., heterogeneity, interaction, scaling) and process, $\mathbf{Z}$ (i.e., stochasticity, asymmetry, adaptation). Associated with this link function are auxiliary sensitivity and validation conditions that must be satisfied associated with the rate of change of the objective function and the feasibility of its constraints, $(I_o, I_1, I_2, ..., I_m)$, whereby,

$$\Delta\left[I_o, I_1, ..., I_m\right] = \Delta\left[R^p, R^s, R^{p \to s}, \epsilon^p, \epsilon^s, \epsilon^{p \to s}, q^p, q^s, q^{p \to s}\right] \qquad (4.40)$$

where the relation, $\boldsymbol{\eta} = \boldsymbol{\eta}(\mathbf{R}, \epsilon, \mathbf{q})$, integrates the monitoring and measurement of system reliability in achieving a given sustainability policy on the basis of statistical metrics of integrated risk $(\mathbf{R})$, sensitivity $(\boldsymbol{\epsilon})$, verification and validation $(\mathbf{q})$; each dependent upon the process $(\mathbf{p})$, structural $(\mathbf{s})$ and transitional (i.e., process to structure, denoted $\mathbf{p} \to \mathbf{s}$) changes or incremental improvements.

While the function, ΔI could be introduced as a penalty term to the optimization function, whereby $I_o \to I_o + \Delta I_o$, this does not take into account the structural change, whereby the set of constraint functions, I_i also change. To account for stochastic choices, nonlinear perturbation or "penalty" functions can also be added to the objective function, I_o, and/or individual or joint *chance constraints* that are functions of risk measures or metrics. For example, an additive, perturbed utility function (i.e., optimization cost or objective function) can be introduced, such that Equation 4.38, becomes,

$$I^* = \min_{\boldsymbol{\xi}} E\left[I_o(\boldsymbol{\pi}, \boldsymbol{\xi})\right] = \int_\Omega I_o(\boldsymbol{\pi}, \boldsymbol{\xi}) p(\boldsymbol{\xi}) d\boldsymbol{\xi} + E\left[\int (q_i^+ \Delta_i^+ + q_i^- \Delta_i^-)\right] \qquad (4.41)$$

where unit penalty costs, q_i^+ and q_i^- are assigned to surpluses, Δ_i^+ or shortages, Δ_i^-, respectively. Alternatively, one may restrict the feasibility region by introducing *individual chance (i.e., stochastic) constraints (ICCs)*,

$$P(\Delta_i^+ > 0) \leq \alpha_i, i = (1, 2, .., m) \qquad (4.42)$$

or *joint chance constraints*,

$$P(\Delta_i^- > 0) \leq \alpha, i = (1, 2, .., m) \tag{4.43}$$

where α_i and α are given risk parameters (normalized or defined within [0,1]) and where the interpretation of such chance constraints is that a solution is feasible only if it is not too risky (Haneveld and Van der Vlerk, 2006). Fudenberg et al. (2015) explore the conditions of *acyclicity* (ordering of choices via preference theory) and *selectivity* (consistency in the ratio of choice probabilities).

The stochastic sustainability problem fundamentally embeds monitoring and measurement, the risk related to a set of possible actions, and the sensitivity of the index to national, regional or sector-based negotiation and compromise. Overarching these elements is the social learning rate. There is also both *internal sensitivity* linked with fundamentally defining, measuring and assessing the index I, as well as *external sensitivity* linked with its real-world, operational application. Verification and validation supporting an indicator explicitly measure the reliability of such application. Ecological indicators need to capture the complexities of the ecosystem yet remain simple enough to be easily and routinely monitored (Dale and Beyeler, 2001). Abrupt event-times and/or more gradual transitions in energy, water, and food-based indicators, for example can be tracked by re-calibrating the stochastic sustainability control trajectory between transitions, or by decomposing it into multiple, interdependent subproblems to facilitate regional, rapid threat and impact appraisals (King and Jaafar, 2015). Peavoy et al. (2015) have just demonstrated that *stochastic* climate models are globally stable under nonlinear (i.e., quadratic form) perturbations. They also provide a Bayesian framework for inferring optimized parameters for stochastic systems. This framework consists of a set of stochastic differential equations or SDEs, which are constrained by the underlying physics of the climate system. A sustainability policy sequence toward a sustainable future objective, therefore, involves the integrated control of SES ecosystem processes and structures, human socio-economic processes and structures, and technological learning processes and structures. Integrated learning is crucial to guide, inform, and maintain support as scientific consensus evolves. As science and society learn from the modeling, assessment and forecasting of integrated risks (hazards, losses, costs), global sensitivity or robustness, and verification and validation metrics, different interrelated sustainability policy sequences can then be adjusted *in tandem* with one another.

From an integrated perspective of sustainability involving causal belief networks, deep learning, a stochastic sustainability indicator function can be defined. Adoption of such an indicator fundamentally re-frames sustainability planning and negotiation from one that focuses on reaching agreements on deterministic, comment policy objectives based on realized constraints, rewards and added incentives as part of an adaptive process (Cabezas and Diwekar, 2012). Moreover, Turnhout et al. (2007) argue that sustainability indicators can be expected to be used or rejected strategically, depending on policy context. Applying an integrated network approach with complementary goals in an objective function enables individual indicators to be selected that ensures that the representation of key information and characteristics of ecosystems are tightly coupled to policy and management objectives (Niemeijer and de Groot, 2008; Fath et al., 2001). Chambers (2008) has derived productivity indicator measures for stochastic technologies and stochastic decision environments, demonstrating their application under the assumptions of risk-neutral firms and a common expectation market-based equilibrium. He also highlights the shortcomings of non-stochastic productivity measures in the agricultural context, indicating the need for a broader application of such measures in the financial and insurance sector, that utilize stochastic payouts and products. Lin et al. (2009) outline an integrated, causal and hierarchical network framework, such that "...the hierarchical framework collects and arranges the relevant ecological elements around the general objective covering the key gradients across the ecological systems, while a causal network then identifies the key factors and connects them as indicators by causality chains." Through a process of continual adaptation, the stochastic sustainability challenge essentially seeks to promote and develop a capacity of "self-awareness" of SES systems, their dynamics and behaviors. This is developed in tandem with harnessing the diversity, heterogeneity, and capacity available to ensure greater possibilities and potential for sustainable learning, meaningful process, and realizable structural change.

4.7 Cumulative Impacts and Sensitive Trade-Offs

There is an increasing need for multi-criteria based decision-making due to more stringent market-driven sustainability requirements and greater public awareness of environmental risks. Many food retailers and manufacturers are looking beyond their own operations to realize improvements in environmental performance as an estimated 90% of the food industry's environmental footprint occurs in commodity production, an area outside their direct control. While there is increasing uncertainty and growing concern for sustainable development internationally, there is also increased availability of higher resolution datasets, together with progress in computation and statistics that continues to increase our ability to forecast ecosystem change. In addition to data and model complexity, informatics complexity also exists because of an ever increasing number of specialized tools, models, and software platforms under development worldwide to measure sustainability (Bezlepkina et al., 2011). Each of these methodologies is based on a different set of assumptions, sustainability measures (i.e., criteria/sets of indicators), and accuracy of input datasets. Currently, it is unclear which of these methodologies are best suited to accurately assess Canadian production and processing systems and enable compliance with emerging sustainability requirements.

Integrated assessment models (IAMs) help to address the above complexities, as they offer a system-level perspective of interrelationships between different ecosystem components, such as: ecosystem productivity, land capability/use/change, soil and air quality, water management, energy use, biodiversity (invasive species, habitat capacity), and socio-economic indicators (urban/rural population change, labor availability, and consumer food prices). Many IAMs lack the ability to forecast, and do not incorporate geospatial context and trends, because high-resolution, geospatial information for assessing regional-scale impacts, vulnerabilities, risks and uncertainties is often sparse and difficult to obtain (van Vuuren et al., 2011; Parson and Fisher-Vanden, 1997). Addressing such challenges the development of IAMs parallels similar advances in developing Earth System Models of Intermediate Complexity (EMICs) for exploring ways to bridge gaps between 3D comprehensive and simpler models. These models facilitate evaluation and comparison of historical and auxiliary/proxy observational data. These models need to be enhanced to enable exploration of the regional aspects of climate change, especially to be relevant in addressing extreme event impacts (i.e., droughts/floods).

Stanton et al. (2008) have reviewed 30 existing IAMs under four key aspects of climate change and resource economics. Their comprehensive study compares IAMs (i.e., welfare maximization, general equilibrium (GE), partial equilibrium (PE), simulation and cost minimization-type models). This inter-comparison examines their differing assumptions and model structure. This study also inter-compares the resulting predictions of the IAMs according to the uncertainty in climate change outcomes and future damage projections, equity across time and space, abatement costs, and technological change and transitions). In general, IAM model predictions were found to be highly sensitive to structural assumptions, making the ability to identify and prescribe a set of key model-based policy prescriptions more challenging. Moreover, their is a need to integrate the latest economic and climate change knowledge into the IAMs to keep them scientifically up-to-date and relevant for informing climate and sustainable development policy. This review highlights the need for more extensive sensitivity analyses and a more rigorous examination of risk and uncertainty of IAM models currently being used to inform climate policy. This study reveals that, "none of the models provide adequate information for formulating a policy response to the worst-case extreme outcomes that are unfortunately not unlikely enough to ignore" (Stanton et al., 2008).

An example of an IAM which has gained recognition as a leading agricultural sector model, is the International Model for Policy Analysis for Agricultural Commodities/Trade (IMPACT) model, developed by the International Food Policy Research Institute)[15]. While it was first developed as an agricultural economic model, it has expanded into a network of inter-linked economic, water, and crop models that integrate information from climate models (GCMs), crop simulation models (e.g., the Decision Support System for Agrotechnology Transfer (DSSAT) (Jones et al., 2003)[16], and water models (see Figure 4.11). This is a great example of how IAMs, with their integration perspective, facilitate broader knowledge integration and hierarchical modeling when addressing the complex real-world problems facing resource economic sectors. With its 30-year time horizon, the model generates

---

[15]International Model for Policy Analysis of Agricultural
  Commodities/Trade (IMPACT), www.ifpri.org/program/impact-model
[16]DSSAT, http://dssat.net/

long-term scenarios. It has been used to investigate the international, national- and farm-level production of key food commodities and demand under different climate and global change scenarios, and includes 58 (as of 2014) traded and non-traded agricultural commodities (Rosegrant et al., 2014). See Rosegrant (2012) for further details on IMPACT's equilibrium-based assumptions and the structure of this IAM and its broad application.

IMPACT is a *deterministic* model that specifies different area and crop-specific response functions for irrigated and rainfed crop cultivation. Harvested area changes with crop prices, projected trends in harvested area expansion and water availability, human population pressure and urban growth, land conversion away from farming and soil degradation. Historical and future trends in such factors are specified by combining both data on historical trends and expert judgment on the future projected changes. The model integrates future growth trend factors, alongside area- and yield-commodity price "elasticities." A module that simulates water availability, demand, and uses within different economic sectors at the global to regional scale, is also integrated within the IAM. It comprises two different models—a global hydrological (water-balance) model with a monthly time-step with a spatial resolution of 0.5° latitude by 0.5° longitude, coupled to a water model at the monthly and basic/sub-basin scales simulating and optimizing irrigation, livestock, and domestic and industrial water supply and demand. The model consists of 320 food producing units (FPUs) (across 159 countries). These units result from the intersection of 115 economic (geopolitical) regions and 154 water basins (Rosegrant et al., 2014; Nelson et al., 2010). IAMs are not only being developed and applied across resource sectors, but also within individual sectors. For example, IAMs developed specifically within the agricultural sector to address agricultural production and environmental health problems, include: *SEAMLESS-IF* (System for Environmental and Agricultural Modeling: Linking European Science and Society) to enable quantitative analysis and scenario-based exploration of the impacts of future policies; funded by the EU's 6th Framework Programme for Research Technological Development and Demonstration (Therond et al., 2009), and, *RegIS* (Regional Climate Change Impact and Response Studies in East Anglia and North West England) was developed as a stakeholder-driven methodology for regional climate change impact assessment has evaluated scales of impact, adaptation options, and cross-sector interactions between biodiversity, agricultural land use, water use, and land suitability (Holman et al., 2008)

Despite their increasing application use, many IAMs still do not comprehensively characterize, propagate, and analyze uncertainty—still a major criticism of many IAMs and a primary hallmark of good integrated assessment (Schneider, 1997; Morgan and Dowlatabadi, 1996). In addressing such gaps, a prototype, probabilistic/stochastic (i.e., Bayesian) IAM that treats uncertainty has recently been designed and tested for assessing and forecasting potential impacts of global change scenarios on the Canadian agricultural system at the regional-scale (i.e., scale of "ecodistricts") (Newlands et al., 2013). Ecodistricts are subdivisions of larger eco-regions that are characterized by having distinctive assemblages of relief, landforms, geology,

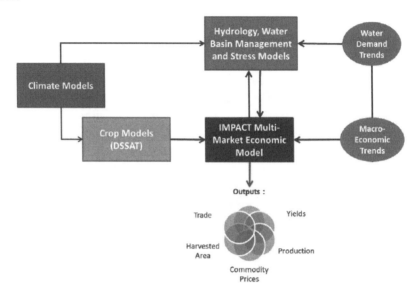

FIGURE 4.11: The IMPACT network system of inter-linked economic, water, and crop models (Rosegrant et al., 2014).

soil, vegetation, water bodies, and fauna[17]. This model's integrated design enables both parameter and structural sensitivity analysis and model cross-validation. In the initial testing of this model prototype, two major Canadian crops (i.e., spring wheat and canola) and production regions (i.e., Western Prairies, South-Eastern Canada/Ontario Great Lakes Region) were considered. Example mid-term forecast scenarios out to 2020 for crop yield (kg/ha) for spring wheat and canola within each of these two test regions are shown, in relation to the historical baseline and its trend (Figure 4.12). Also, the trends and the relative uncertainty associated with agri-environmental indicators (spatial covariates) that are selected by the model are profiled. These are: cropland net greenhouse gas emission (AirGHGIndex), cumulative growing season forcing temperatures for crop growth (i.e., summation of growing degree days, GSsumGDD), growing-season precipitation (GSSumPrecip), soil erosion index (soilErosionIndex) and the water contamination index (Water-ContamIndex).

One of the primary drawbacks is the availability of historical data on agri-environmental indicators—the length of their time record and its regional-scale coverage and resolution. The results of this study identified the importance of considering uncertainty, and particularly its spatial dependence, whereby at the finer ecodistrict scale, the crop response within a single or small set of ecodistricts can have a strong influence on the forecasted ranges at broader, regional scales. This

---

[17]Canadian National Ecological Framework,
  http://sis.agr.gc.ca/cansis/nsdb/ecostrat/index.html

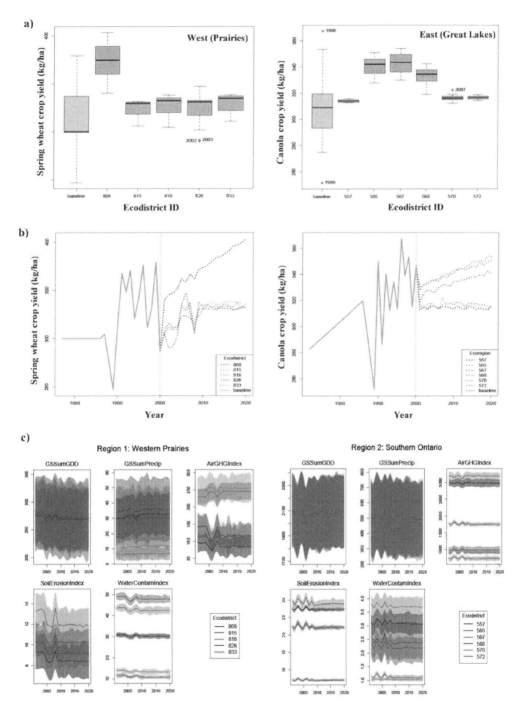

FIGURE 4.12: Regional-scale crop outlook for spring wheat and canola out to 2020: a) forecasted crop yield distribution and b) mean trend by ecodistrict in relation to the historical baseline, c) spatial variation in agri-environmental indicators used to guide the IAM system model forecast (Newlands et al., 2013).

identifies the importance of testing the *robustness* of systems in terms of how they can handle variability and remain effective for estimation and/or prediction. *Robust* statistical methods are not unduly affected by outliers and have good performance (e.g., accuracy) under small departures from the model assumptions (e.g., input parametric distributions), or more generally, they provide insights to a problem, despite having its assumptions altered or violated. Greater objectivity in the identification and selection of the best predictors or explanatory variables, from all possible ones, based on ranking their correlation and sensitivity, proved particularly useful for reducing forecast error. This indicates the importance of considering *cumulative effects or impacts*, often neglected by many IAMs.

Cumulative effects or impacts are the direct and indirect changes to the environment that are caused by an action in combination with other past, present, and future human actions. In this way, no single human activity causes cumulative effects, but instead they are caused by the addition and/or accumulation of impacts from different sources and human activities over time. Examples of cumulative effects are: time crowding (repetitive effects on an ecosystem), time-lags (delayed effects), space crowding (high spatial density on an ecosystem), trans-boundary (effects occurring away or distant from a source), fragmentation (change in landscape patterning), compounding effects (effects arising from multiple sources or pathways), indirect effects (secondary effects), and broader-scale drivers, triggers, and thresholds (fundamental changes in ecosystem structure and behavior) (see CEQ (1997) and Spaling (1995)).

While Environmental Impact Assessments (EIAs) typically focus on local scales and ecological footprints across areas covered by a given action, Cumulative Effects Assessments (CEAs) are applied at the regional-scale over longer time duration and broader spatial scales (Weber et al., 2012; Hegmann et al., 1999). Key gaps in CEAs, which have been recently identified, include: 1) the development of transparent and defensible methods for developing scenarios and applying the approach in data-limited regions, 2) valuing intangible and in-commensurable ecological goods and services, 3) incorporating social values and indicators that explore plausible outcomes, while acknowledging that communities and individuals can choose alternate courses of action, and 5) refining methods to communicate integrated modeling results with non-technical participants (Weber et al., 2012). Many cumulative effects, such as the long-term exposures to harmful environments and air pollution which impairs human health, tend to be nonlinear and to increase more rapidly as the accumulation level increases (Kong et al., 2010). As Kong et al. (2010) identifies, "long-term cumulative effects, although recognized as important, have not been properly modeled or quantified by existing methodologies, and are thus often ignored before they become serious and, by then, it is too late to act."

Stochastic modeling of epidemics as well as the impact of multiple, large-scale teleconnections exemplify the importance of cumulative (i.e., spatial and time-lag effects) in ecological forecasting (Newlands and Stephens, 2015; Bonner et al., 2014; Soubeyrand et al., 2007; Anderson and Britton, 2000). To illustrate how long-term cumulative effects can be integrated into stochastic models: for a response variable, $\mathbf{Y(t)}$, explanatory variables, $\mathbf{Z(t)}$ and covariates, $\mathbf{X(t)}$, a multivariate (i.e., semiparametric additive model) that incorporates cumulative effects, is given by,

$$Y(t_i) = \mathbf{Z}^\mathsf{T}(t)\boldsymbol{\beta}^0 + \sum_{k=1}^{q} g_k\left(\int_0^{\Delta} X_k(t_i - \tau)\theta_k(\tau)d\tau\right) + \epsilon(t_i), i = (1, 2, ..., n), \quad (4.44)$$

where $\boldsymbol{\beta}$ is an unknown p-dimensional parameter vector, g_k are unknown (smooth) link functions, and τ_k are time-lags, where $k = (1, 2, ..., q)$. In this model, cumulative effects of a single covariate, $X_1(\tau)$, can be represented as, $\int_0^{\Delta} X_1(t-\tau)\theta(\tau)d\tau$, defined over an interval $[0, \Delta]$, where $\Delta > 0$ for some weighting function $\theta(\tau) \geq 0$, given by,

$$E\left\{g_k\left(\int_0^{\Delta} X_k(t - \tau)\theta_k(\tau)d\tau\right)\right\} = 0 \quad (4.45)$$

$$\int_0^{\Delta} \theta_k(\tau)d\tau = 1, k = (1, 2, ..., q) \quad (4.46)$$

$$E(\epsilon(t_i)^2|\mathbf{Z}(\mathbf{t}), \mathbf{X}(\mathbf{s})) = \sigma^2 \quad (4.47)$$

This formulation avoids unstable estimation, and difficulties in interpretation when just increasing the number of additive components. Kong et al. (2010) have investigated the impact of cumulative effects of air pollution on respiratory diseases in Hong Kong, and influenza and immunity in France, applying the above model using penalized splines as the smoothing functions. In the first of these case-study applications, their modeling provides new evidence that the effect of air pollution is far more pronounced than that suggested by thresholds of US National Ambient Air Quality Standards (NAAQS). In the second case-study application, they reveal that overall influenza immunity can last as long as three years, although it is relatively weak. These are important insights that can help to improve human health and reduce societal health care costs. Santoro et al. (2015) have also recently shown how water levels of the Egyptian Nile River are strongly linked to variations in climate teleconnections, specifically the El Niño/Southern Oscillation (ENSO), the North Atlantic Oscillation (NAO), and the Pacific Decadal Oscillation (PDO), and how the *cumulative effects* of these three teleconnections are strongly linked to the reoccurrences of famine in Egypt over the last thousand years. Bonner et al. (2014) applied functional principal components analysis (FPCA) to explore variations in the impacts of four major teleconnection indices affecting the Northern Hemisphere, namely the Southern Oscillation Index (SOI), Pacific North American (PNA), PDO, and North American Oscillation (NAO). This study examined regional impacts of multiple teleconnections within British Columbia, Canada on surface air temperature. Similar to the approach of Kong et al. (2010), smooth link functions are specified. A polynomial spline is used to interpolate the climate data to generate climate surfaces, with Bayesian P-splines applied to estimate the impacts of the four teleconnections over time. This study revealed important spatial patterning between different teleconnections across the study region, whereby there are marked differences in the relative magnitude in the monthly temperature effects associated with spatial impact of PNA, PDO, ENSO/SOI and NAO. Cumulative effects of multiple teleconnections, reveals "hot-spots" as regions of increased/decreased temperature variability that may have higher climate risk or be suitable for expansion of agricultural activity (Iizumi et al., 2014). The findings of these recent case studies illustrate the importance of modeling, assessing and forecasting long-term cumulative effects when addressing real-world problems.

While ecosystem models may exhibit similar prediction skill in validation studies, the *sensitivity* of their long-term predictions to climate variability can differ significantly. Such discrepancies often relate to an outstanding dilemma in ecosystem science—how best to balance parameter (i.e., number of parameters) versus structural (hierarchical interactions) complexity with the availability of data (Cariboni et al., 2007). In particular, SA analysis enables one to explore the effect and consequence of more factors than one can control in open field or closed laboratory conditions, as well as their interactions. To help address such challenges, model sensitivity analysis (SA) is of major importance for reliably quantifying the influence of input parameters on a model's outputs. SA addressed many crucial questions when applying models to the real-world world—such as: How is variation in an output to be partitioned between inputs? Which parameters are insignificant and can be held constant or eliminated to form a reduced model form? What leading interactions exist between variables? What input ranges are associated with best/worst outcomes? What are the key controllable sources of reducible uncertainty? SA can identify hidden relationships between model components and can be used to simplify (i.e., reduce the dimensionality of) complex ecosystem models. A suite of methodologies has been developed and tested, but each has its strengths and weaknesses. The range of available techniques include: Differential, Nominal Range, Pearson/Spearman Correlation, Response Surface, Mutual Information Index, Classification and Regression Tree (CART), and variance-based methods like the Fourier Amplitude Sensitivity Test (FAST) and Sobol's Method (Frey and Patil, 2002).

Local sensitivity analysis varies the inputs one at a time, while keeping the remaining parameters fixed, but it cannot be used in applications where the model is nonlinear. For example, local sensitivity indices, S_i, which measure the effect of an output, Y due to perturbation of an input, X_i around a fixed fraction, x_i^*, of the mean of the output value $\overline{Y}$ (or alternatively by a fixed fraction, $\sigma(X_i)$ of X_i's standard deviation, $\sigma(Y)$), is,

$$S_i(x_i^*, \overline{Y}) = \frac{x_i^*}{\overline{Y}} \frac{\partial Y}{\partial X_i} \tag{4.48}$$

$$S_i(\sigma(X_i), \sigma(Y)) = \frac{\sigma(X_i)}{\sigma(Y)} \frac{\partial Y}{\partial X_i}. \tag{4.49}$$

In contrast, *global* sensitivity analysis allows multiple inputs to vary simultaneously across given ranges, and has recently become more favorable due to advances in computational power. As complex ecosystem models usually contain some degree of nonlinearity, it is generally not advised to apply local sensitivity analysis methods, as they can give misleading results.

The law of total variance states that,

$$V(Y) = V[E(Y|X_i)] + E[V(Y|X_i)] \qquad (4.50)$$

Assuming that input factors, X_i are independent of each other, the total variance of a model output, $V(Y)$, can be decomposed as a sum of increasing dimensionality, such that the sensitivity of a given model output Y to a given model input X_i (Saltelli et al., 2000) is given by the unconditional variance,

$$V(Y) = \sum_i V_i + \sum_i \sum_{j>i} V_{ij} + ... + \sum_{i<j<m} V_{ijm} + ... + V_{12\cdots k}, \qquad (4.51)$$

where, $V_i = V[E(Y|X_i = x_i^*)]$ is the first-order *fractional* conditional variance with respect to a fixed factor or i^{th} input variable, x_i^*. This measure of sensitivity is more general than a linear regression coefficient of determination (i.e., $\beta_j^2 = V_i/V(Y)$), for example, as it can be applied for both linear and nonlinear models, but does not account for interactions between input factors (Saltelli et al., 2000). This variance is computed over all possible values of x_i^*. Also, $V_{ij} = V[E(Y|x_i = x_i^*, X_j = x_j^*)] - V[E(Y|x_i = x_i^*)] - V[E(Y|x_j = x_j^*)]$. In addition the requirement that inputs are independent, individual terms in the above expansion must be zero-mean (i.e., $\int_0^1 V_i dX_i = 0, \forall X_i, i = (1, 2, ..., k)$) and 'square-integrable' (i.e., $\int_0^1 \int_0^1 V_{ij}(X_i, X_j) dX_i dX_j = 0, \forall X_i, X_j, i < j$) (Saltelli, 2002).

When inputs are correlated, because of a double loop computational condition that exists when trying to compute the expectation integrals, stochastic sampling methods like MCMC must be used. In such situations, a reduction of variance is only achievable by first, fixing a factor, then, fixing others that are correlated with the first factor selected. In this way, a recursive procedure can be employed that depends on the increments and fixing order of the input variables. For this reason, abrupt shifts in the sensitivity of model outputs arise when there is significant correlation between input variables. A variance-based, global SA method is called FAST (Fourier Amplitude Sensitivity Test) involves defining a set of transformation functions and distinct, incommensurate frequencies for each model input factor, and computing total sensitivity indices (TSIs). Another widely-applied variance-based method, called Sobol's method of global SA, computes TSIs, like the FAST method, but instead of employing a spectral/Fourier method, it uses a unique decomposition into sums of increasing dimensionality that does not rely on a transformation function. This approach propagates uncertainty of input parameters, partitioning the sensitivity of outputs according to main effects (i.e., first-order) and total effects indices. The main effects (first-order) index is obtained by normalizing $V[E(Y|X_i)]$, such that (Sobol, 2001; Saltelli et al., 2000; Sobol, 1993),

$$S_i = \frac{V_i}{V(Y)} = \frac{V[E(Y|X_i)]}{V(Y)}. \qquad (4.52)$$

The total output variance of all input parameters, except X_i is denoted $V[E(Y|X_{-i})]$, with the expected value of variance that remains unexplained if only X_i input varies over its uncertainty range denoted, $E[V(Y|X_{-i})]$. The total effects index with respect to a given input, X_i, is then,

$$S_{T_i} = 1 - \frac{V[E(Y|X_{-i})]}{V(Y)} = \frac{V(Y|X_{-i})}{V(Y)}. \qquad (4.53)$$

The total effects index includes contributions from the main effect and the interactions between inputs, and if $V(Y) - V[E(Y|X_{-i})]$ is negligible, then the input, X_i is non-influential. Alternatively, a significant difference between the main and total effects indices signals an important influential interaction between such inputs. Sobol's method can be applied to both nonlinear and non-monotonic models, but is computationally intensive and its ease of application depends on the complexity of a model. All variance-based SA methods may underestimate the mean and range of the most sensitive inputs, can be computational intensive, and assume all the information on model uncertainty is captured in variance (Cariboni et al., 2007; Frey and Patil, 2002).

While there is currently no formal method for estimating confidence intervals for SA indices, standard error (SE) in the main and total effects indices can be estimated by re-sampling of model output with statistical bootstrap sampling and MC simulation. Yang (2011) have explored the convergence and uncertainty analysis of MC-based SA methods such as Sobol's, and find that such simulation approaches are robust when quantifying sensitivities and ranking parameters, despite a large number of model executions (Yang, 2011). For Sobol's method, the total number of model evaluations $N_{tot} = n(s + 2)$, where n is the base sample size and s is the number of model parameters in the SA analysis. Oakley and O-Hagan (2004) have devised a Bayesian probabilistic approach to SA analysis, showing how this approach is computationally efficient and out performs other based methods. It also can compute a full range of sensitivity measures. However, it may be unrealistic to assume Gaussian-distributed input priors. Glen and Isaacs (2012) have evaluated the performance and reliability of estimating Sobol indices when inputs are correlated. They find that this method is well-suited to high-dimensional stochastic simulation models, such as the US EPA's Multimedia Stochastic Human Exposure Model (SHEDS)[18]. They also highlight how the subscripts i and j in the variance decomposition of Equation 4.51 can refer not only to individual inputs, but also to groups of inputs, as long as the groups are independent of each other (but inputs within the same group can be correlated)(Sobol, 2001). In this way, grouped sensitivity indices can be computed. As is common in many large, complex models, especially if they are hierarchically structured, one group may have just a single input, while another may have thousands.

Sensitivity in the response of outputs of a complex ecosystem model, namely, soil water and crop productivity under different soil and climate conditions was recently explored by Newlands et al. (2012). The model was calibrated and validated using experimental data obtained from long-term agricultural research sites. The widely-applied Biome-BioGeoChemical Cycles (BGCs) (Biome-BGC) model, with

---

[18]EPA SHEDS Model, `www.epa.gov/heasd/research/sheds.html`

a daily time-step, was adapted for application to cropland and used to simulate the mass-balanced dynamics of carbon, water, nitrogen, and energy. This study found that significant changes in the model sensitivity accompany changing climatic regimes (Beamish et al., 1999). If such variability is not corrected for, substantial predictive error can result when simulating across time and space. These findings lend further support (in the case of agricultural croplands) for applying a hierarchical statistical approach when analyzing and modeling agricultural and forestry ecosystems, as previously highlighted by Wang et al. (2009). Moreover, the same domain order holds for sets of input parameters ranked according to their relative sensitivity on model output was found for cropland ecosystems, as Wang et al. (2009) reports for forestry ecosystems, namely: phenology, leaf-area dynamics, light-interception, vegetation morphological, and eco-physiological variables. A vegetation sensitivity index, integrating the enhanced vegetation index of Moderate Resolution Imaging Spectroradiometer (MODIS) with three climate variables (air temperature, water availability, cloud cover), used to identify areas sensitive to climate variability over the past 14 years has been developed (Seddon et al., 2016). Using this index, a global map of climate sensitivity has been generated, showing the Arctic tundra, areas of boreal forest, the tropical rainforest, the Alpine, steppe, and prairie, deciduous forest and other regions with different levels of sensitivity and response amplification to climate variability.

Guided by objective SA methods and analysis results, ecosystem models may be further simplified, re-structured and re-adapted for different applications, contexts, and needs. Such model re-structuring also enables a clearer understanding associated with input parameters and their interaction within complex model structures. Moreover, their prediction power may be significantly increased by ranking all possible variables by their sensitivity.

5

Future Outlook

CONTENTS

> "The world as we have created it is a process of our thinking. It cannot be changed without changing our thinking."

Albert Einstein, 1946, The New Quotable Einstein (2005), edited by Alice Calaprice

5.1   Perspective

Modern society is undergoing a transformative process characterized by an "era of risk"—led by not resources or knowledge alone, but by our ability to identify, understand, anticipate, communicate and take real, meaningful action to minimize known and unknown risks due to climate and global change. Such transformations will require societies around the world to collectively re-configure how they interact with future sustainable ecosystems. Transformative processes must be peaceful, democratic, and be based on sound principles of social justice, resource equity and social welfare, otherwise they may lead to migrant and refugee crises and global uprising, as an inevitable outcome of human societies facing environmental destruction and wealth polarization (Hedges, 2015).

The sustainability challenge of the 21st century requires us, today, more than ever before, to build global adaptive capacity and opportunity. It requires us to construct, apply, adapt, and improve integrated risk assessment frameworks involving statistical-based metrics and indicators (Chapter 1). Despite the current state of our ecosystems, the depletion of resources, the extent to which we have placed our planet under pressure (Chapter 2), and the limits and uncertainties in our current scientific knowledge on complex, adaptive systems (Chapter 3), there is still huge opportunity for integrative approaches to guide and inform our path toward a sustainable future. While improvements and changes in human activity will be necessary, so too, will the way we humans think, design, and apply new scientific knowledge and technology. Integrated, real-world problem solving (Chapter 4) will require not only greater creativity and rapid learning, but also computational thinking as a fundamental competency and skill[1]. Such a fundamental change, within the context of a wide array of competing risks and a large complex set of data and knowledge, will require us to utilize or leverage artificial intelligence/machine-learning concepts as part of fully integrative (i.e., "end to end") thinking, interfacing and adapting with real-world systems (Meinke et al., 2009; Wing, 2006). As human societies transform to a sustainable future, some challenges and associated uncertainties will, realistically, scale faster (i.e., out-pace) immediate (and reliable) scientific knowledge and technological solutions. Integrative solutions may even lie well beyond the insights and capabilities of human intelligence alone. Therefore, in facing such challenges, we will need to extend our collective human intelligence and capabilities, as well as explore their integration with AI and machine learning ML, to inform public policy and drive social action and change. This will create opportunity, but will also introduce new complexities, risks and uncertainties. More and more aspects of our lives will very likely be guided by sustainability decision-support tools having machine-learning intelligence that enables human intervention (so-called "human-in-the loop" principles and processes). As evidenced by the extent that we continue to develop, apply, and rely on technology, it is conceivable that, in the future, our thinking and activities will merge and be integrated with machines to benefit more fully from their complex learning capabilities. This will enable us to venture beyond our current limits in our ability to rapidly and reliably continually integrate an ever increasing amount of interdisciplinary knowledge and data. It will enable us to develop more reliable future predictions (forecasts) well beyond the current technological limits in how we compute and solve real-world complex problems today. Such a symbiosis will, in the future, advance human society, if guided for the public good. The symbiosis will involve highly collaborative learning environments. These social hubs will help to provide a distributed learning and interaction network for monitoring, assessing, and forecasting the long-term sustainability of the ecosystems around our planet Earth (and any other Kepler-like planets humans might inhabit in the future). Nonetheless, we must recognize that crucial transitions and hurdles must be collectively overcome in achieving our international sustainability goals. Transitions of our SES may induce a myriad of anticipated (and unforeseen), multi-scale impacts (measurable and/or observable) alongside longer-term cumulative effects.

5.2　Transformation Paths and Transition Dynamics

General equilibrium theory in economics considers the behavior of supply, demand, and prices in a whole economy with several or many interacting markets and assumes a set of prices exists that results in an overall system equilibrium. In other words, it assumes that all economic markets are in equilibrium. This is an obvious abstrac-

---

[1]Computational Thinking, Googleplex in Mountain View, CA.
www.alumni.ubc.ca/podcasts/computational-thinking-for-the-21st-century/

tion from the real-world economy. The theory further assumes that prices at this equilibrium are long-term prices and actual, real-world prices are just statistical deviations from such an equilibrium. This approach aims to capture broad interactions and feedbacks arising from such finer-scale economic shifts and cycles, changes in the ability to achieve long-term policy objectives, or better understand how resilient the economy is to external shocks considering influential commodities, factors, and decision-making agents (e.g., producers, consumers, governments, industry).

Wiedmann et al. (2007) have reviewed Input-Output (IO) models for the assessment of environmental impacts, including single- and multi-region IO models, and those that include feedback loops and simulation-based analyses. Such models capture interactions between different industries and both forward and backward supply flows and intermediate products. Computable General-Equilibrium Model (CGE) models further extend IO models by considering an endogenous output and price system the optimization behavior of individual agents and a complete treatment of income flows in an economy (Carri, 2008). CGEs based on general equilibrium theory, are widely used in the modeling, assessment and forecasting of sustainable development policies, and to frame trade negotiation strategies associated with fiscal, trade, climate change external shocks and international price shocks. In contrast, Partial-Equilibrium Models (PEs) assess the impact of an economic or policy shock (e.g., increasing tariffs) affecting two or more inter-connected markets, assuming the rest of the economy remains fixed. They may consider individual or single economic markets, for simplicity, in analyzing policy issues arising from a shock. Shocks considered are generally those that have limited effects within a particular sector of the economy or industry and/or where the impact on other industries can be considered to be small enough to be ignored in practice. Such models have, therefore, limited scope to handle issues arising from more general shocks that can affect the outputs and prices of many different industries concurrently.

Barker (2004, 1996) provides a major critique of the general equilibrium theory that serves as the basis for CGE models, arguing, "CGE models are misleading and inadequate tools for policy analysis. For long-run sustainability analysis, they are not only misleading but represent a fundamental misunderstanding about economic systems and how they change." CGE models are static and omit population growth, resource dynamics, and investment, and generate forecasts and projections that are typically based on only a single year of training data. Despite such criticism, others argue, however, that the CGE models are one of the most rigorous, cutting-edge quantitative methods to evaluate the impact of economic and policy reforms and shocks across whole economies, and are a tool that is very useful for empirically oriented analyses in economic and public policy design[2] (Cardenete et al., 2012). Because CGEs predict changes in economic conditions at a particular time, they assume that capital, labor and resource levels are constant. As a result, they face considerable challenges in representing transitional paths, whereby a SES adjusts over time to different dynamic equilibria, and may be less useful in guiding

---

[2]CGE, http://www.iadb.org/en/topics/trade/
understanding-a-computable-general-equilibrium-model,1283.html

transformational-type policy (Ackerman, 2002). It is further argued that they often rely on unfounded transition assumptions and involve marginal costs with highly uncertain assumptions for technological change, industrial organizational and consumer behavioral dynamics. This is a serious limitation when one examines marine, agriculture, and forestry policy issues because they are linked to dynamic, renewable resources that can increase and decrease rapidly, while being exploited by multiple industries and requiring substantial capital (and subsidy) investment and provide broad ecosystem services with benefits that change with development (Xie et al., 1996). Just as the World Model (Chapter 1) and its predictions were criticized for being too complex and intractable, CGE models, at the other end of the spectrum of model complexity, are similarly criticized for relying too heavily on functional equations that are more mathematically tractable, than realistic. Within this spectrum, models may also be disadvantaged by being too data-dependent. Instead, the so-called Space-Time economics (STE) systems approach is recommended that is based on measure theory and numerical analysis, being better suited to make assumptions regarding regional heterogeneity and temporal shifts in investment in response to changing risks and levels of uncertainty. It provides a more flexible and realistic approach that accounts for the indivisibilities, externalities, socio-political valuations and niche economies or economies of specialization (Barker, 2004).

A standardized simulation model developed by the Global Trade Analysis Project (GTAP) (GTAP Version 6.2, Center for Global Trade Analysis, Purdue University, 2007) is a static, multi-region, multi-sector CGE model comprising 57 sectors and 87 countries/regions for capturing interactions between domestic sectors and international trade (Wiedmann et al., 2007)[3]. Model extensions and more dynamic models have been developed and tested, in addition to the standard GTAP model. Addressing the need for better representing the real-world dynamics of SES when forecasting economic metrics and policy impacts, the COmprehensive Model for Policy ASSessment (3E COMPASS) was developed and simulates mid-term dynamics (i.e., up to 2020) of the global economy. This model considers, among many aspects, interactions between energy and the environment. Its main aim is to help guide policy dialog (Uno, 2002). It decomposes the global GDP into 60 regions and 25 annual commodity trade flows, and integrates information on production, trade, balance of payments, saving and investment, energy supply and demand, and environmental impacts. In this way, it can track policy measures dynamically (e.g., the effect of tradeable permits on increasing energy price).

The Global INterindustry FORecasting System (GINFORS), which evolved from the 3E COMPASS, is a dynamic simulation model for forecasting environmental impacts (Meyer et al., 2013). It considers most of the essential components of an impact assessment model (IAM) for the analysis of sustainable economic, social and environmental development, but relies on a comprehensive Multi-Region Input-Output (MRIO) historical time-series database. It also assumes not long-term equilibrium conditions (i.e., long-run simulated equilibria), but decision-making agents that make shorter-term decisions under conditions of bounded rationality. This

---

[3]GTAP, https://www.gtap.agecon.purdue.edu/

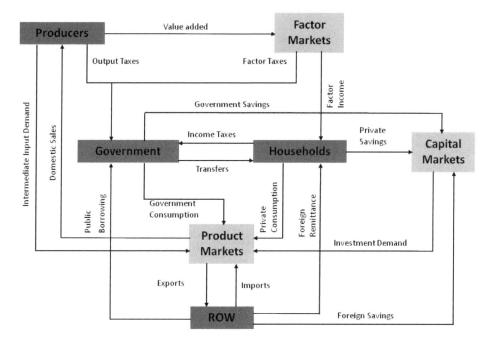

FIGURE 5.1: Simplified structure (components and flows) of CGE models.

model serves as a tool for designing and assessing sustainable development actions and options. As highlighted by Wiedmann et al. (2007), regional trade data, land-use and land-use change, $CO_{2,e}$ emissions and energy-use data across all the industrial sectors, countries and regions represented, means that MRIO-based models, such as GINFORS, are highly data-dependent (Wiedmann et al., 2007). Given such strong dependency on historical conditions, many econometric models must be able to generate additional inferences, to reliably and realistically portray the full range of future possibilities and uncertainties associated with different types of transitional trends and event shocks that can occur. This applies to when they are used to generate plausible short-term forecasts as well as long-term scenario projections (i.e., sustainable "futures" or "outlooks").

Figure 5.1 provides a simplified structure of a CGE model[4]. This structure contains the capital (i.e., land, labor, infrastructure), producer, factor and product markets. In particular, factor markets allocate *services* of the factors of production, particularly for capital goods. Firms within such markets then produce the final goods for various products markets. The interaction between product and factor markets involves the principle of *derived demand*, whereby the demand for productive resources is derived from the final demand for goods and services, and prices are set in relation to price elasticities for resource supply (PERS) and demand (PERD). This structure also includes the government, household, and global (i.e., rest of the world, ROW) key components.

---

[4]Inter-American Development Bank (IDB), www.iadb.org/en/topics/trade/ understanding-a-computable-general-equilibrium-model,1283.html

Von Lampe et al. (2014) have investigated how different models exhibit different behaviors and generate contrasting projections for agriculture over the long-term. They inter-compare the output of 10 global economic models (i.e., six CGE and four Partial-Equilibrium Model (PE) models) with respect to a historical baseline reference and shared, prescribed set of alternate socioeconomic, climate change, and bioenergy scenarios. Income and price elasticities for staple/basic food commodities decrease as income increases. The ensemble of models contains different (i.e., complementary and conflicting) assumptions of land productivity, population growth, and agricultural land-use (intensification versus expansion), and assumptions of agricultural technological change, biophysical and climate change assumptions. CGE models generated smoother price paths with lower price increases (or even decreases) in reference scenarios, and smaller price changes relative to alternative assumptions on exogenous drivers. Global economic models under scenarios of socio-economic development, climate change and bioenergy expansion, project that by 2050, global food demand will increase by 59–98% alone, based on 2005 levels, with increasing disparity between developed and non-developed nations (Valin et al., 2014). Such projections are, however, associated with many underlying assumptions and drivers with significant uncertainty. Some congruent patterns in long-term model projections do emerge. For example, in developed countries, total food consumption declines, attributed to lower income and price elasticities and population decreases. In developing countries, total demand generally increases for crops and decreases for livestock, due to the higher income elasticities for these products, with lower economic growth dominating population growth effects.

What isn't often included in environmental or socio-economic forecasts is innovation. In Chapter 1 we discussed adaptive capacity and its importance linked with science and technology innovation, but here we characterize different innovation types in order to gain insight into how they can drive future, beneficial socio-economic transitions and transformations. A 2012 report by the EU Standing Committee on Agricultural Research (SCAR) elaborates on *innovation cycles* and how they lead to four different kinds of system transitions, namely: innovation novelties, niches, regimes, and landscapes (EU-SCAR, 2012). Novelties are localized "breaks with routines" that are constrained by laws, actors, and norms/status-quo. Niches result from the aggregation of smaller systems with more flexible learning rules, routines, and a higher rate of active learning and embedding of new paradigms linked with a change in normal activities and processes. When niche paradigms evolve into a change in operational practices or are integrated more fully and concretely into SES, then regimes result, as networks of structured and coordinated rules. Evolving from a niche to a regime can involve many contradictions and disruptions, with resistance to the innovation. As regimes of innovation are upscaled (or downscaled) depending on the natural scale or domain size of an innovation, innovation *landscapes* emerge. This level of innovation is radical whereby situations and events emerge that lie outside of national policies and radical innovation is required to address and solve them. Cycles of human innovation take time as they mediate transitions in SES dynamics and regulate our capacity to be proactive or reactive in confronting to social, economic and environmental risks. To strengthen a precautionary and/or proactive capacity to respond to anticipated risks, however, requires developing a fundamental knowledge of real-world causal interactions and using statistics to devise alternative ways to predict (e.g., hind-cast or forecast) risk measures and to evaluate alternative options and actions.

Transitions can be defined as a gradual process of change that transforms the structure or sub-structure of a SES. Niccolucci et al. (2012) further identify four main typologies in reversible and irreversible development paths based on biocapacity (BC, ecological wealth) and ecofootprint (EF) measures: parallel, scissor, wedge, and descent. Countries facing each development path have a very different ordering of priorities and thus must face different sequences of transitions. Other studies on the typology of socio-technical transition pathways also characterize SES system dynamics into *four* main transition pathways (transformation, reconfiguration, technological substitution, de-alignment and re-alignment) that result in regular, hyper-turbulence, specific shock, disruptive, and avalanche environmental impacts based on the attributes of frequency, amplitude/intensity, speed and scope (Geels, 2007, 2004).

Consider a single, stochastic, multi-state process occurring over $[0, C]$ in time with E events that starts in state, X_o at time t_0 and is observed to evolve in steps, t_j, $(j = 1, 2, ..., E)$ or "transition times," into a new state, X_j. Under Markov assumptions it is assumed that a hazard depends only on the current time and the state and is independent of the times of previous transitions. The total, multi-state process likelihood is denoted, $L = L(X_{j-1}, X_j, t_{j-1}, t_j, \lambda_{X_{j-1}, X_j})$ and given by (Hougaard, 1999),

$$L = \prod_{j=1}^{E} \lambda_{X_{j-1}, X_j}(t_j) exp \left\{ -\int_{t_j - 1}^{t_j} \lambda_{X_{j-1}}(u) du \right\} exp \left\{ -\int_{t_E}^{C} \lambda_{X_E}(u) du \right\}. \quad (5.1)$$

The likelihood contains a contribution from each transition and the hazard (λ) from state X_{j-1} to X_j at each transition time. In the real world, one may observe a set of dependent, causally linked processes over the same or different time periods that may be recurrent or non-recurrent, progressive or non-progressive. A progressive process has only one possible transition into each of its states, except the initial state that has no possible transitions into it. A non-progressive process allows for an arbitrary dependence on the number and times of previous state transitions.

Recently, Cai et al. (2014) have devised and tested a general stochastic method (with Markov assumptions) for the prediction of the full distribution of future events involving a single response variable, based on the relationship between its previous phenological events (i.e., event-history) *and* an environmental covariate. For this general model involving a covariate, the likelihood involves a product of conditional probability of the event occurring at a previous time, given values of a covariate Z, and a specified monotonic link function, g, where $g : (0, 1) \to (-\infty, \infty)$ that relates the conditional probability, $\lambda_{j, t_j} = \lambda(X_{j, t_j} = 1 | X_{j, t_j - 1} = 0, Z_{j, t_j})$ of a transition from an initial to the final state, to the state and transition times of a covariate. The likelihood is denoted, $L' = L'(X_{j-1}, X_j, Z_{j-1}, Z_j, t_{j-1}, t_j, \lambda_{X_{j-1}, X_j, Z_{j-1}, Z_j})$, expressed as,

$$L' = \prod_{j=1}^{E} \left\{ g^{-1}g(Z_{j,t_j}) \prod_{k=0}^{t_j-1} \left(1 - g^{-1}g(Z_{j,k})\right) \right\}, \quad (5.2)$$

where g^{-1} is the inverse of the link function, g. Expanding this model to include multiple responses and covariates (i.e., multiple link functions) or more complex spatio-temporal dependencies of state and transition times is possible. Such an extension would enable broader testing of the prediction of multi-state processes. It would also help to explain the observed transitions. It would also be a useful model for selecting metrics having different levels of performance (forecast skill) related to sustainability metrics and indicators.

5.3 Causality and Forecasting

The notion of cause and effect is fundamental to devising testable scientific hypotheses, guiding sound inference, and developing prediction and foresight on the impact, response and adaptive behavior (i.e., resilience) inherent in complex socio-economic and biophysical systems. Causal analysis is concerned with identifying causes and effects of observed phenomena with the purpose of understanding, predicting, and eventually intervening on society and on individuals. In general terms, it involves the *cognitive goal* to relate to explanation and the *action-oriented goal* to relate to inference and decision-making. Causality in the statistical sense, can be defined such that a variable $x(t)$ Granger-causes another variable $y(t)$ (denoted $x(t) \rightarrow y(t)$), if given information of both $x(t)$ and $y(t)$, the variable $y(t)$ can be better predicted in the mean-square-error sense by using only the past values of $x(t)$ than by not doing so. The null hypothesis here is: $x(t)$ does not Granger-cause $y(t)$. In other words, having knowledge of past values of $x(t)$ improves the ability of a model or index to predict $y(t)$. A weaker condition is that of instantaneous causality where not only the past, but also the present values of $x(t)$ improve prediction of $y(t)$. Feedback can occur where $x(t)$ causes $y(t)$ and $y(t)$ causes $x(t)$ (Cromwell et al., 1994). Granger causality statistical tests are sensitive to data availability, random variability, and especially whether the variables arise from a deterministic or stochastic process (Granger, 1996). Uncertainty in causal relationships can amplify when data is sparse, random variability increases and/or due to the mediating effects of hidden (i.e., latent) variables. Granger causality considers the extent to which the lag process in one variable explains the current values of another variable (Bai et al., 2011) .

A detailed review of the importance, need, and consequence of causal inference in solving natural resource problems (spanning a wide range of scientific domains) suggests that a scientific paradigm shift is needed to move from traditional statistical-based methods that infer correlation/association to those that specify a stronger cause-effect or causal relationship (Illari et al., 2011). Such a transformation would: 1) more reliably explain interacting processes using multivariate data, 2) devise more integrative, system-level or multi-scale sustainability assessment frameworks, and, 3) better guide multi-objective stakeholder decision-making and more complex policy evaluation of agricultural or resource economics (Illari et al., 2011). They highlight the critical need for integrated thinking about causality, probability, and mechanisms in scientific methodology. While causality and probability are long-established central concepts in the sciences, different scientific disciplines have developed very different methods, and often, very different understandings and perspectives on mechanisms. Assumptions regarding open versus closed boundaries, degree of modularity and level of complexity in the relationship between interacting components can differ widely between causal and systems-level approaches. Nonetheless, they are highly congruent, compatible, and amenable for advancing both econometric and environmental risk models for guiding resource decision analytics, and to devise operational solutions of real world economic, societal and environmental opportunities and problems involving water, food, energy and climate change aspects (Russo, 2010). Causal and systems analyses both fundamentally embrace broader considerations and assumptions and explore a wider suite of known and unknown observed and/or latent interactions between variables. They also increasingly involve probabilistic-based statistical methods.

Causal statistical methods include: vector autoregressive (VAR) models, parametric and non-parametric decision-tree analysis, stochastic simulation, the state contingent approach and mathematical programming-optimization techniques, statistical risk production functions, and Bayesian belief networks (Pearl, 2009). The availability of Big Data (e.g., sensor-based, remote-sensing based, technological, socio-economic data), computational power, and more efficient statistical methods is driving rapid innovation in these methods and the integrated modeling of risk. *Takens' Theorem* states that the essential information of a CAS is retained in the time-series of any single variable of that system (Takens, 1981). Convergent cross-mapping (CCM) is based on this theorem detects whether two time-series variables belong to the same dynamical system (or subsystem). van Nes et al. (2015) have applied this method in an analysis of the time-lags and causality between global temperature variability and GHG over glacial-inter-glacial timescales.

High economic uncertainty and volatile food prices are key drivers now and will continue to be in the future. They are anticipated to have an increasing influence on food availability and demand, given the anticipated impacts of climate change. Yet, consumer (CPI) and producer price (PPI) indices, are often used as "core" inflation measures (i.e., the long-run trend in price levels) without the impact of food and energy taken into account (Hall, 2009). Recently, Amano and Murchison (2005) devised and tested a stochastic VAR-type causal model for inflation under the so-called New Keynesian Phillips curve (NKPC) to improve the understanding of inflation dynamics. This model has far more realistic assumptions, namely: a non-constant historical inflation target, significant costs when adjusting labor and firm-specific capital (i.e., more than just a single rental market for capital). In the case of firm-specific capital, the capital stock of each firm is invariant to its relative price (i.e., a price dampening effect). This new NKPC stochastic model with dynamic indexation is validated against Canadian inflation data that is calibrated with a discount rate of 0.99 or 4% annual real interest rate. This model represents both price persistence and price deviations well, estimating the average duration between price corrections (re-balancing or re-optimization) is eight months.

The degree, for example, to which the increasing volatility in food commodity markets is transferred to energy markets, and whether the linkage between both non-renewable, crude oil and renewable (e.g., bioenergy, biofuel) energy markets has a *causal* structure, is central to understanding the energy transition to renewables Saghaian (2010). Supporting earlier findings, Saghaian (2010) has determined that, while there are significant correlations between energy, agriculture, and exchange rate markets, there is only weak evidence of a significant causal, predictive link (i.e., Granger causality) between crude oil and commodity prices. Instead, rising price volatility within commodity markets is attributed primarily to rising speculation in hedging integrated risks when futures-contract type investments seek to profit from such price fluctuations within the derivative markets. When such activity is so high, it can destabilize entire markets and become, both a cause and effect of increased volatility. Higher price volatility also translates into higher costs of managing such integrated risks, such as increased crop insurance premiums (e.g., weather index-based insurance) and margins on futures-contracts, which may over the longer-term also contribute to higher consumer prices. As well, available market instruments (e.g., futures and forwards contracts, commodity swaps, call and put options, commodity-indexed bonds and long-term contracts) may not be sufficient (Tothova, 2011). Commodity markets are, however, directly affected when governments intervene by devaluing/depreciating their currency (e.g., China devaluation of the Yuan in August 2015) to stimulate investment and the attractiveness of their exports during economic slowdowns driven by reduced fossil fuel production due to weakening demand and over supply from other oil producing nations. Despite international currency competition, according to the IMF, "Greater exchange rate flexibility is important for China as it strives to give market forces a decisive role in the economy."[5]

To help solve "boom-bust cycles" typified by markets driven by one resource (i.e., fossil-fuel), structural change in economic markets is needed to both diversify marketed goods and services, as well as the integration of non-market goods and services. It is important to realize that our current market allocation system excludes most non-marketed natural and social capital assets and services, yet they are critical contributors to sustainability and human well-being (Costanza, 2012)[6]. There is a clear need to reduce the risk of national, regional and global debt crises, to repair the damage that debt crises incur, and to fund economic growth, boom-bust cycles must be avoided with greater stability built into the global economic system (MGI, 2015). It is risky to transition the global economy and difficult for societies to fully recognize that there can be compatibility of sustainability within modern society,

---

[5]The Guardian,
  www.theguardian.com/business/2015/aug/12/china-yuan-slips-again-after-devaluation
[6]Capital is defined as a stock that has a flow of services over time

if our economic system does not reliably and broadly track ecosystem goods and services. While a price and market can be opened for carbon as a way of mitigating climate change due to rising GHGs, this doesn't lessen the need for more broad structural change in the global economic system to ensure the long-term sustainability of ecosystems. Robert Costanza, along with other ecological economists, argues that we need a new, sustainable ecological economic model to better measure real-world economic efficiency (Costanza, 2012). It must include the contributions of natural and social capital and an integration of ecosystem services: direct provisioning, regulating, and cultural, including other indirectly supporting services. This will, undoubtedly, involve complex decision-making over shared public and private property rights in sustainably managing resources. Most natural and social capital assets are public goods. Open rights and common resources that have neither public nor private property rights, or privatizing them entirely, is not an answer, but property rights can be allocated to resources without privatizing them, by, for example, civic or common property trusts[7]. Brunckhorst and Marshall (2007); Brunckhorst (2005) outline robust common property right regimes and collaborative, adaptive ecosystem management arrangements and have discussed the necessity of an integration research at the regional-scale for ensuring ecosystem sustainability. Walrut (2002) provides a legislative framework for sea ranching in the open ocean that integrates aspects of common law.

Costanza and Liu (2014) have compared China and the US in terms of ecosystem services and their governance, and while their approach to governance is very different (i.e., monocentric in China versus multi-level in the US), significant transitioning in their management of natural resources and the environment is taking place, indicating that there is adaptive capacity for both to adopt an integrated, adaptive regional governance approach (interdisciplinary, stakeholder groups and multi-generational). Figure 5.2 outlines such a bottom-up community-driven ecosystem enterprise. Whether there is sufficient adaptive policy, science, and institutional capacity at the national-scale necessary to support regional-scale arrangements over the long-term, remains to be determined. Well-informed and bold civic, national, and international political leadership will be essential to design, test, and implement such a new sustainable/ecological economic system.

Brede and de Vries (2013) have explored investment decisions within a competitively and cooperatively future managed world. The degree that a nation perceives its integrated risk in making energy (or other) transitions, mitigates and adapts to climate change and other SES risks is key. Speculation, confusion and misinformation only postpones the inevitable—that we must transition economies driven by finite non-renewables like fossil fuels, to ones that are driven by renewable energy resources. Economic model simulations reveal that there is a critical threshold of competition whereby large economies can suffer large damages, despite their own attempts to prevent serious impacts of climate and other global SES changes. Furthermore, joint recommendations and their implementation can often be naively

---

[7]UK National Trust, US Nature Conservancy, Australian Landscape Trust, Australian Civic Trust/Water Action Coalition (WAC)

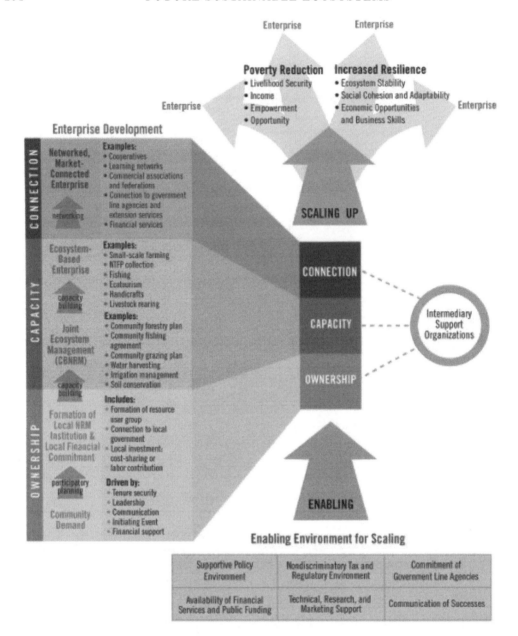

FIGURE 5.2: Scaling up community-driven ecosystem enterprise. The table highlights the key ingredients for successfully *enabling* the scaling up ecosystem-based enterprises to reduce poverty and build resilience, whereby these ingredients interact to generate ecosystem enterprises and drive them to scale, distinguished as: *Ownership*: a local stake in development, and enterprise, *Capacity*: social, technical, and business skills to manage resources and establish enterprises, and *Connection*: links to learning, support, and commercial networks and associations (WRI, 2008, World Resources Report 2013-2015: Creating a Sustainable Food Future, http://www.wri.org/resources/charts-graphs/scaling-community-driven-ecosystem-enterprise/)

assumed to be incentive compatible, whereby the intended or anticipated outcomes of a new policy is automatically seen to establish an equilibria because it is supported by consensus (Conradt and List, 2009). Moreover, the robustness and reliability of shared, transboundary or international policy objectives do not necessarily imply that all parties can or will transition at the same rate. This is because of significant time-lags, varying degrees of causality, and cumulative effects. Despite such lags, minimum, critical threshold targets (i.e., strict climate policy) are still needed to avoid tipping points, to guide our transformation path, and spur beneficial transitions through mitigation and adaption consensus, and collective action (Lontzek et al., 2015). Furthermore, unanticipated consequences are possible, if sufficient regional context (needs, incentives, risks) are not statistically measured and integrated well when setting shared/common mitigation and adaptation targets (Brede and de Vries, 2013; Sidle et al., 2013).

Rapid appraisals and disaster early-warning systems must increasingly become regionally-specific, and rely on a global monitoring network of rapid, automated seasonal forecasts or outlooks that integrate near-real time, satellite remote-sensing data, model sensitivity, and confidence statistical measures, rather than only longer-term scenario projections. Rapid appraisals could then better inform the increasingly complex decision-making that will be required to identify food productivity and production gaps and ensure global food security (Kouadio and Newlands, 2014; Lobell, 2013; Lobell et al., 2009). A third of crop yield variability has been attributed to global climate variation alone (Ray et al., 2013). The strong influence of climate teleconnections and an increasing potential future risk of crop diseases (that can complement and counteract each other) are also threatening agricultural food production around the globe (Iizumi et al., 2014; Juroszek and von Tiedemann, 2013). To counteract such threats, it is crucial to forecast metrics of interest and the associated risks at the regional-scale, and to determine causal-based forecasting time-windows as part of the advance warning and risk outlooks (Newlands and Stephens, 2015; Newlands et al., 2014). While leading covariates can be integrated into a model for improved prediction, at the regional-scale a different approach is often warranted. In addition to included covariates in modeling subregions that comprise a larger region of interest, additional auxiliary variables that can inform a forecast or help gauge its uncertainty can be extremely helpful. Auxiliary indices may come directly from data, another model scenario, expert-judgment or be a probabilistic measure. Figure 5.3 depicts an ensemble-based forecasting approach that integrates structural uncertainty across different models and model-types, as well as parameter uncertainty across spatial and temporal scales. Ideally, an ensemble-based approach applied at a regional-scale should include: 1) spatio-temporal upscaling/downscaling uncertainty from local to regional-scales, 2) regional-scale variability and 3) regional to global-scale variability. This assumes that the forecasting starts at a specific point in time (x-axis) for a relevant sustainability metric. For example, the performance of state-of-the-art global climate models participating in the Coupled Model Intercomparison Project Phase 5 (CMIP5) in simulating climate extremes shows that while multi-model ensembles are generally able to simulate climate extremes and their trend patterns, substantial discrepancies may exist in reanalysis (i.e., observational,

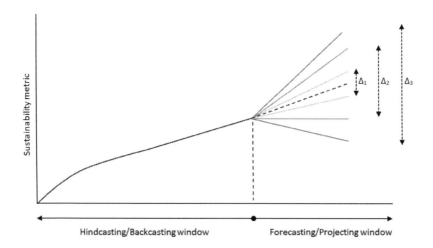

FIGURE 5.3: Simplified depiction of a tertiary ensemble-based forecasting approach that integrates structural uncertainty across different models and model-types, as well as, parameter uncertainty across spatial and temporal scale, shown partitioned by: 1) spatio-temporal upscaling/downscaling uncertainty from local to regional-scales, 2) regional-scale variability and 3) regional to global-scale variability. This depiction assumes that forecasting (linear) starts at a specific point in time (x-axis) for a relevant sustainability metric.

or observationally constrained) datasets (Sillmann et al., 2013a). While reanalysis data may have adequate spatial and temporal coverage to aid in model evaluation, they also can include uncertainties due to inhomogeneities caused by climate station instrumentation/location/exposure and/or the inaccurate representation of surface, boundary layer and convective processes. Also, future projected changes in climate extreme indices, as defined by the Expert Team on Climate Change Detection and Indices (ETCCDI), have recently been analyzed, showing that for precipitation-based indices, there is substantial disagreement between models (sign of change), requiring further regional-based investigation (Sillmann et al., 2013b). A complete set of climate extremes indices calculated from the latest CMIP5 ensembles are available on the ETCCDI indices archive[8].

Next, we outline such an ensemble-based stochastic approach that is able to integrate across models, and datasets and to integrate multi-scale variability (i.e., contributions 1-3 above). We refer readers to several studies involving the spatio-temporal upscaling and downscaling of climate variables (Kouadio and Newlands, 2015; Newlands et al., 2014), prediction of regional and global-scale teleconnection variability (Newlands and Stephens, 2015; Bonner et al., 2014), and seasonal-scale forecasting using machine-learning (Newlands et al., 2014).

If we observe data, $\mathbf{y}$, from a sampling density (i.e., distribution of the data), $p(\mathbf{y}|\boldsymbol{\theta})$, where $\boldsymbol{\theta}$ is a vector of parameters, and assign $\boldsymbol{\theta}$ a prior $p(\boldsymbol{\theta})$, and further

---

[8]ETCCDI indices, http://www.cccma.ec.gc.ca/data/climdex/climdex.shtml

denote underlying state variables by $\mathbf{q}$, then the joint posterior probability that integrates the $\boldsymbol{\theta}$ parameters with such state variables, is,

$$p(\boldsymbol{\theta}, \mathbf{q}|\mathbf{y}) \propto p(\mathbf{y}|\mathbf{q}, \boldsymbol{\theta})p(\mathbf{q}|\boldsymbol{\theta})p(\boldsymbol{\theta}). \tag{5.3}$$

The posterior of model parameters is then,

$$p(\boldsymbol{\theta}|\mathbf{y}) = \int_{\mathbf{q}} p(\boldsymbol{\theta}, \mathbf{q}|\mathbf{y})d\mathbf{q}. \tag{5.4}$$

In the case of sequential-based forecasting within a state-space modeling framework of a dynamic system, the probability densities above are expressed as functions of a set of discrete observed states, $p(y_t|x_t, \boldsymbol{\theta})$, and discrete states of an underlying process, $p(x_t|x_{t-1}, \boldsymbol{\theta})$. These states evolve across a sequence of times $t = (1, ..., T)$ from the initial state, $p(x_1, \boldsymbol{\theta})$. Introducing auxiliary explanatory variables, denoted as z_t, then Equation 5.3 can be re-expressed as (King, 2012; Tanner and Wong, 1987),

$$p(\boldsymbol{\theta}, \mathbf{z}|\mathbf{y}, \mathbf{x}) \propto p(\mathbf{y}|\mathbf{x}, \boldsymbol{\theta}, \mathbf{z})p(\mathbf{x}|\boldsymbol{\theta}, \mathbf{z})p(\mathbf{z}|\boldsymbol{\theta})p(\boldsymbol{\theta}) \tag{5.5}$$

where,

$$p(\mathbf{y}|\mathbf{x}, \boldsymbol{\theta}, \mathbf{z}) = \prod_{t=1}^{T} p(y_t|x_t, \boldsymbol{\theta}, z_t) \tag{5.6}$$

$$p(\mathbf{x}|\boldsymbol{\theta}, \mathbf{z}) = \prod_{t=1}^{T} p(x_t|z_t); \ p(\mathbf{z}|\boldsymbol{\theta}) = p(z_1) \prod_{t=2}^{T} p(z_t|z_{t-1}). \tag{5.7}$$

The x_t denote environmental states and z_t denote states of additional auxiliary explanatory variables. One can further identify different spatial regions, $\mathbf{s}$, such that $x_t \leftarrow x_{s,t}$ and $z_t \leftarrow z_{s,t}$. The posterior predictive or forecast density is then either the replication of $\mathbf{y}$ given the model (denoted $\mathbf{y}^{rep}$) or the prediction of a new and unobserved $\mathbf{y}$, (denoted $\mathbf{y}^{new}$), given the model. This is the likelihood of the replicated or predicted data, averaged over the posterior distribution $p(\boldsymbol{\theta}|\mathbf{y})$, given by;

$$p(\mathbf{y}^{rep}|\mathbf{y}) = \int_{\boldsymbol{\theta}} p(\mathbf{y}^{rep}|\boldsymbol{\theta})p(\boldsymbol{\theta}|\mathbf{y})d\boldsymbol{\theta}; \ or \ p(\mathbf{y}^{new}|\mathbf{y}) = \int_{\boldsymbol{\theta}} p(\mathbf{y}^{new}|\boldsymbol{\theta})p(\boldsymbol{\theta}|\mathbf{y})d\boldsymbol{\theta}. \tag{5.8}$$

We assume a linear model that, in matrix notation, for a general response vector, $\mathbf{Y}$, is given by,

$$\mathbf{Y} = \mathbf{D}\boldsymbol{\beta} + \boldsymbol{\epsilon}, \tag{5.9}$$

under the distributional assumptions,

$$\mathbf{Y}|\boldsymbol{\beta}, \sigma^2, \mathbf{D} \sim N_n(\mu, \boldsymbol{\Sigma}) \sim N_n(\mathbf{D}\boldsymbol{\beta}, \sigma^2\mathbf{I}) \tag{5.10}$$

where $\mathbf{D}$ is the *design matrix*, $\boldsymbol{\beta}$ is a vector of unknown model parameters, and $\boldsymbol{\epsilon}$ is a vector of independently and identically-distributed normal random errors with mean zero and variance σ^2. Our design matrix, $\mathbf{D}$, involves two sets of explanatory variables, such that $\mathbf{D}=(\mathbf{X}|\mathbf{Z})$, having columns $x_1, ..., x_{n_p}$ augmented with columns $z_1, ..., z_{n_q}$. N_n denotes the multivariate normal distribution (MVN) of dimension, n, with mean μ, variance-covariance matrix $\boldsymbol{\Sigma}$ and $\mathbf{I}$ the identity matrix.

Within a given sub-region (i.e., area of interest), a multivariate regression equation with spatially-varying coefficients can be specified (Banarjee et al., 2005),

$$E(\mathbf{Y}|\beta, \mathbf{X}, \mathbf{Z}) = \hat{y}_{i,j} = (\gamma_{i,0} + \gamma_{i,1} \times j) + \alpha_i y_{i,j-1} + \sum_{l=1}^{n_p} \beta_{i,j}^{(l)} x_{i,j}^{(l)} + \sum_{l=n_{p+1}}^{n} \beta_{i,j}^{(l)} z_{i,j}^{(l)} \quad (5.11)$$

where $\hat{y}_{i,j}$ denotes the estimated or expected value of $y_{i,j}$, the forecast metric for year j (i.e., to distinguish calibration yearly time-step from the forecast monthly time-step, denoted by t), where $j = (2, \ldots, T)$, within a given sub-region, i, where $i = (1, \ldots, C)$. $x_{i,j}^{(l)}$ and $z_{i,j}^{(l)}$ denote the l predictor variables for i at time j. The total number of predictors (i.e., n_p covariates and n_q auxiliary indices) is $n=n_p + n_q$. The coefficients, $\beta_{i,j}^{(l)}$ are spatially and temporally-varying. Uncertainty, $\varepsilon_{s,i}$ is independent and normally distributed (random error) with mean zero and variance σ_i^2. The regression coefficients, $\gamma_{i,0}$ (yield intercept), and $\gamma_{i,1}$ (technology trend coefficient) are used to de-trend the forecast metric and α is a lag-1 autoregressive term. The technology trend accounts for historical increases in a forecast metric and is assumed to be linear. Inter-annual autocorrelation was assumed to vary across sub-regions. For the ith sub-region, the design matrix, for fixed i, is expressed as,

$$\mathbf{D}_i = \left(X_i^1, X_i^2, ..., X_i^{n_p} | Z_i^{n_p+1}, Z_i^{n_p+2}, ..., Z_i^n \right), \quad (5.12)$$

associated with the model parameter vector, $\boldsymbol{\Theta}_i = \left(\gamma_0, \gamma_1, \alpha, \beta_i, \sigma_i^2 \right)$. In matrix notation, the integrated design matrix is then,

$$\mathbf{D}_i = \begin{pmatrix} x_{i,2}^{(1)} & x_{i,2}^{(2)} & \cdots & x_{i,2}^{(n_p)} & z_{i,2}^{(n_p+1)} & z_{i,2}^{(n_p+2)} & \cdots & z_{i,2}^{(n)} \\ x_{i,3}^{(1)} & x_{i,3}^{(2)} & \cdots & x_{i,3}^{(n_p)} & z_{i,3}^{(n_p+1)} & z_{i,3}^{(n_p+2)} & \cdots & z_{i,3}^{(n)} \\ \vdots & \vdots & & \vdots & \vdots & \vdots & & \vdots \\ x_{i,T}^{(1)} & x_{i,T}^{(2)} & \cdots & x_{i,T}^{(n_p)} & z_{i,T}^{(n_p+1)} & z_{i,T}^{(n_p+2)} & \cdots & z_{i,T}^{(n)} \end{pmatrix} \quad (5.13)$$

Additional statistical algorithms can be integrated to improve the robustness of regional-scale forecasts. Robust regression is a statistical technique that ensures that a model is less sensitive to outliers and may be applied to situations of unequal variance (Khan et al., 2010, 2007). This technique was applied to account for heteroscedasticity and outliers in the historical data during model training and calibration. Heteroscedasticity occurs when the variance of an explanatory or predictor variable is dependent on its value. Robust regression also provides a flexible and general technique for modeling based on residuals, because it is less influenced by the presence of outliers. Variable-selection is then employed for prediction in the case where there are no significant outliers. Robust regression is a compromise between excluding outliers entirely from the analysis and treating all the data points equally, as it is done in ordinary least squares (OLS) regression. Robust regression weighs observations differently depending on how well-behaved they are. Several approaches to robust estimation have been proposed, including R-estimators and L-estimators. However, M-estimators are more widely used because of their generality, high breakdown point, and efficiency. M-estimators are a generalization of maximum likelihood estimators (MLEs). The standard MM-type regression estimates use a bi-square re-descending score function and returns a highly robust and efficient estimator (with 50 % breakdown point and 95 % asymptotic efficiency for normal errors). For example, the tuning parameters of lmrob, the R Project for Statistical Computing (R) package that implements the robust regression technique, comprises an MM-type robust linear regression estimator that consists of an initial S-estimate, followed by an M-estimate via a regression that enables one to specify a breakdown point and asymptotic efficiency consistent with normal distributional assumptions (Koller and Stahel, 2011).

Bootstrap robust least angle regression (B-RLARS) can also be applied to select a leading set of $m \leq n_q$ potential predictors from $z^1, z^2, \ldots, z^{n_q}$ after adjusting for the effects of $x^1, x^2 \ldots, x^{n_p}$. Let $z^{1,*}, z^{2,*}, \ldots, z^{m,*}$ denote the leading n_q ranked variables. A final, reduced model, with $m^* \leq n_q$ predictors, is then obtained by robust leave-one-out cross validation (LOOCV) involving the selection of auxiliary variables/predictors from $z^{1,*}, z^{2,*}, \ldots, z^{n_q,*}$, after adjusting for the influence of $x^1, x^2 \ldots, x^{n_p}$ (Khan et al., 2010, 2007). Let the complete model design matrix be denoted as $\mathbf{D}_{C,T,(n_p+m^*)}$, based on the selection of best-fit predictors. Let $\boldsymbol{\Psi}$ be the vector of model parameters corresponding to the joint distribution of $\mathbf{D}_{C,T,(n_p+m^*)}$ (recall the model parameter vector is $\boldsymbol{\Theta}_i = \left(\gamma_0, \gamma_1, \alpha, \beta_i, \sigma_i^2\right)$). The conditional likelihood function of $(y_2, y_3, \ldots, y_n)$ given y_1 becomes,

$$f\left(y_2, \ldots, y_n | y_1, \Theta, \mathbf{D}_{C,T,(n_p+m^*)}\right) = f\left(y_2|y_1\right) f\left(y_3|y_1, y_2\right) \cdots f\left(y_n|y_1, y_2, \ldots, y_{n-1}\right)$$

$$= N\left(\gamma_0 + 2\gamma_1 + \sum_{l=1}^{n_p} \beta_2^{(l)} x_2^{(l)} + \sum_{l=n_{p+1}}^{n} \beta_2^{(l)} z_2^{(l)} + \alpha y_1, \sigma^2\right)$$

$$\times N\left(\gamma_0 + 3\gamma_1 + \sum_{l=1}^{n_p} \beta_3^{(l)} x_3^{(l)} + \sum_{l=n_{p+1}}^{n} \beta_3^{(l)} z_3^{(l)} + \alpha y_2, \sigma^2\right) \qquad (5.14)$$

$$\times \cdots \times N\left(\gamma_0 + n\gamma_1 + \sum_{l=1}^{n_p} \beta_n^{(l)} x_n^{(l)} + \sum_{l=n_{p+1}}^{n} \beta_n^{(l)} z_n^{(l)} + \alpha y_{n-1}, \sigma^2\right),$$

where the sub-region subscript, i is here omitted. Given that $(y_2, \ldots, y_n)$ is independent of $\boldsymbol{\Psi}$ given y_1, Θ, and $\mathbf{D}_{C,T,(n_p+m^*)}$, it follows that,

$$p\left(y_2, \ldots, y_n, \mathbf{D}_{C,T,(n_p+m^*)} | y_1, \Theta, \boldsymbol{\Psi}\right) = \qquad (5.15)$$

$$p\left(y_2, \ldots, y_n | y_1, \Theta, \mathbf{D}_{C,T,(n_p+m^*)}\right) p\left(\mathbf{D}_{C,T,(n_p+m^*)} | \boldsymbol{\Psi}\right).$$

Under a fully Bayesian approach, prior distributions for both $\boldsymbol{\Theta}$ and $\boldsymbol{\Psi}$ must be specified. Assuming a separation of variables, such that, $p\left(\boldsymbol{\Theta}, \boldsymbol{\Psi}\right) = p\left(\boldsymbol{\Theta}\right) p\left(\boldsymbol{\Psi}\right)$, an empirical prior distribution for $\boldsymbol{\Theta}$ can be constructed by residual bootstrapping the data from neighboring subregions. This bootstrapping procedure cross-validates model forecasts for a given subregion with model forecasts of its neighboring subregions. The top k ranked subregions that minimize the cross-validation error are selected. The *predictive* joint posterior density function can be expressed as;

$$p(\mathbf{y}^{new}|\mathbf{y}) = \int p(\mathbf{y}^{new}|\boldsymbol{\beta}, \sigma^2)p(\boldsymbol{\beta}, \sigma^2|\mathbf{y})d\boldsymbol{\beta}d\sigma^2 \qquad (5.16)$$

The random forests algorithm generates multiple bootstrapped regression trees without pruning and averages the outputs, and has proved very effective in reducing variance and error in high dimensional datasets (Chen et al., 2012; Breiman, 2011). This machine-learning algorithm has also recently been applied in generating national crop yield forecasts or yield outlooks for Canada (all major crops) in order to provide enhanced agriculture decision-support within the growing season (Kouadio and Newlands, 2014; Newlands and Zamar, 2012). Specifying non-parametric Bayesian priors makes this approach more flexible and more broadly applicable, as it can integrate a wide set of variables, indices and metrics.

5.4   Adaptive Science, Policy, and Institutions

What does the future hold for science and its role for informing policy? What kind of institutional changes will accompany how we sustainably manage our ecosystems and the goods and services they provide? As Grasso and Roberts (2014) highlight, "In terms of international cooperation to adequately address the climate crisis, the most urgent and complicated coordination problem is the development of an inclusive, concerted framework for promptly abating GHG. Such a framework must meet three criteria: effectiveness, feasibility, and fairness." They recommend that *production-based* GHG emissions accounting, upon which international agreements have been based and that have encouraged a displacement of carbon-intensive products and economic activity, be replaced by *consumption-based* accounting. This would effectively divide countries into either carbon exporters or carbon importers. This accounting measures emissions derived from the *final* use of goods and services (i.e., national production-based inventory emissions+ emissions imported – emissions exported), reduces "carbon-leakage" arising from the displacement of people and industrial production activities.

Keohane and Victor (2010), as part of the Harvard Project on International Climate Agreements, identify key design elements of what they argue is a scientifically sound, economically rational, and politically pragmatic international policy architecture for global climate change. They argue that a comprehensive, fully integrated regime for mitigating and adapting to climate change is not realistic[9]. Instead, they envision that a "regime complex" for climate change is "likely to persist and that

---

[9]Harvard Project on International Climate Agreements,
http://belfercenter.ksg.harvard.edu/climate

efforts to build an effective, legitimate, and adaptable comprehensive regime will likely not succeed in the long-term." This is because of current international climate change commitments being highly interdependent and with nations differing widely in their motivations and ability to comply with both voluntary and legislative/mandatory commitments. As they highlight, bilateral and multilateral agreements add greater complexity, and as long as they involve *non-binding* commitments and may further enhance adaptive capacity, without constraining international negotiations and agreements. Similarly, Vasconcelos et al. (2013) find that a polycentric approach involving multiple institutions is more effective than that associated with a single, global one, in finding a real world solution for complex and global dilemmas.

The Common But Differentiated Responsibilities (CBDR) principle of international environmental law establishes that all states are responsible for addressing global environmental destruction, yet not equally responsible. The principle balances the need for all states to take responsibility for global environmental problems, with the need to recognize the wide differences in levels of economic development between states. These differences are, in turn, linked to the states' contributions to, as well as their abilities to address, these problems. This CBDR Principle was formalized within international law at the 1992 United Nations Conference on Environment and Development (UNCED) in Rio de Janeiro. Countries have agreed to publicly outline what post–2020 climate actions they intend to take under a new international agreement—termed INDC. They will determine whether (or not) the world will achieve an ambitious 2015 agreement and whether we decide to collectively follow a sustainable development path toward a low-carbon, climate-resilient future[10]. Consistent with the CBDR Principle, the INDC structure enables a self-differentiating approach, whereby constructive feedback loops (or transition steps) are taken by a country at the national-scale in relation to international decision-making on climate change. Integrated sustainability assessment frameworks that provide scientific support such a multi-tier policy-making process will play an increasing role, alongside model-based evaluation of multi-functional beneficial/negative impacts, new opportunities, sector, and multi-sectoral policy options (Paracchini et al., 2011). All of this relies on our ability to model, assess, and forecast sustainability within national-scale policy cycles involving their design, evaluation, and implementation (see Figure 5.3).

Lackey (2006) proposes 9 general axioms of contentious and transformative ecological policy problems, in addition to recognizing that their unique features (i.e., not universally true, but universally applicable):

- The policy and political dynamic is a zero-sum game
- The distribution of benefits and costs is more important than the ratio of total benefits to total costs
- The most politically viable policy choice spreads the benefits to a broad majority with the costs limited to a narrow minority of a population
- Potential losers are usually more assertive and vocal than potential winners and are, therefore, disproportionately important in decision making

---

[10]INDC: www.wri.org/indc-definition

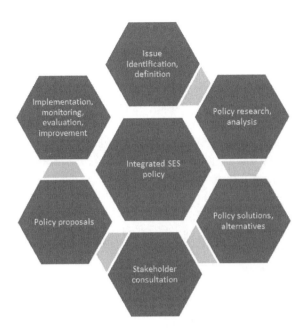

FIGURE 5.4: General overview of steps in the policy cycle, involving design, evaluation, and implementation steps.

- Many advocates will cloak their arguments as science to mask their personal policy preferences and ideology
- Even with complete and accurate scientific information, most policy issues remain divisive
- Demonizing policy advocates who support competitive policy options is often more effective than presenting rigorous analytical (e.g., statistical) arguments
- If something can be measured accurately and with confidence, it is probably not particularly relevant in policy decision-making (e.g., negotiation)
- The meaning of words matters greatly and arguments over their precise meaning are often surrogates for debates over values

Lackey (2006) further cautions us to be constantly aware of normative science in both our scientific language and thinking, as it has subtle, loaded or "built-in" policy preferences and biases. Science's role is thus to identify, evaluate the feasibility, and compare options as scientifically-derived decisions. Choice preferences and their selection, however, is a societal enterprise of shared responsibility. In our sustainable future, this traditional model of policy decision-making will change dramatically and scientific, social, and political operational enterprises (i.e., not just their knowledge realms) will merge as sustainable policy decision making necessitates. Nonetheless, Lackey (2006) indicates there is often an ambiguous role for science, where science is often not pivotal in evaluating policy options, but science often ends up serving inappropriately as a surrogate for debates over values and preferences. Without a meaningful and more complete integration of science in policy decision-making, negotiation and policy decision-making runs the risk of becoming irrelevant. There is also the increased possibility of broad-based social uprising, unless complex decision-making is disentangled and representatives at negotiation tables are replaced by voting (i.e., referendums/plebiscites). These public votes can, however, be socially and economically disruptive, especially without a well-developed, tested and broadly-communicated science and policy-based integrated risk framework.

5.5 Public Good Science

The role of science and technology (ST) and its networks in decision making within future SES is crucial, as scientific knowledge and appropriate technologies are central to resolving the economic, social, and environmental problems that make current development paths unsustainable. Bridging the development gap between the developed and developing countries, and alleviating poverty to provide a more equitable and sustainable future for all, requires novel integrated approaches that fully incorporate existing and new scientific knowledge. A clear implication of this is that international cooperation in Science and Technology (ST), from small cross-border research projects to global-scale coordination, must be considered as a key tool for enhancing sustainable development. Globalization of ST has been dominated by primarily "bottom-up" approaches, driven largely by purely scientific community dynamics or commercial considerations. This is not particularly well-suited to defining or addressing sustainability objectives. In international terms, ST is primarily a matter for the public sphere, and cooperation in science and technology provides the most important vehicle for implementing a sustainable development agenda globally. Effective dissemination of knowledge about science, technology, and sustainable development is also critically important for the application of solutions at the *local* and *regional* level or scale. This implies a need for an interdisciplinary knowledge infrastructure, bridged by statistical concepts, methods and applications, to enable a fair, consistent and reliable assessment of impacts and actionable responses. "Ecosystem resilience" decision-making needs to extend to all parts of the world—both the highly industrialized as well as the underdeveloped (Ahmed and Stein, 2004).

Shawn Otto, a science advocate, producer of US Presidential Science Debates, winner of the prestigious IEEE-USA's National Distinguished Public Service Award and founder of sciencedebate.org, in his book, *Fool Me Twice: Fighting the Assault on Science*, details the rise of anti-science beliefs and the current, anti-science revolution crisis (Otto, 2011). He raises critical questions of what will happen in a world dominated by complex science and whether people will continue to be well informed and trust their own governments. With less than 2% of the US Congress having any professional knowledge and background of science, one can ask, how can governments be trusted to lead us in the right direction? Because of misinformation and trust breakdown, at some point in the future, the science, social, and political enterprises may even become fully integrated with AI guided tools under human-intervention (e.g., under a hierarchy of international, national organizations and regional/local, civic trust collectives).

Ensuring the vitality and shared/common benefits of public good science means ensuring that public trust in the experts is not eroded, and ensuring a central role of government scientists who are dedicated to upholding public trust, confidence, and ethics for framing, consolidating, and integrating expert knowledge to support innovation in government policy. It is risky to rely solely on the advice of specially selected experts from private corporations or academic institutions; the independence of their motivation can always be questioned when relying on them to fully integrate knowledge and information across all stakeholder interests. The removal of key national science advisors appointed by elected officials further erodes public trust in their governments (and their scientists) in undertaking open, transparent and fair evidence-based decision-making and in representing and speaking for the often under-represented, public good at decision-making tables. Pidgeon and Fischhoff (2011) discuss the role of social and decision sciences in communicating uncertain risks (which government scientists are exposed to and provided professional training and work experience in), in addition to the subject-matter experts. The scientific integrity of government scientists and advisors must be actively promoted, maintained and upheld by academic, corporate and citizens alike for the benefits it offers and promotes to all. The premise that government science and evidence-based government decision-making necessarily competes with external interests, whether they be academic, corporate and/or public, is fundamentally wrong. Instead, they are crucial for elucidating evidence-informed options, rather than simply advocating a single course of action. Such expert panels should always involve government scientists, otherwise it subverts and erodes the role of government scientists in their role in ensuring the public good and government as a whole in upholding their responsibility for developing and implementing policy instruments that achieve effective, feasible, and fair transitions and outcomes. The use of advisors, advisory councils and academia need not be mutually exclusive (Gluckman, 2014). But it is crucial that science advice to government, whether from external, internal, or a mixture of external/internal experts and panels, integrate sound principles for building not only independence and trust, but also engagement and influence. The latter requires the role of municipal, provincial, and federal government to be effective and feasible (i.e., implementable/operational). From the perspective, knowledge, and experience as a national science advisor, there are ten general principles for achieving trust, independence, engagement, and influence (Gluckman, 2014):

- Maintain the trust of many
- Protect the independence of advice
- Report to the top
- Distinguish science for policy from policy for science
- Expect to inform policy, not make it
- Give science privilege as an input into policy
- Recognize the limits of science
- Act as a broker not an advocate

- Engage the scientific community
- Engage the policy community

In the future, governments will have an increasing role to lead and/or promote citizen-centric change in the face of increasing global-scale integrated risks, and creating and broadening the public value that they generate. As detailed by the World Economic Forum (WEF) in a report entitled, "Future of Government - Fast and Curious," a global agenda for transforming governments into being "fast, agile, streamlined and tech-enabled" (FAST) organizations is outlined (WEF, 2012). This can only be accomplished if people and institutions external to government do not lose trust in their public institutions and public servants and their role and integrity, while transforming to meet the changed needs of citizens, does not become eroded, diminished, and subjugated by power and/or ideology of a few (i.e., a niche group or elite).

One key aspect at the interface between science for decision-making and developing evidence-based, sound policy, is *reproducibility and selection bias*. Statistical science best addresses the controversial domain between science and policy, as it integrates both the knowledge (evidence support) and operational tools (actions, choices) required for framing policy decisions. Operational models that support sound policy decision-making, for example, must be able to detect and minimize the selection bias between positive reported findings of different stakeholders, assumptions and considerations that may have a competitive advantage over false of negative results, findings and discoveries (ISI, 2014)[11],[12]. Reproducibility has a subtle distinction from replication—it can be difficult or impossible to replicate an experiment, but data and methods can be independently verified and validated under different conditions and contexts (Kenall et al., 2015).

5.6 Super-Intelligent, Sustainability Tools

Lipsey et al. (2005) argue that long-term economic growth and economic transformations are predominately driven by pervasive technologies that can affect entire economies (of which they identify only 24 since the Neolithic Agricultural Revolution (9000–8000 BC) termed General Purpose Technologies (GPTs) (e.g., electricity, automobiles, automation, computers, Internet). GPTs are defined by: 1) a core, recognizable technology, 2) widely used across an economy with improvement potential, 3) has broad application and uses, 4) creates spill-over technology effects. In this way, new GPTs may decrease productivity before increasing it because of learning costs, the time and costs associated with tranforming infrastructure and labor

---

[11]Statistics and Science: A Report of the London Workshop on the Future of the Statistical Sciences: http://www.worldofstatistics.org/wos/pdfs/Statistics&Science-TheLondonWorkshopReport.pdf

[12]Focus on Reproducibility in Science: www.nature.com/nature/focus/reproducibility/index.html

re-adjustment. Super-intelligent tools may represent a future GPT that develops to solve our global, real world sustainability problems, but what specific advances would they have, beyond the way machines compute and learn today?

In pursuit of an artificial machine that could describe nature and intelligence, Turing devised a universal computational model known as the *Turing machine* that distinguishes the structure of a computational machine from its program or set of operations, along with an accessible memory of external storage (Russell and Norvig, 2010; Turing, 1936). The Turing machine follows the transition $f : (\alpha, s) \rightarrow (\beta, m, s')$, whereby at each recursion, an input $\alpha \in [0, 1, blank]$ is cross-referenced to a state $s \in [1, 2, ..., S]$ defining three output rules: 1) a new binary value $\beta \in [0, 1]$, a left or right shift/move $m \in [L, R]$, or new state $s' \in [1, 2, ..., S]$ until a halting state is reached. This generates finite, binary sequences as input and output of the computation with state values as rational numbers, not real-valued (Siegelmann, 1995). The *Church–Turing theorem* or *Turing Limit* associated with the Turing model states that no realizable computing device can be more powerful than a Turing machine. This realizable or real-world Turing limit, alongside Gödel's 1931 Incompleteness Theorems stating that there are fundamental limits to proof and computation, may mean that such formally-defined systems cannot ever be fully complete in describing the real world. Nonetheless, one needs to recognize that the Turing machine only computes what it has been programmed for, and possesses no ability to learn or adapt to new situations. For this reason, Siegelmann (2013, 1999, 1995); Siegelmann and Sontag (1991) have challenged the Turing model and proposed an alternative model with computational power that surpasses the Turing model and is realizable in highly chaotic dynamical systems (subset of CAS). Nonetheless, Edward N. Lorenz in *The Essence of Chaos* shows that certain deterministic dynamic systems also have formal predictability limits (Lorenz, 1995).

How does one then mimic real-world adaptability? Levin (2002) characterizes adaptability as requiring the development of an appropriate statistical mechanics that allows one to separate the knowable and unknown from the truly unknowable dealing with heterogeneous ensembles of interacting agents, continually refreshed by novel and unpredictable types. Markose (2004) combines the Turing limit and Gödel's Incompleteness Theorems into a computational theory of novelty in CAS, whereby adaptive novelty or strategic innovation can only emerge if agents, operating as universal Turing Machines, are able to make *self-referential* calculations of hostile behavior. In this context, agents possess behavior according to finitely encodable algorithms rather than a machine that can learn intelligent behavior directly from data and not have its behavior explicitly programmed. This requires focusing on a computational model of machine-learning with behaviors that are learned, not pre-programmed, and with outcomes that are not optimal and may fail, mimicking how real world CASs learn and adapt. Siegelmann (2013) describes this adaptability that achieves Super-Turing; "Super-Turing computation is not precise in the sense of Turing, it makes mistakes and tries new avenues. This does not mean that Super-Turing computation is inaccurate, rather, it constantly tries things and makes small corrections to achieve an overall accuracy supported by experience and learning." Super-Turing breaks down, because of limits on computer processing (a technological limit), and not because of any technological singularity associated with real-valued computation itself.

In general, there are two main types of artificial neural networks (ANNs), namely: feedforward (acyclic) (FNNs) and recurrent (cyclic) (RNNs) types (see Chapter 4, Section 4.3.5). Recurrent neural networks (RNNs) contain dynamic states that evolve depending on the input the system receives and on their current state's distribution. This gives them a significantly larger and richer memory and computational capacity to learn and carry out complicated transformations of data over extended periods of time (Graves et al., 2014). Such capabilities of RNNs can be further enriched by coupling the system of dynamic states (termed the neural network *controller*) to a large, addressable memory bank, analogous to the working memory in human cognition, and simulating the ability of humans to rapidly create sets of variables through a recursive processing procedure involving *causal time-lags* between various inputs and outputs (i.e., structures of varying lengths termed probabilistic or asynchronous computation) including every component being *differentiable* and able to be trained by gradient-descent. These enhancements form a Neural Turing Machine (NTM) (Graves et al., 2014). Interested readers are referred to an application of the gradient-decent algorithm in optimizing marine ecosystem network models (Kavanagh et al., 2004).

Robust and highly complex analog computing occurs within all living systems, from single cells to complex organisms (Hermans et al., 2014). Analog Recurrent Neural Network (ARNN) models have dynamic internal states that are analog (i.e., continuous rather than discrete). Such systems consist of a finite number of neurons (N) having M total external input signals with X_j activations, based on u_j external inputs and a_{ij}, b_{ij}, c_i real-valued weight coefficients (i.e., analog), such that the internal state of each neuron processor, $i = (1, 2, ..., N)$ is updated according to,

$$X_i(t + 1) = \sigma \left(\sum_{j=1}^{N} a_{ij} x_j(t) + \sum_{j=1}^{M} b_{ij} u_j(t) + c_i \right) \tag{5.17}$$

with σ a sigmoidal function. The surprising finding has been that when analog networks assume real-valued weights, their power encompasses and transcends that of digital computers as they compute a super-Turing class of variable-length input sequences (i.e., nonuniform P/poly class) in polynomial time and all binary functions in exponential time (Siegelmann, 1995). One can thus advance from Turing to *Super-Turing* computation via real-valued weights and stochastic computation.

Deep learning neural networks develop novelty by artificial agents learning successful policies directly from high-dimensional sensory inputs using end-to-end reinforcement learning, as a form of statistical-based learning (Mnih et al., 2015; Hastie et al., 2008; Vapnik, 2000). In 2013, the first *deep learning* model employing ANNs that could successfully learn control policies directly from high-dimensional raw sensory input based on reinforcement learning (i.e., modified *Q-algorithm*) was identified. Deep Learning using the *Q-learning* algorithm minimizes a sequence of loss functions, which change at each minimization step/iteration, typically using a stochastic gradient descent approach (Mnih et al., 2015, 2013). This signifies the high prediction power that network models have in helping humans to solve complex real world problems, by devising realistic loss functions and learning by integrating high-dimensional data input and model output efficiently. Fouskakis and Draper (2002) inter-compare the sensitivity and performance of leading stochastic optimization methods, namely simulated annealing, genetic algorithms and tabu heuristic search, concluding that the winning optimization method is highly context-specific, and the integration or hybridization of competing algorithms often narrows the performance gap between methods. Also, an inter-comparison of optimization methods can result in a novel approach that is superior to any of the algorithms being hybridized.

A necessary condition for optimality, the *Bellman's dynamical programming equation* breaks dynamic optimization problems into simpler subproblems, whereby the value of a decision problem at a certain point in time as a function with two contributions—a payoff from some initial choices *and* the value of the remaining decision problem resulting from such initial choices. In this way, Bellman's equation requires an optimal cost-to-go function, $v(x, t)$ which is the sum of immediate costs $l(x, y)$ accumulated by initializing the system at state x and time t and controlling it, in a globally optimal way, until a final time T, involving the computation of v everywhere—but this is numerically infeasible when the dimensionality n is large. Mnih et al. (2013) provide an outline of the *Q-algorithm* and its associated loss function based on the Bellman equation. Let rewards or payoffs be discounted by a learning parameter or factor γ, $0 \leq \gamma \leq 1$, per time-step, i (where T is the final time). If γ is closer to zero, a learning *network or agent* will tend to consider only immediate rewards, whereas if γ is closer to one, an agent will consider future rewards with greater weight and be willing to delay a reward. This is expressed by the *discounted return* at time or step i,

$$d_i = \sum_{t'=i}^{T} \gamma^{t'-i} d_{t'}. \tag{5.18}$$

The maximum (optimum) expected return from a strategy (i.e., sequence of previous actions, s, and then taking some action a), is given by,

$$Q^*(s, a) = \max_{\pi} E[d_i | s_i = s, a_i = a, \pi], \tag{5.19}$$

where π is a *policy* that maps sequences to actions or action distributions. If, at the step, i, the optimal value $Q^*(s', a')$ is known for all possible actions, and return, d_i, a', then the reinforcement strategy,

$$Q_{i+1}(s, a) = E\left[d_i + \gamma \max_{a'} Q_i(s', a') | s, a\right] \tag{5.20}$$

converges to the optimum (i.e., $Q \to Q^*$) for $i \to \infty$. The Q function can be approximated by ANN networks, by *minimizing* a sequence of loss functions at each time step, i, $L_i(\theta_i)$, where θ is given by network weights,

$$L_i(\theta_i) = E_{s,a \sim \rho(s,a)} \left[(y_i - Q(s, a, \theta_i))^2\right] \tag{5.21}$$

where the ith target distribution, y_i is,

$$y_i = E_{s' \sim \epsilon} \left[d_i + \gamma \max_{a'} Q(s', a', \theta_{i-1}|s, a) \right] \tag{5.22}$$

associated with $\rho(s, a)$, the *behavioral* distribution as a probability distribution over all sequences s and actions a. The gradient of this *network-based loss function*, is then given by Mnih et al. (2013),

$$\nabla_{\theta_i} L_i(\theta_i) = E_{s, a \sim \rho(s,a), s' \sim \epsilon} \left[\left(r + \gamma \max_{a'} Q(s', a', \theta_{i-1}) - Q(s, a, \theta_i) \right) \nabla_{\theta_i} Q(s, a, \theta_i) \right] \tag{5.23}$$

where $\nabla(\cdot)$ denotes the gradient operator, and network weights, θ_i are updated after each ith step, and single samples of previous sequences and actions, $\rho(s, a)$ and *emulator* ϵ, of future or next sequences, then the loss function can be optimized in a computationally-efficient way using stochastic gradient-descent, termed *Q-learning*. (Mnih et al., 2013; Watkins and P., 1992).

Self-awareness can be defined as, "the capacity to become the object of one's own attention" (Lewis et al., 2012; Moran, 2006). Self-awareness involves creating and testing alternative models of each process based on the real (sensory) or imagined information when making predictions, and may be measurable as *self-improvement*, whereby machines are observed to learn by themselves and discover errors in algorithms, and fix them in real-time without human guidance or intervention. This extends from adapting their own algorithms to adapting their systems more broadly involving *self-replication* as machines (i.e., highly intelligent computers, agents or robots) teach themselves how to adapt to different events, conditions and situations. In this sense, self-awareness is "limitless". As Lewis et al. (2012) describe, "For a system to adapt itself effectively, it is important that it has the ability to be self-aware. Self-awareness is concerned with the availability, collection and representation of knowledge about something, by that something. A self-aware node has knowledge of itself, permitting reasoning and intelligent decision making to support effective autonomous adaptive behavior. While some work in self-aware computing exists, there is no general methodology for engineering self-aware systems or for validating and benchmarking their behavior." In studying microscopic systems, England (2013) explores the statistical limit (i.e., lower bound) of self-replicating processes (i.e., bacterial cell division and the emergence of self-replicating nucleic acids). This study has found that the process sensitivity is governed by internal entropy, durability (i.e., inverse of the growth rate of a replicating system or population), and the heat that is dissipated.

Entropy (i.e., Shannon metric of entropy) of a finite sample of a stochastic variable, X, is given by,

$$H(X) = E[I(X)] = -\sum_i P(X_i) log(P(X_i)) = E[-ln(P(X))], \tag{5.24}$$

where the entropy is the expected (or average) value of *self-information* as a stochastic variable. Here, *self-information* is the negative of the probability function P of *self-information* I. Thus, the more uncertain an event, the more information its occurrence contributes to our understanding of a process. (Mayer et al., 2014). Entropy measures emergent ecosystem behavior, as the concept of Maximum Entropy Production (MEP) postulates that states or configurations that maximize entropy production (degradation of gradients) are selected because they are compatible with a greater number of environments (Mayer et al., 2014). A deep connection between statistical learning from diverse information, entropy maximization and intelligence may exist. Exploring this question further, Wissner-Gross and Freer (2013) propose a generalization of "causal entropic forces" that spontaneously induce sophisticated adaptive behaviors (i.e., closely associated with human "innovation niches"). This proposition attempts to account for the statistical bias of dynamical system to maximize instantaneous entropy production and evolve toward states of higher-entropy involving *entropic forces, known as the Maximum Entropy Principle* (Chakrabarti and Ghosh, 2013). In general, an entropic force, $\mathbf{F}$ is associated with a macro-state (here considered analogous to a CAS network state partition) $\mathbf{X}$ according to,

$$F(X_0) = T\nabla_{\mathbf{X}}H(\mathbf{X})|_{\mathbf{X_0}}, \qquad (5.25)$$

where T is the "reservoir temperature," $S(\mathbf{X})$ is the entropy, H, of a final network partition $\mathbf{X}$ from an initial partition, X_0.

Topsoe (2002) has explored aspects of game theory underlying the Maximum Entropy Principle, showing that maximizing entropy, H matches minimizing integrated risk, R. As highlighted by Grünwald and Dawid (2004), the close relationship between maximizing entropy and minimizing risk (as worst-case expected loss) has only recently be recognized. Let $\mathbf{A}$ be an alphabet (i.e., a countable set) and $\mathbf{M(A)}$ be a set of probability distributions over $\mathbf{A}$.. Let $\mathbf{K(A)}$ be the set of codes, algorithms or mappings (i.e., $\kappa : \mathbf{A} \to [0, \infty]$ with $\sum_{a \in \mathbf{A}} exp(-\kappa(a)) = 1$). Then, (κ, P) is a *matching pair* or κ *is adapted to* P if $\kappa(a) = -ln(P(a))$ for each $a \in \mathbf{A}$.

For entropy, H, and information divergence, D, for any probability distribution $P \in M(\mathbf{A}$ and $\kappa \in K(\mathbf{A})$, an *integration identity* can be defined as,

$$< \kappa, P >= H(P) + D(P|Q), \qquad (5.26)$$

where Q is the probability distribution that matches the set of codes, κ. In the above identity, $D(P|Q)$ is equal to the $D(P|\kappa)$, defined as the *redundancy* associated with the set of codes, κ in relation to a true distribution P. Divergence and redundancy are inherent properties of real-world CAS because agent behaviors may be imperfect, and their adaptation algorithms are approximate—they are not complete descriptions of all possible dynamical states that completely mimic a complete distribution of behavior/s. Furthermore, if P is in equilibrium, (P_n) is asymptotically optimal, and κ^* is a minimum risk code and P^* the corresponding maximum entropy distribution, stated mathematically as (Topsoe, 2002),

$$\lim_{n\to\infty} H(P_n) = H_{max} = R_{min} = R(\kappa^*) = H(P^*). \qquad (5.27)$$

Although originally defined in terms of eliminating logarithmically divergent integrals, *renormalization* in the context of systems is defined as the existence of multiple length and energy scales (Wilson, 1975). Here, renormalization group methods break up or decomposes computations into separate stages, with a stage for each length or energy scale, and each stage involves a matrix diagonalization (or branching of a regression tree etc.) such that each sub-problem involves fewer degrees of freedom than the complete or full multi-scale problem. In the limit of an infinite number of iterations or recursions, there is no difference at all except for the change in scale and the approaches a fixed point, although there is no guarantee that a given renormalization group computation will lead to a fixed point. Also, renormalization breaks down due to the interaction strength/correlation length scaling limits.

Real world optimization problems are often non-convex, having an exponential number of local optima. Here, AI can help design, build, test, and deploy robust solutions to complex, nonlinear sustainability problems that require optimization; whether involved in interactivity, for distributed control and automation, data mining and machine learning, or solutions that embed all of these functionalities (Fisher, 2011). Robust solutions achieve required levels of performance under a broad range of realizations (e.g., different structures, behaviors, future adaptations) of a real-world system. Standard convex techniques to find only local optima and convex methods like gradient-search, fail to find the global optimum in a reasonable amount of time. (Friesen and Domingos, 2015) demonstrate how by repeatedly decomposing a problem into independently solvable subproblems one can solve problems involving continuous variables in polynomial time problems, not exponential time. This optimization solver or learning algorithm technique is called, (R)ecursively (D)ecomposing the function into locally (I)ndependent (S)ubspaces or RDIS. It recursively defines sets of variables so as to simplify and decomposes (or constructs from parts) an objective function into approximately independent sub-functions, until the remaining functions are simple enough to be optimized by standard techniques like gradient-descent. The variables to set are chosen by graph partitioning, ensuring decomposition whenever possible.

The RDIS algorithm recursively chooses a subset of the variables $\mathbf{X_C} \in \mathbf{X}$ that forms a partition $(\mathbf{x_C}, \mathbf{X_U})$ of $\mathbf{X}$ and assigns them values ρ_C in the problem objective function, $f|_{\rho_C}(\mathbf{X_U})$ such that it decomposes into multiple independent sub-functions, $f_i|_{\rho_C}(\mathbf{X_U})$, $(\mathbf{X_{U_1}}, ..., \mathbf{X_{U_k}})$ is a partition of $\mathbf{X_U}$, $1 \leq k \leq n$. This procedure is repeated for each partition/sub-function until the objective function is globally optimized, conditioned on the assignment of variables and their objective function values, $\mathbf{X_C} = \rho_C$. After each recursion, optimal values of $\mathbf{X_U}$ are used to choose new subset values for $\mathbf{X_C}$ until a heuristic stopping criterion is satisfied.

One can further postulate that an algorithm that maximizes entropy or minimizes integrated risk associated with different *behavioral modes* or *internal states*, which themselves may be decomposable according to a fundamental set of rules for creating (graph or network) partitions—termed *switch, jump and link (SJL)* rules. Coupling learning via recursive decomposition with reinforcement provided by a memory bank of learned decomposition sequences (i.e., synonymous with "K-lines" hypothesized by Marvin Minsky) of associated with such SJL rule-based partitions may enable self-awareness to emerge in systems with a Super-Turing computational capacity. A system that can update its memory on previously decomposed partitions following decomposition sequences comprising the *SJL* rule types, may then adapt such sequences further in response to external feedbacks and cues. Such updating enables a self-awareness algorithm to fail and to allow for redundancy that decomposition can't capture. However, updating can emulate self-awareness whereby a smart machine makes a decision based on internal decision making and memory, and then updates its decision after receiving new external inputs. True self-awareness requires the smart machine to innovate and construct its own updating rules and decomposition sequences.

Gorbenko et al. (2012) define *self-awareness* as the capability of an agent to focus attention on the representation of internal states. There are also other types of awareness such as situational, contextual, and uncertainty awareness. Finite State Machines (FSMs) have multiple internal states, each with an associated action or set of actions and set of references (i.e., memory and/or observations). FSMs learn the rules for switching between internal states. They have been applied in machine device controllers (i.e., sequential logic circuits). Grush (2004) has developed and explored an *emulation theory of representation* on how the human brain functions, involving statistical models that can be run off-line that produce imagery, estimate outcomes of different actions, and evaluate and develop motor plans.

Reggia (2014) discusses the *computational explanatory gap* in creating conscious machines using AI, as due to our current inability to explain and/or implement high-level cognitive algorithms in terms of neurocomputational computing. More specifically; "it is the current lack of understanding of how high-level cognitive information processing can be mapped onto low-level neural computations. Through an integrated, stochastic process on a large enough neural network, with multi-scale activation cascades (i.e., requiring a high-level cognitive transitional process that renormalizes signal patterning), a SJL rule-based dynamical decomposition of internal states or modes and system updating (i.e., re-enforcement learning) based on memory and perception/external sensory inputs, self-awareness and consciousness may spontaneously emerge as a self-organizing process. Repetitive tasks may rely on low-level re-enforcement learning with reduced multi-scale cascading. Here, signal cascades are generated through a complementary micro-columnar organization mechanisms involving so-called "double bouquet cells" (i.e., distinguished from Large Basket and Candelier Cells) as long descending, vertically bundled axons, which cross several sets or bundles of cortical layers and migrate radially (DeFelipe, 2011). In this system, self-awareness would essentially be an internal neural response that is generated through repetitive stochastic switching between modes, whereby a re-correction or re-adjustment is made when dynamically switching between sets of concurrent SJL-rule decomposition processes having extended time-lags. So, self-awareness is generated through an agent's focused attention to this internal state representation and re-adjustment process. Activation sensitivities would apply in general through a biological network. Consciousness is an emergent phenomenon involving perception, learning and building associative memories (Starzyk and Prasad, 2011). In the above model context, this would, hypothetically, then imply that it could emerge from an interaction-feedback loop between the stochastic dynamics of mode-switching and the process that renormalizes multi-scale cascades of the system as a whole. This interaction with feedback would establish a link between behavior and risk perception—in essence a system's own risk emulator, controller, and adapter. Such whole-system (e.g., whole-brain) modeling and simulation exploration linking with cognitive and systems biology, however, remains controversial (*Nature*, 2015a).

Nick Bostrom, in *Superintelligence: Paths, Dangers, Strategies* raises the possibility of an AI revolution coming that will give machines "superintelligence" and cognitive capabilities (including self-awareness), beyond that of humans (Bostrom, 2014). Ray Kurzweil (AI expert, inventor, thinker, and futurist) envisions the human-machine symbiosis will occur in the 2030s, whereby the human brain will merge with computer networks to form a hybrid artificial intelligence, connecting directly from our neocortex (containing 300 million pattern processes responsible for human thought) to the Internet's cloud (Kurzweil, 2013)[13,14]. Others envision it will take place not in the 2030s, but later, in the 2100s. Marvin Minsky in *The Emotion Machine: Commonsense Thinking, Artificial Intelligence, and the Future of the Human Mind* emphasizes how increased complexity and the integration of different ways of thinking and AI, despite a fragmented and incomplete perspective on how our minds work (Minsky, 2006).

What happens to humanity, whether humanity would even survive, would then depend on the goals of the superintelligence (Bostrom, 2014). Faced with such uncertainty regarding the consequences of an upcoming AI-driven "technological singularity", Elon Musk (inventor and futurist, CEO of Tesla Motors, SpaceX) has called for the establishment of international regulations on AI development, considering it to be "the human race's biggest existential threat."[15] Stephen Hawking also envisions that "the development of full artificial intelligence could spell the end of the human race," calling for researchers to do more to protect humans from the risk of AI[16]. He has also emphasized that "whereas the short-term impact of AI depends on who controls it, the long-term impact depends on whether it can be controlled at all."[17] Bill Gates (Cofounder of Microsoft, philanthropist) envisions that "First the machines will do a lot of jobs for us and not be super intelligent. That should be positive if we manage it well," but that, "A few decades after that though, the intelligence is strong enough to be a concern."[18] Such concerns have prompted AI experts around the globe to recently sign an Open Letter prepared by *The Future of Life Institute* pledging to that pledges to "safely and carefully coordinate progress in the field to ensure it does not grow beyond humanity's control." To the contrary, Noam Chomsky has argued that AI's intensive use of statistical techniques to pick regularities in masses of data is unlikely to yield the explanatory insight that science ought to offer, and that statistical learning techniques to better mine and predict

---

[13]Creating a Mind:http://www.howtocreateamind.com/

[14]Human-Machine Symbiosis, http://www.cbc.ca/news/technology/ artificial-intelligence-human-brain-to-merge-in-2030s-says-futurist-kurzweil-1. 3100124

[15]Elon Musk on AI,http://www.bbc.com/news/technology-30290540

[16]Stephen Hawking on AI,http://www.bbc.com/news/technology-30290540

[17]Stephen Hawking on AI,http://www.independent.co.uk/news/science/

[18]Bill Gates on AI,http://www.forbes.com/sites/ericmack/2015/01/28/ bill-gates-also-worries-artificial-intelligence-is-a-threat/

data are unlikely to yield general principles about the nature of intelligent beings or about cognition.[19,20]

But AI already achieves and may outperform human intelligence in specific applications and domains (e.g., Internet searches, robot control, remote sensing, games, medical diagnoses, financial trading etc.) and will continue to be integrated into all aspects of our society. In 2014 alone, there have been some major breakthroughs that are facilitating its rapid advancement: inexpensive parallel computation, Big Data (massive databases and storage, advanced data-mining tools) and better learning algorithms (e.g., deep learning)[21]. The use of AI in our sustainable future is likely critical. A broad examination and evaluation of the scientific evidence and support for forecasts of an AI singularity and its potential consequences is provided in *The Singularity Hypothesis: A Scientific and Philosophical Assessment* from the perspective of computer scientists, physicists, philosophers, biologists, economists and others (Eden et al., 2012). For example, Omohundro (2012) of *Self-Aware Systems*, outlines a Safe-AI Scaffolding development strategy embedding long-term strategies that ensure AI technology contributes to the greater human good. Cortés et al. (2000) provide a historical overview of the impact of AI techniques on the definition and development of Environmental Decision Support Systems (EDSS). Gomes (2009) explains how we must develop new computational models, methods, and tools to help balance environmental, economic, and societal needs for a sustainable future. New research is now exploring Gödel machines—general problem-solvers that are capable of self-optimization. Such machines are self-referential, machine-dependent, and contain self-modifying algorithms (e.g., computer code) as they react with real-world system responses (Schmidhuber, 2007). Research in *structured machine-learning*, which involves testing learning hypotheses from data with rich internal structure is also rapidly advancing (Dieterich et al., 2008). Moreover, Villa et al. (2014); Villa (2001) have demonstrated how a robust and adaptable, integrated modeling methodology for sustainability (i.e., ecosystem goods and services) can be designed. Their methodology is called ARIES (ARtificial Intelligence for Ecosystem Services)[22] and addresses the diversity of real-world problems and situations within SES and more rapid Ecosystem Service Assessment and Valuation (ESAV).

Statistics is able to reconnect fragmented knowledge, as an interdisciplinary knowledge bridge. The importance of a statistical understanding and the critical role of statistical inquiry is to address knowledge gaps, identifying "blind spots," while ensuring the reliability of metrics to guide sustainability problem-solving. Statistics also offers an efficient way for devising methods that are capable of integrating diverse and large sets of data, obtained using different technologies and observation platforms to enable and facilitate an integrated approach to sustainability problem

---

[19]Noam Chomsky on AI,http://www.theatlantic.com/technology/archive/2012/11/ noamchomskyonwhereartificialintelligencewentwrong/261637/

[20]2011 MIT Symposium on Brains, Minds and Machines: http://mit150.mit.edu/symposia/brains-minds-machines

[21]AI Breakthroughs: http://www.technologyreview.com/news/533686/ 2014-in-computing-breakthroughs-in-artificial-intelligence/

[22]ARIES, http://www.ariesonline.org/

solving across spatial and temporal scales (e.g., global, national, regional, local). The development and use of statistical learning algorithms embedded within AI tools may better enable the reproduction of experimental and analytic findings within both real or virtual worlds and domains. It also can quantify selection bias and its significance across multiple, completing or interlinked sets of scientific findings. The broader use and integration of statistical thinking, techniques, frameworks, and decision-support tools can address our increasing need for greater reproducibility, and reduce selection bias. This helps to increase the public trust needed in science to support sound, integrated sustainable policies for the future. One cannot rely on AI alone. We must evolve ourselves, adapting, and capitalizing better on the tremendous power of our human creativity and ability for integrated learning that emerges when fully engaged in real world problem solving (Nagel, 1996). Moreover, machine-learning algorithms that use Bayesian or Minimum Description Length (MDL) metrics for classifying may not predict well, even though they have large, complex information (and even infinitely many samples) to learn/train from (Grünwald and Langford, 2007).

Collaborative models and highly collaborative learning environments, between real and virtual, will be increasingly needed to develop, harness, and exploit human ingenuity to solve real world problems. Sir Ken Robinson, who led the National Commission on creativity, education and the economy for the UK Government and its report, "Our Futures: Creativity, Culture and Education (The Robinson Report)"[23] raises some of the most important issues facing government, business, and the broader society in the 21st century and urges a "radical rethink of our school systems, to cultivate creativity and acknowledge multiple types of intelligence."[24] He advocates for systems to educate creative thinkers and more fully embrace the gift of human imagination and innovation. In a recent talk regarding creativity in education, Sir Robinson concludes, "I believe our only hope for the future is to adopt a new conception of human ecology, one in which we start to reconstitute our conception of the richness of human capacity. Our education system has mined our minds in the way that we strip-mine the Earth: for a particular commodity. And for the future, it won't serve us. We have to rethink the fundamental principles on which we're educating our children."[25] Mooney et al. (2013) discuss how emerging research programs for sustainability research, such as *Future Earth*,[26] are building upon lessons-learned from past knowledge integration efforts. Such efforts include the International Geosphere-Biosphere Programme (IGBP), NASA, the US Global Change Research Program, MEA, and Global Environmental Change Programme (GEC) of the International Council for Science (ICSU). The ICSU involves The International Programme of Biodiversity Science (DIVERSITAS), The International Geosphere-Biosphere Programme on Global Change and Terrestrial Ecosystems, The Int'l Human Dimensions Programme on Global Envir. Change (IHDP)

---

[23]Our Futures (UK Report): http://sirkenrobinson.com/pdf/allourfutures.pdf
[24]Ken Robinson: www.ted.com/speakers/sir_ken_robinson
[25]www.ted.com/talks/ken_robinson_says_schools_kill_creativity/transcript?language=en
[26]A new global research platform to accelerate our transformations to a sustainable world, www.futureearth.org/

and The World Climate Research Programme (WCRP).[27] *Future Earth* aims to be an international hub that coordinates new, interdisciplinary, approaches to research under three main themes: Dynamic Planet, Global Sustainable Development, and Transformations toward Sustainability.

Advances in AI will continue to transform our lives, especially education and employment in our sustainable future[28] (PEW, 2014). Advances may have smart machines undertake repetitive tasks, so that humans can invest their time and energy on more creative work. Human consciousness, creativity/imagination, empathy, and compassion, that we all have a capacity to express and share, may be human attributes that AI smart machines may never fully emulate or duplicate. Future AI machines might, however, at the time of the future technological singularity (assuming this happens before any global change tipping point catastrophe), still want to ask us why we did not do all we could to protect our Earth and sustain it, when all humans collectively depended on it for their long-term survival.

---

[27] Global Environmental Change Programmes (GEC) of the International Council for Science (ICSU), www.icsu.org/about-icsu/structure/interdisciplinary-bodies-1

[28] PEW, http://www.pewinternet.org/2014/08/06/future-of-jobs/

Acronyms

Glossary

Adaptation: Adaptation is a process by which individuals, communities, and countries seek to cope with the consequences of climate or other changes. Adaptation measures describe those activities and responses that assist in coping or living with adverse effects.

Adaptive Capacity: Adaptive capacity of an ecosystem is the capacity of a system to adapt if the environment where the system exists is changing. It is a property of an ecosystem or CAS linking change to system stability with functional and structural resilience. The adaptive capacity of a SES also encompasses the ability of institutions and networks to learn, and store knowledge and experience, creative flexibility in multifunctional decision-making and problem solving, and the existence of power structures that are responsive and consider the needs of all stakeholders.

Agroecosystem: A spatially and functionally coherent unit of agricultural activity that includes both living and nonliving components and their interactions. Agroecosystems are managed systems and provide humans worldwide with food, energy, fiber and other ecosystem services.

Artificial Intelligence: The theory and development of computer systems able to perform tasks normally requiring human intelligence, such as visual perception, speech recognition, decision-making, and translation between languages (*Oxford Dictionaries*); The simulation of intelligent behavior in computers and the capability of a machine to imitate intelligent human behavior (*Merriam-Webster Dictionary*).

Baseline: The baseline (or reference) is any datum against which change is measured. It might be a "current baseline," in which case it represents observable, present-day conditions. It might also be a "future baseline," which is a projected future set of conditions excluding the driving factor of interest. Alternative interpretations of the reference conditions can give rise to multiple baselines.

Bifurcation: A bifurcation point is a point in parameter space where the behavior of a dynamical system changes qualitatively, such that an equilibrium changes stability, an equilibrium appears or disappears, or a non-equilibrium attractor appears or disappears.

Biogeochemical Cycle: A pathway whereby a chemical substance moves through both biotic (biosphere) and abiotic (lithosphere, atmosphere, and hydrosphere) compartments of the Earth and involves interrelated biological, geological and chemical factors, and undergoes a series of changes involving storage and flow, whereby recycling occurs.

Carrying Capacity: The carrying capacity is the level of population size that the resources of the environment can just maintain or carry without a tendency to either increase or decrease (De Vries, 2013). Alternatively, it is the maximum population size of a biological species that the environment can sustain indefinitely, given the food, habitat, water, and other necessities available in the environment. What the environment can sustain is prescribed by biogeochemical cycles, biophysical limits, and thresholds.

Causality: In a statistical sense, causality is defined such that a variable $x(t)$ Granger-causes another variable $y(t)$ (denoted $x(t) \rightarrow y(t)$), if given information of both $x(t)$ and $y(t)$, the variable $y(t)$ can be better predicted in the mean-square-error sense by using only past values of $x(t)$ than by not doing so. The null hypothesis here is: $x(t)$ does not Granger-cause $y(t)$. In other words, having knowledge of past values of $x(t)$ improves the ability of a model or index to predict $y(t)$. A weaker condition is of instantaneous causality where not only past, but also present values of $x(t)$ improve the prediction of $y(t)$. Feedback can occur where $x(t)$ causes $y(t)$ and $y(t)$ causes $x(t)$.

Chaos: Chaos is the inherent/controlled randomness within a CAS, defined as, "unpredictable order, whereby there is inherent unpredictability in the state and state dynamics/behavior of a CAS given its high sensitivity to initial conditions," instead of its more common definition of, "complete confusion and disorder a state in which behavior and events are not controlled by anything" (*Merriam-Webster*) (Manson, 2001; Lorenz, 1995).

Cognitive Map: The mental image or representation made by human individuals and groups of their environment and their relationship to it, involving not only the rational aspects of attitudes and behaviors, but also the values and belief components that shape human perception (Laszlo and Krippner, 1998).

Complex, adaptive systems/theory: A system that exhibits collective or "emergent" properties from small-scale interactions. When adaptation is included in systems behavior, complexity arises. As one cannot describe, measure, or model all the variables in a complex system at once, complex, adaptive theory describes how theoretical insights and data from small-scale experiments may be combined to better understand and predict large-scale patterns and processes.

Copula: A function that links univariate marginal distributions to form a joint multivariate distribution

Consilience: The synthesis of knowledge from different specialized fields of human endeavor based on the principle that evidence from independent, unrelated sources can converge to strong conclusions and can emerge across different lines of scientific enquiry leading to "unity through diversity" epistemologically, ontologically, and ethically (Hofkirchner, 2005).

Crowdsourcing: A collective intelligence process whereby individuals gather and analyze information and complete tasks using web technology, often using mobile devices such as cellular phones, transferring data to integrated databases within computing environments. Individuals with these devices form interactive, scalable sensor networks that enable professionals and the public to gather, analyze, share, and visualize local knowledge and observations and to collaborate on the design, assessment, and testing of devices and results (Adapted from NAP (2013)).

Culture: A property of the group or of the individual's relationship to the group.

Cumulative effects/impacts: These are direct and indirect changes to the environment that are caused by an action in combination with other past, present, and future human actions.

Degradation: The reduction or loss in the biological and economic productivity and complexity of terrestrial ecosystems, as well as in the ecological, biochemical and hydrological processes that operate in them.

Domain of attraction: Is the domain of stability whereby a steady steady state remains under external perturbation/s.

Ecofootprint: The biologically productive area that is required to produce the resources required by the human population and to absorb humanity's carbon dioxide emissions (Ecological Footprint Standards).

Ecosystem: A dynamic complex of plant, animal, and micro-organism communities and the non-living environment interacting as a "functional unit" (UN Convention on Biological Diversity) (De Vries, 2013).

Ecosystem Services: The benefits people obtain from ecosystems. These include provisioning services such as food and water; regulating services such as flood and disease control, cultural services such as spiritual, recreational, and cultural benefits, and supporting services, such as nutrient cycling, that maintain the conditions for life on Earth (MEA, 2005). Many ecosystems are both providers and consumers of ecosystem services. Agro-ecosystems, covering 40% of terrestrial land globally, are intensively managed ecosystems designed to provide key *provisioning services* that include food, forage, fiber, bioenergy and pharmaceuticals, but critically depend on *supporting services* that include genetic biodiversity, soil formation, structure, fertility, nutrient cycling and carbon sequestration, and water quantity and quality. Ecosystem *disservices* include loss of biodiversity, nutrient runoff, pesticide contamination, GHG, loss of critical habitat (Power, 2010). Crop pathogens (i.e., viruses, bacteria, fungi, Oomycetes, nematodes) can have potentially beneficial as well as harmful effects for provisioning services, so can be considered both a regulating service and/or a disservice.

Eigenvalues, Eigenvectors, Eigenfunctions, Eigenspaces: Solutions to many problems (i.e., eigenproblems) involving systems of equations require transforming matrices, operators or spaces to "diagonal" representations/forms. In n-dimensional space, involving linear transformations and $n \times n$ square matrices (A), if a non-zero vector v exists such that $Av = \lambda v$, then λ is a scalar quantity that is termed the eigenvalue of the matrix A and v the eigenvector. An eigenvector does not change direction (only reversed) under linear transformations, as it can be stretched by the factor λ or reversed if its eigenvalue is negative. Eigenvalues are also termed characteristic values or roots of a linear system of equations. Similarly, linear operators as infinitely differentiable real functions (e.g., D as a differential operator, d/dt), satisfy the relation $Df = \lambda f$ and are termed eigenfunctions. An eigenbasis of A consists of linearly independent eigenvectors of A (i.e., vector space V).

Emergence: Emergence can be defined as a process whereby larger entities, patterns, and regularities or stability arise through interactions among smaller or simpler entities that themselves do not exhibit such properties.

Endogenous and Exogenous risk: The risk from shocks that are generated and amplified within the system. It stands in contrast to exogenous risk, which refers to shocks that arrive from outside the system. Financial markets are subject to both types of risk (Danielsson, 2013; Danielsson and Shin, 2002)

Environmental Risk: The chance of harmful effects to human health or to ecological systems resulting from exposure to an environmental stressor. A stressor is any physical, chemical, or biological entity that can induce an adverse response. Stressors may adversely affect specific natural resources or entire ecosystems, including plants and animals, as well as the environment with which they interact.

Forecasting: An operational research technique used to anticipate outcomes, trends, or expected future behavior of a system, using statistics and modeling. It is used as a basis for management planning and decision making and is stated in less certain terms than a prediction (NAP, 2013, 2010).

Geography: The study of human use and interaction with the Earth and the identification of spatial-temporal variation in natural and human processes, applying both natural and social science principles to analyze and interpret change in the environment such as globalization and cultural diversity, resource management, climate change, and environmental hazards.

Geocomputation: Complex, integrated geospatial techniques that combine geographical information science (GIS), spatio-temporal modeling, ML and AI computer techniques and are closely linked with geography and its integration across the social sciences (Paegelow and Olmedo, 2008).

Geospatial Intelligence and Planning: The ability to describe, understand, and interpret processes (data, information), products (knowledge), and organizational aspects (functions, decision requirements), to anticipate human impacts of an event or action within a spatio-temporal environment.

Geospatial Intelligence (GEOINT) Fusion or Integration: The aggregation, integration, and conflation of geospatial data across time and space with the goal of removing the effects of data measurement systems and facilitating spatial analysis and synthesis across information sources (NAP, 2013).

Geostatistics: A branch of applied statistics that focuses on providing a quantitative description of observations of geological and ecological variables that are distributed in space, or in time and space. It refers to a specific set of models and techniques stemming from its early development and application in mining, meteorology and the theory of stochastic processes and random fields. It offers the practitioner a probabilistic methodology for quantifying spatio-temporal uncertainty (Chilès and Delfiner, 2012).

General Systems Theory: The concepts, principles, and models that are common to all kinds of systems and the isomorphisms between and among various types of systems (Straussfogel and von Schilling, 2009).

Green Growth: Environmental sustainability economic growth as a strategy for achieving sustainable development, which focuses on on overhauling or adapting the economy in a way that integrates economic growth and environmental protection, and one in which investments in resource savings as well as sustainable management of natural capital are drivers of growth.

Greenhouse Effect: The effect of heat generated from sunlight at the Earth's surface being trapped by certain gases and prevented from escaping through the atmosphere.

Heteroskedasticity: is defined for a finite sequence of random variables, Y_t and X_t, associated with the conditional expectation, $E(Y_t|X_t)$, the sequence Y_t is termed heteroscedastic if the conditional variance, $\sigma^2(Y_t|X_t)$, changes with t.

Hierarchy Theory: A theory explaining self-organization in real-world ecosystems (as in complex dynamic systems) arising from the hierarchical structure of these systems in space and time, whereby complexity built upon complexity.

Hotelling's Rule: If the market is perfectly competitive, the net price (price minus marginal extraction cost) must rise at a rate equal to the rate of interest. This is known as Hotelling's rule. It is at the heart of the economics of natural resources (Long, 2000)

Hyperparameter: A hyperparameter is a parameter of a prior distribution; the term is used to distinguish them from parameters of the model for the underlying system under analysis. The probability distribution of a hyperparameter is termed a "hyperprior."

Hysteresis: A nonlinear response that occurs within a complex dynamic system involving a single variable that drives system dynamics after a delay or time-lag (i.e., "memory"), whereby its influence persists. It involves a cause and effect or "causal" lag in a given variable or property of a system with respect to its effect, such that its effect is generated as its magnitude varies.

Industrial Ecology: An emerging scientific field within sustainability science that seeks to bridge industry standards, processes, and practices with sustainable development. It specifically addresses the need to further develop and advance industry or sector-specific requirements, goals, or targets for environmental performance and sustainability beyond current international standards. It investigates water, energy, land resource use patterns, efficiencies, and environmental risks and uncertainties associated with human activity and industrial production systems. It aims to develop and optimize new models of highly efficient resource use and production, to guide and improve the environmental performance of different industries, with strong collaboration and communication with a broad set of stakeholders.

Integrated: The interdisciplinary combination or coupling of theory, empirical data, techniques, methods, and tools in devising, testing, and deploying problem solving research and development (R& D) scientific and technological solutions for real-world operational application. For example, within the field of statistical ecology, "Integrated models are designed to analyze multiple data sources simultaneously within a single robust analysis, permitting information to be borrowed (or pooled) across datasets. Thus, integrated models can be viewed as multivariate observation processes that correspond to the different forms of data collected, with the system process describing the different underlying biological processes. The different observation processes may observe either the same underlying states or different states that evolve simultaneously over time but that depend on the same demographic parameters." King (2014).

Interdisciplinary: Interdisciplinary research is a mode of research by teams or individuals that integrates information, data, techniques, tools, perspectives, concepts, and/or theories from two or more disciplines or bodies of specialized knowledge to advance fundamental understanding or to solve problems whose solutions are beyond the scope of a single discipline or area of research practice" (NAP, 2015; NRC, 2010).

Intractable: A problem or system is intractable if its complexity grows exponentially with the size of input.

Latitude: The maximum amount (threshold) that a system can be reversibly or irreversibly changed.

Markov Random Field (MRF): A random field superimposed on a graph or network structure, or more specifically, an undirected graphical model in which each node corresponds to a random variable or a collection of random variables, and its edges represent conditional probability dependencies. The great attraction of these models is that they leverage the so-called "knock-on effect," whereby explicit short-range conditional dependencies give rise to extended, long-range correlations, providing solutions to many problems in vision science, including image reconstruction, image segmentation, 3D vision, and object labeling (Blake and Kohli, 2011).

Mitigation: A series of activities (e.g., technologies and social behaviors) which contribute to reducing climate change, such as activities that mitigate the emission of greenhouse gases and their sequestration within marine and terrestrial ecosystems.

Multifunctionality: The concept of multifunctionality recognizes agriculture as a multi-output activity producing not only commodities (food, fodder, fibers and biofuels), but also non-commodity outputs such as ecosystem services, landscape amenities, and cultural heritages. This definition of multifunctionality is the one adopted by the OECD.

Natural Resource: Natural resources are derived from the environment. Some of them are essential for our survival, while most are used for satisfying our needs. They may be biotic (living, organic) or abiotic (non-living and non-organic), and they may exist as separate entities or comprise many entities that are aggregated and inseparable. Some are ubiquitous (e.g., sunlight, air), existing everywhere around us, while others are partitioned (e.g., land, water).

Panarchy: The degree that a certain hierarchical level of an ecosystem is influenced by other levels.

Persistence: The ability of a stable steady state to stay within a domain of attraction under external perturbations.

Phase Diagram: A geometric representation of the trajectories of a dynamical system in the phase plane, whereby each set of initial conditions is represented by a different point or curve.

Precariousness: A measure of how close a system is to a limit or threshold.

Projection: A "what-if" scenario whereby not all assumptions follow boundary conditions that are expected to hold continuously. This compares with "future predictions" or "forecasts," where initial conditions are specified and all assumptions are expected to hold and approximate reality.

Regime/Regime shift: A regime is a characteristic behavior of a system which is maintained by mutually reinforced processes or feedbacks. Regimes are considered persistent relative to the time period over which the shift occurs. Regime shifts (ecology) are large, abrupt, persistent changes in the structure and function of a system. Switching between regimes occurs when a smooth change in an internal process (feedback) or a single disturbance (external shocks) triggers a completely different system behavior.

Renormalization: Originally defined in terms of eliminating logarithmically divergent integrals, *renormalization* in the context of systems is defined as the existence of multiple length and energy scales. Renormalization group methods break up or decompose computations into separate stages, with a stage for each length or energy scale, such that each sub-problem involves fewer degrees of freedom than the complete or full multi-scale problem. In the limit of an infinite number of iterations or recursions there is no difference at all except for the change in scale and the approaches to a fixed point, although there is no guarantee that a given renormalization group computation will lead to a fixed point. Also, renormalization breaks down due to the interaction strength/correlation lengths as scaling limits.

Representative Concentration Pathway (RCP): Four different evolution patterns for atmospheric greenhouse gas emissions and concentrations, land-use changes and emission of air pollutants (e.g., ozone and aerosols), but do not consider future natural forcings of the Earth system, such as volcanic eruptions, changes in some natural sources (e.g., methane CH_4 and nitrous oxide N_2O emission), total solar irradiance or adaptive responses by the Earth system to potential inter-related cascading impacts. They were developed using IAM's and a wide range of climate model simulations to project future consequences of the climate system that consists of five main interacting components: the atmosphere, the hydrosphere, the cryosphere, the land surface, and the biosphere. These latest IPCC (AR5) scenarios compare to previous Special Report on Emissions Scenarios (SRES) used in previous IPCC assessments (e.g., AR4), except that RCP's, include a consideration of climate policy (IPCC, 2014b; van Vuuren et al., 2011; Moss et al., 2010). RCP8.5 is comparable to SRES A2/A1F1 as a very high GHG emissions scenario. The two stabilization scenarios are RCP6.0 and RCP4.5, comparable to SRES B2 and B1, respectively. There is no equivalent scenario for the low GHG forcing scenario RCP2.6 in SRES.

Resilience: Resilience is defined as the capacity of a system to absorb disturbance, undergo change, and still retain essentially the same function, structure, identity, and feedbacks. Accordingly, one metric of resilience of a system is the full range of perturbations over which the system can maintain itself (De Vries, 2013).

Resistance: The ease or difficulty of a structural or process change in a system.

Robustness: A *robust* system can handle variability and remain effective. Robust statistical methods are not unduly affected by outliers and have good performance (e.g., accuracy) under small departures from model assumptions (e.g., input parametric distributions), or more generally, provide insights to a problem, despite having its assumptions altered or violated. Robust solutions therefore achieve the required levels of performance under a broad range of realizations (e.g., different structures, behaviors, future adaptations) of a real-world system.

Risk Discount and Tolerance: Individual or organization deciding to receive less of a return on an investment in exchange for less risk (i.e., risk adverse). This is the opposite of a risk premium. The degree that one chooses different discount amounts varies, depending on their risk tolerance.

Scenario: A projection may serve as the raw material for a scenario, but scenarios often require additional information (e.g., about historical baseline or benchmark conditions). A set of scenarios is often adopted to reflect, as well as possible, the range of uncertainty in projections. Other terms that have been used as synonyms for scenario are "characterization," "storyline" and "construction" [29].

Self-Awareness: The capability of an agent to focus attention on the representation of internal states (Gorbenko et al., 2012).

Self-Organized Criticality (SOC): A large, interactive system that evolves toward a self-organized critical state, whereby a minor event acts as a tipping point and can cause a cascading catastrophe. In this state, the frequency and magnitude of events follow a power-law distribution that is statistic all stable and has no characteristic scale such that events are correlated across all scales, termed scale invariance. (Bak, 1996; Bak and Chen, 1991; Bak et al., 1988).

Self-Protection and Self-Insurance: Self-protection is the increasing probability of favorable outcomes from investment that reduces risk, while reducing risk of the consequences or severity of risk is termed self-insurance.

Shadow Price: The instantaneous change, per unit of the constraint, in the objective value of the optimal solution of an optimization problem obtained by relaxing the constraint. In other words, it is the marginal utility of relaxing the constraint, or, equivalently, the marginal cost of strengthening the constraint.

Social Discount Rate: The amount one is willing to spend less than a dollar now to prevent a dollar's worth of damage in a year, or a decade in the future, to avert climate change.

Stochastic System: A stochastic system (i.e., pure or mixed) having a sufficient degree of random variability, whereby the state of such a system changes with a degree of randomness—i.e., changes of state have multiple possible outcomes, each having varying degrees of certainty or uncertainty.

Sustainability: Sustainability is an act, a process or a situation, which is capable of being upheld, continued, maintained or defended.

Sustainable Development: Sustainable development is development that meets the needs of the present generation without compromising the ability of future generations to meet their own needs (United Nation's UNWCED Report, "Our Common Future" (1987).

Sustainability Science: Sustainability science aims to better understand the dynamics, resilience, and ways to maintain evolving, coupled with applying an integrated, interdisciplinary approach to problem solving. It focuses on interactions between resources, its users and the governance required to sustain ecosystems while also delivering what people need and value.

---

[29]*www.ipcc − data.org/guidelines/pages/definitions.html*

Threat (or Hazard): The frequency of potentially adverse events, or cumulative probability of the persistence of adverse conditions.

Tipping Point or Critical Transition: An abrupt change in a dynamical system that occurs rapidly in comparison to past system dynamics whereby the system crosses a threshold near a transition and the new state of the system is far away from its previous state (Kuehn, 2011; Bentley et al., 2014).

Tipping Elements: Leading candidate variables or indicators linked with human activities that are likely to cause future critical transition in the form of a bifurcation inducing tipping in Earth system dynamics between steady states, and thus have the highest relevance for political decision making regarding global change.

Transition: Transitions can be defined as a gradual process of change that transforms the structure or sub-structure of a SES (EU-SCAR, 2012).

Vulnerability: A measure of the likelihood of success of a particular threat. It is a function of exposure of a system to hazards, its intrinsic sensitivity to that exposure, and its adaptive capacity.

Bibliography

Ackerman, F. (2002). Still dead after all these years: Interpreting the failure of general equilibrium theory. *Journal of Economic Methodology 9*(2), 119–139.

Ahi, P. and C. Searcy (2013). A comparative literature analysis of definitions for green and sustainable supply chain management. *Journal of Cleaner Production 52*, 329–341.

Ahi, P. and C. Searcy (2014). A stochastic approach for sustainability analysis under the green economics paradigm. *Stochastic Environmental Research and Risk Assessment 28*, 1743–1753.

Ahi, P. and C. Searcy (2015). An analysis of metrics used to measure performance in green and sustainable supply chains. *Journal of Cleaner Production 86*, 360–377.

Ahmad, S. and A. Yu (2014). *A socially aware Bayesian model for competitive foraging*, Volume 1 of *Proceedings of the 36th Cognitive Science Society Conference (Quebec City, Canada, July 23-26)*. Red Hook, NY: Cognitive Science Society, Inc. and Curran Associates, Inc.

Ahmed, A. and J. Stein (2004). Science, technology and sustainable development: A world review. *World Review of Science, Technology and Sustainable Development 1*(1), 5–24.

Akaeva, A. and V. Sadovnichiib (2010). Mathematical model of population dynamics with the world population size stabilizing about a stationary level. *Doklady Mathematics 82*(3), 978–981.

Akiyama, E. and K. Kaneko (2000). Dynamical systems game theory and dynamics of games. *Physica D 147*, 221–258.

Albertine, J., W. Manning, M. DaCosta, K. Stinson, M. Muilenberg, and C. Rogers (2014). Projected carbon dioxide to increase grass pollen and allergen exposure despite higher ozone levels. *PLoS ONE 9*(11), e111712.

Aleklett, K., H. Mikael, K. Jakobsson, M. Lardelli, S. Snowden, and B. Soderbergh (2010). The peak of the oil age - Analyzing the world oil production Reference Scenario in World Energy Outlook 2008. *Energy Policy 38*, 1398–1414.

Alessa, L. and F. Chapin III (2008). Anthropogenic biomes: A key contribution to earth-system science. *Trends in Ecology and Evolution 23*, 529–531.

Alexander, L., X. Zhang, T. Peterson, J. Caesar, B. Gleason, A. Tank, M. Haylock, D. Collins, B. Trewin, F. Rahimzadeh, A. Tagipour, A. P., K. Rupa Kumar, J. Revadekar, G. Griffiths, L. Vincent, D. Stephenson, J. Burn, E. Aguilar, M. Brunet, M. Taylor, M. New, P. Zhai, M. Rusticucci, and J. Vazquez-Aguirre (2006). Global observed changes in daily climate extremes of temperature and precipitation. *Journal of Geophysical Research: Atmospheres 111*, D05109.

Alkema, L., A. Raftery, P. Gerland, S. Clark, F. Pelletier, T. Buettner, and G. Heilig (2011). Probabilistic projections of the total fertility rate for all countries. *Demography 48*(3), 815–839.

Allesina, S. and S. Tang (2012). Stability criteria from complex ecosystems. *Nature 483*, 205–208.

Almeida-Neto, M., P. Guimaraes, P. R. Guimaraes, R. D. Loyola, and W. Ulrich (2008). A consistent metric for nestedness analysis in ecological systems: Reconciling concept and measurement. *Oikos 117*, 227–1239.

Amano, R. and S. Murchison (2005). *Factor-Market Structure, Shifting Inflation Targets, and the New Keynesian Phillips Curve*. Proceedings of a Conference held by the Bank of Canada: Issues in Inflation Targeting (Ottawa, Canada, April). Bank of Canada.

Amsler, C., M. Doser, A. M., D. Asner, K. Babu, H. Baer, H. Band, R. Barnett, E. Bergren, J. Beringer, W. Bert, H. Bichsel, O. Biebel, P. Bloch, E. Blucher, S. Blusk, R. Cahn, M. Carena, C. Caso, A. Ceccucci, D. Chakraborty, M.-C. Chen, R. Chivukula, G. Cowan, O. Dahl, G. D'Ambrosio, T. Damour, d. Gouvéa, T. DeGrand, B. Dobrescu, M. Drees, D. Edwards, S. Eidelman, V. Elvira, J. Erler, V. Ezhela, J. Feng, W. Fetscher, B. Fields, B. Foster, T. Gaisser, L. Garren, H.-J. Gerber, G. Gerbier, T. Gherghetta, G. Giudice, M. Goodman, and C. Grab (2008). Review of particle physics. *Physics Letters B 667*(1–5), 1–1340.

Anderies, J., M. Janssen, and B. Walker (2002). Grazing management, resilience, and the dynamics of a re-driven rangeland system. *Ecosystems 5*, 23–55.

Anderson, B., P. Armsworth, F. Eigenbrod, C. Thomas, S. Gillings, A. Heinemeyer, D. Roy, and K. Gaston (2009). Spatial covariance between biodiversity and other ecosystem service priorities. *Journal of Applied Ecology 46*(4), 888–896.

Anderson, H. and T. Britton (2000). *Stochastic Epidemic Models and their Statistical Analysis*. Springer Lecture Notes in Statistics. Berlin: Springer-Verlag.

Andres, R., T. Boden, F.-M. Bréon, P. Ciais, S. Davis, D. Erickson, J. Gregg, A. Jacobson, G. Marland, J. Miller, T. Oda, J. Olivier, M. Raupach, P. Rayner, and K. Treanton (2012). A synthesis of carbon dioxide emissions from fossil-fuel combustion. *Biogeosciences 9*, 1184–1871.

Apanosovich, T. and M. Genton (2010). Cross-covariance functions for multivariate random fields based on latent dimensions. *Biometrika 97*(1), 15–30.

Aral, M. (2014a). Climate change and human population dynamics. *Water Quality Exposure and Health 6*, 53–62.

Aral, M. (2014b). Climate change and persistent high temperatures: Does it matter? *Frontiers in Environmental Science 2*, 1–8.

Arnell, N. (1999). Climate change and global water resources. *Global Environmental Change 9*, S31–S49.

Arnell, N. (2015). *A Short Guide to Climate Change Risk.* Surrey, UK and Burlington, VT, USA: Gower.

Arneth, A., C. Brown, and M. Rounsevell (2014). Global models of human decision-making for land-based mitigation and adaptation assessment. *Nature Climate Change 4*, 550–557.

Austin, J., N. Butchart, and K. Shine (1992). Possibility of an arctic ozone hole in a doubled-CO_2 climate. *Nature 360*, 221–225.

Avery, J. (2014). *Malthus' Essay on the Principle of Population.* Introduction to Sustainable Development. Paris, France: UNESCO-Encyclopedia of Life Support Systems (EOLSS) Publishers.

Bai, Z., H. Li, W. Wong, and B. Zhang (2011). Multivariate causality tests with simulation and application. *Statistics and Probability Letters 81*, 1063–1071.

Bak, P. (1996). *How Nature Works: The Science of Self-Organized Criticality.* New York, NY, USA: Copernicus (Springer-Verlag Inc.).

Bak, P. and K. Chen (1991). Self-organized criticality. *Scientific American 264*, 46–53.

Bak, P., C. Tang, and K. Weisenfeld (1988). Self-organized criticality. *Physical Review A 38*(1), 364–374.

Bale, C., L. Varga, and T. Foxon (2015). Energy and complexity: New ways forward. *Applied Energy 138*, 150–159.

Banarjee, S., B. Carlin, and A. Gefland (2005). *Hierarchical Modeling and Analysis for Spatial Data* (2nd. ed.). Monographs of Applied Statistics and Probability. Baton Rouge, FL, USA: CRC Press/Taylor and Francis Group, LCC.

Bannister, R. (2008). A review of forecast error covariance statistics in atmospheric variational data assimilation. I: Characteristics and measurements of forecast error covariances. *Quarterly Journal of the Royal Meteorological Society 134*, 1951–1970.

Barbarossa, S., S. Sardellitti, and P. Di Lorenzo (2013). Distributed detection and estimation in wireless sensor networks. *CoRR: Computer Science: Distributed, Parallel, and Cluster Computing abs/1307.1448*, 92.

Bardgett, R. (2011). Plant-soil interactions in a changing world. *F1000 Biology Reports 3*(16), 6.

Barker, T. (1996). *Space-Time Economies.* Cambridge Econometrics. Cambridge, UK: Cambridge University Press.

Barker, T. (2004). *Economic theory and the transition to sustainability : a comparison of general-equilibrium and space-time-economics approaches.* Tyndall Centre for Climate Change Research, Working Paper 62, Norwich, UK, November, http://www.tyndall.ac.uk/sites/default/files/wp62.pdf.

Barnard, J., R. McCulloch, and X.-L. Meng (2000). Modeling covariance matrices in terms of standard deviations and correlations with application to shrinkage. *Statistica Sinica 10*, 1281–1311.

Barnosky, A., E. Hadly, J. Bascompte, E. Berlow, J. Brown, M. Forelius, W. Getz, J. Harte, A. Hastings, P. Marquet, N. Martinez, A. Mooers, P. Roopnarine, G. Vermeij, J. Williams, R. Gillespie, J. Kitzes, C. Marshall, N. Matzke, D. Mindell, E. Revilla, and A. Smith (2012). Approaching a state shift in Earth's biosphere. *Nature (Review) 486*(7 June), 52–58.

Barrett, S. and S. Dannenberg (2013). Sensitivity of collective action to uncertainty about climate tipping points. *Nature Climate Change (Letters) 4*(January), 36–39.

Barton, D., S. Kuikka, O. Varis, L. Uusitalo, H. Henriksen, M. Borsuk, A. de la Hera, R. Farmani, S. Johnson, and J. Linnell (2012). Bayesian networks in environmental and resource management. *Integrated Environmental Assessment and Management 8*(3), 418–429.

Bast, E., S. Makhijani, S. Pickard, and S. Whitley (2014). *Fossil Fuel Bailout: G-20 Subsidies for Oil, Gas and Coal Exploration.* London, UK: Overseas Development Institute (ODI) and Oil Change International, November, http://www.odi.org/publications/8678-fossil-fuel-bailout-g20-subsidies-oil-gas-coal-exploration.

Batjes, N. (2011). *Global distribution of soil phosphorus retention potential.* Wageningen, The Netherlands: International Soil Reference and Information Centre (ISRIC) - World Soil Information (with dataset) (ISRIC Report 2011/06), http://www.isric.org/sites/default/files/ISRIC_Report_2011_06.pdf.

Batt, R., S. Carpenter, J. Cole, M. Pace, and R. Johnson (2013). Changes in ecosystem resilience detected in automated measures of ecosystem metabolism during a whole-lake manipulation. *Proceedings of the National Academy of Sciences 110*(43), 17398–17403.

Beamish, R., D. Noakes, A. McFarlane, G, L. Klyashtorin, V. Ivanov, and V. Kurashov (1999). The regime concept and natural trends in the production of Pacific salmon. *Canadian Journal of Fisheries and Aquatic Sciences 56*(3), 516–526.

Beer, T. (2003). *Environmental Risk and Sustainability*, pp. 39–61. Dordrecht, The Netherlands: Springer Science & Business Media.

Beirlant, J., Y. Goegebeur, J. Segers, and J. Teugels (2004). *Statistics of Extremes. Theory and Applications.* Wiley Series in Probability and Statistics. Chichester, UK: John Wiley & Sons Ltd.

Belyaev, A., M. Brown, R. Foadi, and M. Frandsen (2014). Technicolor Higgs boson in the light of LHC data. *Physical Review D: Particles, Fields, Gravitation, and Cosmology 90*, 035012.

Bennett, E., S. Carpenter, and N. Caraco (2001). Human impact on erodable phosphorus and eutrophication: A global perspective. *BioScience 51*(3), 227–234.

Bentley, R. A., E. Maddison, P. Ranner, J. Bissell, C. Caiado, P. Bhatanacharoen, T. Clark, M. Botha, F. Akinbami, M. Hollow, R. Michie, B. Huntley, S. Curtis, and P. Garnett (2014). Social tipping points and Earth system dynamics. *Frontiers in Environmental Science 2*(35), 1–7.

Berger, J. (2013). *Statistical Decision Theory and Bayesian Analysis* (2nd. ed.). Springer Series in Statistics. New York, NY, USA: Springer Science & Business Media Inc.

Berger, T. (2001). Agent-based spatial models applied to agriculture: A simulation tool for technology diffusion, resource use changes and policy analysis. *Agricultural Economics 25*, 245–260.

Besag, J. (1974). Spatial interaction and the statistical analysis of lattice systems. *Journal of the Royal Statistical Society, Series B 36*(2), 192–236.

Bezlepkina, I., P. Reidsma, S. Sieber, and K. Helming (2011). Integrated assessment of sustainability of agricultural systems and land use: Methods, tools and applications. *Agricultural Systems 104*, 105–109.

Biggs, R. (2009). Turning back from the brink: Detecting an impending regime shift in time to avert it. *Proceedings of the National Academy of Sciences 106*, 826–831.

Biggs, R., M. Schluter, D. Biggs, E. Bohensky, S. BurnSilver, G. Cundill, V. Dakos, T. Daw, L. Evans, K. Kotschy, A. Leitch, C. Meek, A. Quinlan, C. Raudsepp-Hearne, M. Robards, M. Schoon, L. Schultz, and P. West (2012). Toward principles for enhancing the resilience of ecosystem services. *Annual Review of Environment and Resources 37*, 421–448.

Binnemans, K., P. Jones, B. Blanpain, T. Gerven, Y. Yang, A. Walton, and M. Buchert (2013). Recycling of rare earths: A critical review. *Journal of Cleaner Production 51*, 1–22.

Blake, A. and P. Kohli (2011). *Introduction to Markov Random Fields.* Markov Random Fields for Vision and Image Processing. Boston, MA, USA: MIT Press.

Blaschke, T., G. J. Hay, Q. Weng, and B. Resch (2011). Collective sensing: Integrating geospatial technologies to understand urban systems - an overview. *Remote Sensing 3*, 1743–1776.

Boccara, N. (2010). *Modelling Complex Systems* (2nd. ed.). Graduate Texts in Physics. Berlin: Springer-Verlag.

Bochner, S. (1933). Monotone funktionen: Stieltjes integrale und harmonische analyse. *Mathematische Annalen 108*, 378–410.

Boden, T., G. Marland, and R. Andres (2010). *Global, Regional, and National Fossil-Fuel CO_2 Emissions.* Oak Ridge, TN, USA: Carbon Dioxide Information Analysis Center, Oak Ridge National Laboratory, US Department of Energy, http://cdiac.ornl.gov/trends/emis/overview.html.

Boer, E., K. de Beurs, and A. Hartkamp (2001). Kriging and thin plate splines for mapping climate variables. *International Journal of Applied Earth Observation and Geoinformation 3*(2), 146–154.

Bolin, B., P. Crutzen, P. Vitousek, R. Woodmansee, E. Goldberg, and R. Cook (1983). *SCOPE 21 -The Major Biogeochemical Cycles and Their Interactions.* Chichester, UK: John Wiley & Sons Ltd.

Bonabeau, E. (2002). Agent-based modeling: Methods and techniques for simulating human systems. *Proceedings of the National Academy of Sciences 99*(3), 7280–7287.

Bonhommeau, S., L. Dubroca, O. Le Papec, J. Bardeb, D. Kaplan, E. Chassot, and A.-E. Nieblas (2013). Eating up the world's food web and the human trophic level. *Proceedings of the National Academy of Sciences 110*(51), 20617–20620.

Bonner, S., N. Newlands, and N. Heckman (2014). Modeling regional impacts of climate teleconnections using functional data analysis. *Environmental and Ecological Statistics 21*(1), 1–26.

Bornn, L., G. Shaddick, and J. Zidek (2012). Modelling nonstationary processes through dimension expansion. *Journal of the American Statistical Association 107*, 281–289.

Bostrom, N. (2014). *Superintelligence: Paths, Dangers, Strategies.* Oxford, UK: Oxford University Press.

Bostrom, N. and M. Cirkovic (2011). *Global Catastrophic Risks.* Oxford, UK: Oxford University Press.

Bousquet, F. and C. Le Page (2004). Multi-agent simulations and ecosystem management: A review. *Ecological Modelling 176*, 313–332.

Boyle, R. (2014). *Forget the Higgs, neutrinos may be the key to breaking the Standard Model History says neutrinos are where to look for new physics, so current research obliges.* arstechnica, `http://arstechnica.com/science/2014/04/`.

Bradshaw, C. and B. Brook (2014). Human population reduction is not a quick fix for environmental problems. *Proceedings of the US National Academy of Sciences (PNAS) 111*(46), 16610–16615.

Brännström, Å. and D. Sumpter (2005). Coupled map lattice approximations for spatially explicit individual-based models of ecology. *Bulletin of the American Mathematical Society 67*, 663–682.

Brede, M. and B. de Vries (2013). The energy transition in a climate-constrained world: Regional vs. global optimization. *Environmental Modelling & Software 44*, 44–61.

Breeden, D. (1979). An intertemporal asset pricing model with stochastic consumption and investment opportunities. *Journal of Financial Economics 7*, 265–296.

Breiman, L. (2011). Random forests. *Machine Learning 45*, 5–32.

Brook, B., E. Ellis, M. Perring, A. Mackay, and L. Blomqvist (2013). Does the terrestrial biosphere have planetary tipping points? *Trends in Ecology and Evolution 28*(7), 396–401.

Brown, J. and M. Hovmöller (2002). Aerial dispersal of pathogens on the global and continental scales and its impact on plant disease. *Science 297*, 537–541.

Brown, P., N. Le, and J. Zidek (1994). *Inference for a covariance matrix.* Aspects of Uncertainty: A Tribute to D. V. Lindley (Freeman, P.R., Smith, A.F.M., Eds.). Chichester, UK: John Wiley & Sons, Inc.

Brunckhorst, D. (2005). Integration research for shaping sustainable regional landscapes. *Journal of Research Practice 1*(7), 16–28.

Brunckhorst, D. and G. Marshall (2007). *Designing Robust Common Property Regimes for Collaborative Arrangements towards Rural Resource Sustainability.* Sustainable Resource Use: Institutional Dynamics and Economics. London, UK: Earthscan.

Brundtland, G. (1987). *Our Common Future.* Oxford, UK: World Commission on Environment and Development (WCED) and Oxford University Press, `http://www.un-documents.net/our-common-future.pdf`.

Bruno, F., P. Guttorp, P. Sampson, and D. Cocchi (2009). A simple non-separable, non-stationary spatiotemporal model for ozone. *Environmental and Ecological Statistics 16*, 515–529.

Burrows, M., D. Schoeman, A. Richardson, J. Molinos, A. Hoffmann, L. Buckley, P. Moore, C. Brown, J. Bruno, C. Duarte, B. Halpern, O. Hoegh-Guldberg, C. Kappel, W. Kiessling, M. O'Connor, J. Pandolfi, C. Parmesan, W. Sydeman, S. Ferrier, K. Williams, and E. Poloczanska (2014). Geographical limits to species-range shifts are suggested by climate velocity. *Nature 507*, 492–495.

Burton, I., E. Malone, S. Huq, and UNDP (2004). *Adaptation Policy Frameworks for Climate Change: Developing Strategies, Policies and Measures.* New York, NY, USA: UNDP and Cambridge University Press, `http://www.preventionweb.net/files/7995_APF.pdf`.

Butchart, S., H. Akcakaya, J. Chanson, J. Baillie, B. Collen, S. Quader, W. Turner, R. Amin, S. Stuart, and C. Hilton-Taylor (2007). Improvements to the Red List Index. *PLoS ONE 2*(1), e140.

Byrd, D. and R. Cothern (2000). *Introduction to Risk Analysis: A Systematic Approach to Science-Based Decision Making.* New York, NY, USA: Government Institutes.

Cabezas, H. and U. Diwekar (2012). *Sustainability Indicators and Metrics.* Sustainability : Multi-Disciplinary Perspectives. Bentham Science (e-Book), `http://ebooks.benthamscience.com/book/9781608051038/chapter/102790/`.

Cadenasso, M., S. Pickett, and J. Grove (2006). Dimensions of ecosystem complexity: Heterogeneity, connectivity, and history. *Ecological Complexity 3*, 1–12.

Cai, S., J. Zidek, N. Newlands, and D. Neilsen (2014). Statistical modeling and forecasting of fruit crop phenology under climate change. *Environmetrics 25*(1), 621–629.

Calov, R. and A. Ganopolski (2005). Multistability and hysteresis in the climate-cryosphere system under orbital forcing. *Geophysical Research Letters 32*(L21717), 4.

Cameron, D., J. Marty, and R. Holland (2014). Wither the rangeland? Protection and conversion in California's rangeland ecosystems. *PLoS ONE 9*(8), e103468.

Canadell, J. (2010). Carbon sciences for a new world. *Current Opinion in Environmental Sustainability 2*, 209.

Canadell, J., P. Ciais, S. Dhakal, H. Dolman, P. Friedlingstein, K. Gurney, A. Held, R. Jackson, C. Le Quéré, E. Malone, D. Ojima, A. Patwardhan, G. Peters, and M. Raupach (2010). Interactions of the carbon cycle, human activity, and the climate system: A research portfolio. *Current Opinion in Environmental Sustainability 2*, 301–311.

Candiago, S., F. Remondino, M. De Giglio, M. Dubbini, and M. Gattelli (2015). Evaluating multispectral images and vegetation indices for precision farming applications from UAV images. *Remote Sensing 7*, 4026–4047.

Cannon, A. (2009). A flexible nonlinear modelling framework for nonstationary generalized extreme value analysis in hydroclimatology. *Hydrological Processes 24*(6), 673–685.

Caramia, M. and P. Dell-Olmo (2008). *Multiobjective Management in Freight Logistics: Increasing Capacity, Service Level and Safety with Optimization Algorithms.* London, UK: Springer.

Cardell-Oliver, R., K. Smettem, M. Kranz, and K. Mayer (2004). *Field testing a wireless sensor network for reactive environmental monitoring.* IEEE Intelligent Sensors, Sensor Networks and Information Processing Conference (ISSNIP) (14-17 December, Melbourne, Australia), http://ieeexplore.ieee.org/xpl/mostRecentIssue.jsp?punumber=9710. IEEE Xplore Digital Library.

Cardenete, M., A.-I. Guerra, and F. Sancho (2012). *Applied General Equilibrium.* Springer Texts in Business and Economics. Berlin: Springer-Verlag.

Cardinale, B., J. Duffy, A. Gonzalez, D. Hooper, C. Perrings, P. Venail, A. Narwani, G. Mace, D. Tilman, D. Wardle, A. Kinzig, G. Daily, M. Loreau, J. Grace, A. Larigauderie, D. Srivastava, and S. Naeem (2012). Biodiversity loss and its impact on humanity. *Nature 486*, 59–67.

Cariboni, J., D. Gatelli, R. Liska, and A. Saltelli (2007). The role of sensitivity analysis in ecological modelling. *Ecological Modelling 203*(1-2), 167–182.

Carlson, J. and J. Doyle (2002). Complexity and robustness. *Proceedings of the National Academy of Sciences 99*(1), 2538–2545.

Carpenter, S., S. Chisholm, C. Krebs, D. Schindler, and R. Wright (1995). Ecosystem experiments. *Science 269*(21 July), 324–327.

Carri, C. (2008). *Computable General Equilibrium (CGE) Approaches to Policy Analysis in Developing Countries: Issues and Perspectives*, Volume 2 of *Working Paper*. Centre for Economic Development, Health and the Environment (SPERA), http://dse.univr.it/spera.

CEQ (1997). *Considering Cumulative Effects Under the National Environmental Policy Act (NEPA).* Washington, DC, USA: Council of Environmental Quality, Executive Office of the President of the United States, http://energy.gov/sites/prod/files/nepapub/nepa_documents/RedDont/G-CEQ-ConsidCumulEffects.pdf.

Cervantes-De la Torre, F., J. González-Trejo, C. Real-Ramirez, and L. Hoyos-Reyes (2013). Fractal dimension algorithms and their application to time series associated with natural phenomena. *Journal of Physics: Conference Series 475*, 012002.

Chakrabarti, C. and K. Ghosh (2013). Dynamical entropy via entropy of non-random matrices: Application to stability and complexity in modelling ecosystems. *Mathematical Biosciences 245*, 278–281.

Chambers, R. (2008). Stochastic productivity measurement. *Journal of Production Analysis 30*, 107–120.

Champagne, C., A. Berg, J. Belanger, H. McNairn, and R. De Jeu (2010). Evaluation of soil moisture derived from passive microwave remote sensing over agricultural sites in Canada using ground-based soil moisture monitoring networks. *International Journal of Remote Sensing 31*, 3669–3690.

Champagne, C., A. Davidson, P. Cherneski, J. L-Heureux, and T. Hadwen (2014). Monitoring agricultural risk in Canada using L-band passive microwave soil moisture from SMOS. *Journal of Hydrometeorology 16*(1), 5–18.

Chandrasekar, V., J. Sheeba, B. Subash, M. Lakshmanan, and J. Kurths (2014). Adaptive coupling induced multi-stable states in complex networks. *Physica D. 267*, 36–48.

Chang, H., A. Fu, N. Le, and J. Zidek (2007). Designing environmental monitoring networks to measure extremes. *Environmental and Ecological Statistics 14*, 301–321.

Charmantier, A. and P. Gienapp (2013). Climate change and timing of avian breeding and migration: Evolutionary versus plastic changes. *Evolutionary Applications Special issue on Climate change, adaptation and phenotypic plasticity 7*(1), 15–28.

Chen, J., M. Li, and W. Wang (2012). Statistical uncertainty estimation using random forests and its applications to drought forecast. *Mathematical Problems in Engineering 2012*, 1–12.

Chen, L., M. Fuentes, and J. Davis (2006). *Spatial-temporal statistical modeling and prediction of environmental processes, Chap. 7*. Hierarchical Modeling and Analysis for Spatial Data (Clark, J.S., Gelfand, A.E., Eds.). Oxford, UK: Oxford University Press.

Chen, W.-K., H. Sun, X. Zhang, and D. Korošak (2010). Anomalous diffusion modeling by fractal and fractional derivatives. *Computers and Mathematics with Applications 59*, 1754–1758.

Chen, W.-K. and P. Wang (2013). *Fuzzy Forecasting with Fractal Analysis for the Time Series of Environmental Pollution*, Volume 47 of *Time Series Analysis, Modeling and Applications: A computational intelligence perspective*. Berlin and Heidelberg: Springer-Verlag.

Chen, X. and K. Tung (2014). Varying planetary heat sink led to global-warming slowdown. *Science 345*(6199), 897–903.

Chertow, M. (2001). The IPAT equation and its variants: Changing views of technology and environmental impact. *Journal of Industrial Ecology 4*(4), 13–29.

Chickering, D. (2002). Optimal structure identification with greedy search. *Journal of Machine Learning Research Policy 3*, 507–554.

Chilès, J. and P. Delfiner (2012). *Geostatistics - Modeling Spatial Uncertainty* (2nd. ed.). Wiley Series in Probability and Statistics. New York, NY, USA: John Wiley & Sons Ltd.

Chipanshi, A., N. Newlands, P. Cherneski, H. Hill, and A. Howard (2015). *Mitigation and Adaptation Strategies for Reducing Agricultural Greenhouse Gases in Canada - A Review.* Agriculture Management for Climate Change (Chapter 11). New York, NY, USA: Nova Science Publishers, Inc.

Choi, T., K. Dooley, and M. Rungtusanatham (2001). Supply networks and complex adaptive systems: Control versus emergence. *Journal of Operations Management 19*, 351–366.

Christakos, G. (2000). *Modern spatiotemporal geostatistics.* Oxford, UK: Oxford University Press.

Chu, D., R. Strand, and R. Fjelland (2003). Theories of complexity: Common denominators of complex systems. *Complexity 8*(3), 19–30.

Chum, H., A. Faaij, J. Moreira, G. Berndes, P. Dhamija, H. Dong, B. Gabrielle, A. Goss Eng, W. Lucht, M. Mapako, O. Masera Cerutti, T. McIntyre, T. Minowa, and K. Pingoud (2011). *Bioenergy,* pp. 1075. United Kingdom and New York, NY, USA: Cambridge University Press, Cambridge, `http://srren.ipcc-wg3.de/report`.

Cipra, B. (1987). An introduction to the Ising model. *Journal American Mathematical Monthly 94*(10), 937–959.

Clark, J. and A. Gefland (2006). *Hierarchical Modelling for the Environmental Sciences: Statistical Methods and Applications.* Oxford Biology. Oxford, UK: Oxford University Press.

Clark, W. and N. Dickson (2003). Sustainability science: The emerging research program. *Proccedings of the National Academy of Sciences 100*(14), 8059–8061.

Clarvis, M., M. Halle, I. Mulder, and M. Yarime (2015). Towards a new framework to account for environmental risk in sovereign credit risk analysis. *Journal of Sustainable Finance and Investment 4*(2), 147–160.

Clauset, A., C. Shalizi, and M. E. J. Newman (2009). Power-law distributions in empirical data. *SIAM Review 51*, 661–703.

Cline, A. and I. Dhillon (2006). *Computation of the Singular Value Decomposition.* Discrete Mathematics and Its Applications. Chapman and Hall/CRC, `https://www.crcpress.com/Handbook-of-Linear-Algebra/Hogben/9781420010572`.

CMS (2012). Observation of a new boson at a mass of 125 GeV with the CMS experiment at the LHC. *Physics Letters B 716*, 30–61.

CMS (2014). Evidence for the direct decay of the 125 GeV Higgs boson to fermions. *Nature Physics 10*, 557–560.

Cobb, Jr., J. (1995). *Is It Too Late? A Theology of Ecology* (2nd. ed.). Florida, USA: Environmental Ethics Books.

Cocchi, D., F. Greco, and C. Trivisano (2007). Hierarchical space-time modelling of pm_{10} pollution. *Atmospheric Environment 41*, 532–542.

Cole, H., C. Freeman, M. Jahoda, and K. Pavitt (1973). *Thinking about the Future: A Critique of the Limits to Growth.* London: Chatto & Windus for Sussex University Press.

Coles, S. (2001). *An Introduction to Statistical Modeling of Extreme Values.* Springer Series in Statistics. London, UK: Springer-Verlag.

Coles, S., L. Pericchi, and S. Sisson (2003). A fully probabilistic approach to extreme rainfall modeling. *Hydrology 273*, 35–50.

Collins, S., L. Bettencourt, A. Hagberg, R. Brown, D. Moore, G. Bonito, K. Delin, S. Jackson, D. Johnson, S. Burleigh, R. Woodrow, and J. McAuley (2006). New opportunities in ecological sensing using wireless sensor networks. *Frontiers in Ecology and the Environment 4*(8), 402–407.

Conrad, R. (2009). The global methane cycle: Recent advances in understanding the microbial processes involved. *Environmental Microbiology Reports 1*(5), 285–292.

Conradt, L. and C. List (2009). Group decisions in humans and animals: A survey. *Philosophical Transactions of the Royal Society B 364*(1518), 719–742.

Cook, D. (2004). *The Natural Step: Towards a Sustainable Society.* Schumacher Briefings. Devon, UK: Green Books.

Cook, W., D. Casagrande, D. Hope, P. Groffman, and S. Collins (2004). Learning to roll with the punches: adaptive experimentation in human-dominated systems. *Frontiers in Ecology and the Environment 2*(9), 467–474.

Cooper, O., D. Parrish, J. Ziemke, N. Balashov, M. Cupeiro, I. Galbally, S. Gilge, L. Horowitz, N. Jensen, J.-F. Lamarque, V. Naik, S. J. Oltmans, J. Schwab, D. Shindell, A. Thompson, V. Thouret, Y. Wang, and R. Zbinden (2014). Global distribution and trends of tropospheric ozone: An observation-based review. *Elementa: Science of the Anthropocene 2*, 000029.

Cordell, D., J.-O. Drangert, and S. White (2009). The story of phosphorus: global food security and food for thought. *Global Environmental Change 19*, 292–305.

Correa, M., C. Bielza, and J. Pamies-Teixeira (2009). Comparison of Bayesian networks and artificial neural networks for quality detection in a machining process. *Expert Systems with Applications 36*(3), 7270–7279.

Cortés, U., M. Sanchez-Marré, and L. Ceccaroni (2000). Artificial intelligence and environmental decision support systems. *Artificial Intelligence 13*, 77–91.

Costanza, R. (2012). *Sustainability Science: The Emerging Paradigm and the Urban Environment*, pp. 444. New York, NY, USA: Springer Science+Business Media, LCC.

Costanza, R. and S. Liu (2014). Ecosystem services and environmental governance: comparing China and the US. *Asia and the Pacific Policy Studies 1*(1), 160–170.

Craven, P. and G. Wahba (1979). Smoothing noisy data with spline functions. *Numerische Mathematik 31*, 377–403.

Cressie, N., C. Calder, J. Clark, J. Ver Hoef, and C. Wikle (2009). Accounting for uncertainty in ecological analysis: the strengths and limitations of hierarchical statistical modeling. *Ecological Applications 19*(3), 553–570.

Cressie, N. and S. Lele (1992). New models for Markov random fields. *Journal of Applied Probability 29*(4), 877–884.

Cressie, N. and C. Wilke (2011). *Statistics for Spatio-Temporal Data*. Wiley Series in Probability and Statistics. Hoboken, NJ, USA: John Wiley & Sons Ltd.

Creswell, J. (2013). *Research Design: Qualitative, Quantitative and Mixed Methods* (4th ed.). Washington, DC, USA: Sage Publications, Inc.

Cromwell, J., M. Hannan, W. Labys, and M. Terraza (1994). *Multivariate Tests for Time Series Models*, Volume 07-100. Thousand Oaks, CA, USA: Sage Publications, Inc.

Cuddington, K. and P. Yodzis (2002). Predator-prey dynamics and movement in fractal environments. *The American Naturalist 160*(1), 119–134.

Cuéllar, A. and M. Webber (2010). Wasted food, wasted energy: The embedded energy in food waste in the United States. *Environmental Science and Technology 44*(16), 6464–6469.

Dale, V. and S. Beyeler (2001). Challenges in the development and use of ecological indicators. *Ecological Indicators 1*, 3–10.

Daly, C. (2000). High-quality spatial climate data sets for the United States and beyond. *Transactions of the American Society of Agricultural Engineers (ASAE) 43*(6), 1957–1962.

Daly, C. (2006). Guidelines for assessing the suitability of spatial climate data sets. *International Journal of Climatology 26*(6), 707–721.

Daly, C., M. Halbleib, J. Smith, W. Gibson, M. Doggett, G. Taylot, J. Curtis, and P. Pasteris (2008). Physiographically sensitive mapping of climatological temperature and precipitation across the conterminous United States. *International Journal of Climatology 28(15)*(December 2008), 2031–2064.

Daly, H. and J. Cobb, Jr. (1994). *For the Common Good: Redirecting the Economy Toward Community, Environment, and a Sustainable Future* (2nd. ed.). Boston, MA, USA: Beacon Press.

Danielsson, J. (2013). *Global Financial Systems: Stability and Risk.* London, UK.: Pearson.

Danielsson, J. and H. Shin (2002). *Endogeneous Risk.* London, UK: Systemic Risk Centre (SRC), London School of Economics (LSE).

Davies, E. (2011). Endangered elements. Critical thinking. *Chemistry World January*, 50–54.

Davis, R., A. Aden, and P. Pienkos (2011). Techno-economic analysis of autotropic microalgae for fuel production. *Applied Energy 88*, 3424–3531.

de Fatima Andrade, M., R. Ynoue, E. Freitas, E. Todesco, A. Vela, S. Ibarra, L. Martins, J. Martins, and V. Carvalho (2015). Air quality forecasting system for Southeastern Brazil. *Frontiers in Environmental Science (Interdisciplinary Climate Studies) 3*, 1–14.

de Iaco, S., S. Maggio, M. Palma, and D. Posa (2012). *Advances in spatio-temporal modeling and prediction for environmental risk assessment.* Air Pollution - A Comprehensive Perspective (Haryanto, B. Ed.) (Chapter 14), `http://www.intechopen.com/books/air-pollution-a-comprehensive-perspective`. InTech.

de Marchi, S. and S. Page (2014). Agent-based models. *Annual Review of Political Science 17*, 1–20.

de Ponti, T., B. Rijk, and M. van Ittersum (2012). The crop yield gap between organic and conventional agriculture. *Agricultural Systems 108*, 1–9.

De Vries, B. (2013). *Sustainability Science* (1st ed.). Cambridge, UK: Cambridge University Press.

DeAngelis, D. and V. Grimm (2014). Individual-based models in ecology after four decades. *F1000 Prime Reports 6*, 39.

DeFelipe, J. (2011). The evolution of the brain, the human nature of cortical circuits, and intellectual creativity. *Frontiers in Neuroanatomy 5*(29), 1–17.

Deinlein, U., A. Stephan, T. Horie, W. Luo, G. Xu, and J. Schroeder (2014). Plant salt-tolerance mechanisms. *Trends in Plant Science 19*(6), 371–379.

Deo, R. and M. Sahin (2015). Application of extreme learning machine algorithm for the prediction of monthly Effective Drought Index in eastern Australia. *Atmospheric Research 153*, 512–525.

Desjardins, M., P. Rathod, and L. Getoor (2008). Learning structured Bayesian networks: Combining abstraction hierarchies and tree-structured conditional probability tables. *Computational Intelligence 24*(1), 1–22.

Devarajan, S. and A. Fisher (1981). Hotelling's Economics of Exhaustible Resources: Fifty Years Later. *Journal of Economic Literature 19*(1), 65–73.

Diaz, H. and R. Bradley (1997). Temperature variations during the last century at high elevation sites. *Climate Change 36*, 253–279.

Diaz, R. J. and R. Rosenberg (2008). Spreading dead zones and consequences for marine ecosystems. *Science 321*, 926–929.

Diaz-Solis, H., M. Kothmann, W. Grant, and R. De Luna-Villarreal (2006). Use of irrigated pastures in semi-arid grazing lands: A dynamic model for stocking rate decisions. *Agricultural Systems 88*, 316–331.

Dietterich, T., P. Domingos, L. Getoor, S. Muggleton, and P. Tadepalli (2008). Structured machine learning: The next ten years. *Machine Learning 73*, 3–23.

D'Odorico, P., F. Laio, A. Porporato, and I. Rodriguez-Iturbe (2003). Hydrological controls on soil carbon and nitrogen cycles, II: A case study. *Advances in Water Resources 26*, 59–70.

Dorner, S., J. Shi, and D. Swayne (2007). Multi-objective modelling and decision support using a Bayesian network approximation to a non-point source pollution model. *Environmental Modelling and Software 22*, 211–222.

Duda, R., P. Hart, and D. Stork (2001). *Pattern Classification* (2nd. ed.). New York, NY, USA: John Wiley & Sons, Inc.

Dühr, S., C. Colomb, and V. Nadin (2010). *European Spatial Planning and Territorial Cooperation*. New York, NY, USA: Routledge (Taylor & Francis).

Duke, G., D. Neilsen, B. Taylor, A. Cannon, T. Van der Gulik, N. Newlands, G. Frank, and C. Smith (2009). Climate surfaces for the Okanagan Basin Water Supply Demand Project. *One Watershed - One Water Conference*, 73.

Eberhardt, L. and J. Thomas (1991). Designing environmental field studies. *Ecological Monographs 61*(1), 53–73.

Ecofiscal (2015). *Practical Solutions for Growing Prosperity: The Way Forward*. Montreal, QC, Canada: Canada's Ecofiscal Commission (Ecofiscal).

Edelstein-Keshet, L. (2005). *Mathematical Models in Biology*, Volume 46 of *SIAM Classics in Applied Mathematics*. New York, NY, USA: Random House.

Eden, A., J. Moor, J. Soraker, and E. Steinhart (2012). *The Singularity Hypothesis: A Scientific and Philosophical Assessment*. Berlin: Springer-Verlag.

Edwards, M., G. Beaugrand, G. Hays, J. Koslow, and A. Richardson (2010). Multi-decadal oceanic ecological datasets and their application in marine policy and management. *Trends in Ecology and Evolution 25*(10), 602–610.

Egan, M. (2009). Hubble, trouble, toil and space rubble: The management history of an object in space. *Management and Organizational History 4*(3), 263–280.

Ehrlich, P. and J. Holdren (1971). Impact of population growth. *Science 171*(3977), 121–1217.

EIA (2013). *International Energy Outlook 2013*. Washington, DC, USA: US Energy Information Administration (EIA), http://www.worldenergyoutlook.org/media/weo2010.pdf.

Ekins, P. (1999). *Economic Growth and Environmental Sustainability: The Prospects for Green Growth*. London, UK and New York, NY, USA: Routledge (Taylor & Francis).

Elad, Y. and I. Pertot (2014). Climate change impacts on plant pathogens and plant diseases. *Journal of Crop Improvement 28*, 99–139.

Elkington, J. (1997). *Cannibals with Forks: The Triple Bottom Line of 21st Century Business*. Oxford, UK: Capstone.

Elliott, J., C. Müller, D. Deryng, J. Chryssanthacopoulos, K. Boote, Bücher, I. Foster, M. Glotter, J. Heinke, T. Iizumi, R. Izaurralide, N. Mueller, D. Ray, C. Rosenzweig, A. Ruane, and J. Sheffield (2015). The Global Gridded Crop Model Intercomparison: data and modeling protocols for Phase 1 (v1.0). *Geoscience Model Development 8*, 261–277.

Ellis, E. and N. Ramankutty (2008). Putting people in the map: Anthropogenic biomes of the world. *Frontiers in Ecology and the Environment 6*, 439–447.

Elton, S. (2013). *Consumed: Food for a Finite Planet*. Toronto, Canada: HarperCollins Publishers LLC.

England, J. (2013). Statistical physics of self-replication. *Journal of Chemical Physics 139*, 121923.

Enting, I., P. Rayner, and P. Ciaia (2012). Carbon cycle uncertainty in REgional Carbon Cycle Assessment and Processes (RECCAP). *Biogeosciences 9*, 2889–2904.

EPA (2012). *Study of the Potential Impacts of Hydraulic Fracturing on Drinking Water Resources: Progress Report*, Volume EPA 601/R-12/011. Washington, DC, USA, December, http://www.epa.gov/hfstudy/: US Environmental Protection Agency (EPA).

Epstein, J. (2007). *Generative Social Science: Studies in Agent-Based Computational Modeling*. Princeton, NJ, USA: Princeton University Press.

Epstein, S. (2015). Wanted: Collaborative Intelligence. *Artificial Intelligence 221*, 36–45.

Erickson, T. (1989). Proper posteriors from improper priors for an unidentified errors-in-variables model. *Econometrica 57*(6), 1299–1316.

Esty, D., M. Levy, T. Srebotnjak, and A. de Sherbinin (2005). *2005 Environmental Sustainability Index: Benchmarking National Environmental Stewardship.* New Haven, CT, USA: Yale University (Yale Center for Environmental Law and Policy), Columbia University (Center for International Earth Science Information Network), Joint Research Centre (JRC) European Commission and World Economic Forum, http://www.yale.edu/esi/.

EU-SCAR (2012). *Agricultural Knowledge and Innovation Systems in Transition - A Reflection Paper.* Brussels, Belgium: European Union (EU), Standing Committee on Agricultural Research (SCAR), Collaborative Working Group on Agricultural Knowledge and Innovation Systems (AKIS), http://ec.europa.eu/research/bioeconomy/pdf/ki3211999enc_002.pdf.

Eviner, V. and F. Chapin, III (2003). *Biogeochemical Interactions and Biodiversity (Chapter 8)*, pp. 151–173. Washington, DC, USA: Island Press.

Ewald, B., C. Penland, and R. Temam (2004). Accurate integration of stochastic climate models with application to El Niño. *Monthly Weather Review 132*, 154–164.

Falkowski, P., R. Scholes, E. Boyle, J. Canadell, D. CanÞeld, J. Elser, N. Gruber, K. Hibbard, P. Hogberg, S. Linder, F. Mackenzie, B. Moore, T. Pedersen, Y. Rosenthal, S. Seitzinger, V. Smetacek, and W. Steffen (2000). The global carbon cycle: A test of our knowledge of earth as a system. *Science 290*, 290–296.

Falloon, P., A. Challinor, S. Dessai, L. Hoang, J. Johnson, and A.-K. Koehler (2014). Ensembles and uncertainty in climate change impacts. *Frontiers in Environmental Science 2*(33), 1–7.

FAO (2010). *Bioenergy and Food Security: the BEFS Analytical Framework.* Environment and Natural Resources Management Series. The Bioenergy and Food Security Project (BEFS), FAO, http://www.fao.org/3/a-i1968e.pdf.

FAO (2011a). *Climate change, water, food security*, Volume 36. Rome: FAO, http://www.fao.org/docrep/014/i2096e/i2096e.pdf.

FAO (2011b). *The State of the World's Land and Water resources for Food and Agriculture (SOLAW) - Managing Systems at Risk.* Rome, Italy and Earthscan, London: FAO, http://www.fao.org/docrep/017/i1688e/i1688e.pdf.

FAO (2013). *Tackling Climate Change through Livestock: A Global Assessment of Emissions and Mitigation Opportunities.* Rome, Italy: FAO, http://www.fao.org/3/i3437e.pdf.

Farman, J., B. Gardiner, and J. Shanklin (1985). Large losses of total ozone in Antarctica reveal seasonal ClOx/NOx interaction. *Nature 315*, 207–210.

Farmer, J. and D. Foley (2009). The economy needs agent-based modeling. *Nature (Opinion) 460*(6), 685–686.

Fath, B., B. Patten, and J. Choi (2001). Complementarity of ecological goal functions. *Journal of Theoretical Biology 208*, 493–506.

Fayle, T., E. Turner, Y. Basset, R. Ewers, G. Reynolds, and V. Novotny (2010). Whole-ecosystem experimental manipulations of tropical forests. *Trends in Ecology and Evolution 25*(10), 334–346.

Feldman, D. and J. Crutchfield (1997). Measures of statistical complexity: Why? *Physics Letters A 238*(4-5), 244–252.

Fiksel, J. (2006). Sustainability and resilience: toward a systems approach. *Sustainability: Science, Practice & Policy 2*(2), 14–21.

Filatova, T., P. Verburg, D. Parker, and C. Stannard (2013). Spatial agent-based models for socio-ecological systems: Challenges and prospects. *Environmental Modelling and Software 5*, 1–7.

Finus, M. and Caparroós (2015). *Game Theory and International Environmental Cooperation*. The International Library of Critical Writings in Economics Series. Edward Elgar Publishing, `http://www.e-elgar.com/shop/game-theory-and-international-environmental-cooperation-11778`.

Fischer, G., H. Van Velthuizen, M. Shah, and F. Nachtergaele (2002). *Global Agro-Ecological Assessment for Agriculture in the 21st Century: Methodology and Results*. FAO and IIASA, `http://www.iiasa.ac.at/publication/more_RR-02-002.php`.

Fisher, D. (2011). Computing and AI for a Sustainable Future. *IEEE Intelligent Systems 26*(6), 1–8.

Flanagan, E. (2015). *Crafting an Effective Canadian Energy Strategy: How Energy East and the Oilsands Affect Climate Change*. Calgary, AB, Canada: Pembina Institute.

Floyd, J. and S. Shieh (2001). *Future Pasts : The Analytic Tradition in Twentieth-Century Philosophy: The Analytic Tradition in Twentieth-Century Philosophy*. New York, NY, USA: Oxford University Press.

Foerster, H. v., P. Mora, and L. Amiot (1960). Doomsday: Friday, 13, November, A.D. 2026. *Science 132*, 1291–1295.

Folke, C. (2004). Regime shifts, resilience, and biodiversity in ecosystem management. *Annual Review in Ecology and Evolutionary Systematics 35*, 557–581.

Folke, C. (2006). Resilience: The emergence of a perspective for social–ecological systems analyses. *Global Environmental Change 16*, 253–267.

Folke, C., S. Carpenter, T. Elmqvist, L. Gunderson, C. Holling, and B. Walker (2002). Resilience and sustainable development: Building adaptive capacity in a world of transformations. *Ambio 31*(5), 437–440.

Folke, C. and L. Gunderson (2012). Reconnecting to the biosphere: A social-ecological renaissance. *Ecology and Society 17*(4), 55.

Forkuor, G., C. Conrad, M. Thiel, T. Ullmann, and E. Zoungranna (2014). Integration of optical and synthetic aperture radar imagery for improving crop mapping in Northwestern Benin, West Africa. *Remote Sensing 6*, 6472–6499.

Fouskakis, D. and D. Draper (2002). Stochastic optimization: A review. *International Statistical Review 70*(3), 315–349.

Fowler, M. and L. Ruokolainen (2013). Confounding environmental colour and distribution shape leads to underestimation of population extinction risk. *PLoS ONE 8*(2), e55855.

Freedman, D. (2001). Ecological inference and the ecological fallacy. *International Encyclopedia for the Social and Behavioral Sciences 6*, 4027–4030.

Freedman, D. (2009). *Statistical Models: Theory and Practice* (2nd. ed.). Cambridge, UK: Cambridge University Press.

Freedman, J. and D. Noakes (2002). Why are there no really big bony fishes? A point-of-view on maximum body size of teleosts and elasmobranchs. *Reviews in Fish Biology and Fishes 12*, 403–416.

Frei, C. and C. Schar (2000). Detection probability of trends in rare events: Theory and application to heavy precipitation in the Alpine Region. *Journal of Climate 14*, 1568–1584.

Frey, C. and S. Patil (2002). Identification and review of sensitivity analysis methods. *Risk Analysis 22*(3), 553–578.

Friedlingstein, P., R. Houghton, G. Marland, J. Hackler, T. Boden, T. Conway, J. Canadell, M. Raupach, P. Ciais, and C. Le Quéré (2010). Update on CO_2 emissions. *Nature Geoscience 3*, 811–812.

Friedman, N., D. Geiger, and M. Goldszmidt (1997). Bayesian network classifiers. *Machine Learning 29*(2-3), 131–163.

Friesen, A. and P. Domingos (2015). Recursive Decomposition for Nonconvex Optimization. *24th International Joint Conference on Artificial Intelligence (July 25-31, Buenos Aires, Argentina)*, 1–7.

Frigg, R. (2003). Self-organised criticality—what it is and what it isn't. *Studies in History and Philosophy of Science 34*, 613–632.

Fritz, S., I. McCallum, C. Schill, C. Perger, R. Grillmayer, F. Achard, F. Kraxner, and M. Obersteiner (2009). Geo-wiki.org: The use of crowdsourcing to improve global land cover. *Remote Sensing 1*, 345–354.

Fudenberg, D., R. Iijima, and T. Strzalecki (2015). Stochastic choice and revealed perturbed utility. *Econometrica 83*(6), 2371–2409.

Fuentes, M., L. Chen, and J. Davis (2008). A class of nonseparable and nonstationary spatial temporal covariance functions. *Environmetrics 19*, 487–507.

Fuglestvedt, J., B. Samset, and K. Shine (2014). Counteracting the climate effects of volcanic eruptions using short-lived greenhouse gases. *Geophysical Research Letters 41*(23), 8627–8635.

Fung, I., S. Meyn, I. Tegen, S. Doney, J. John, and J. Bishop (2000). Iron supply and demand in the upper ocean. *Global Biogeochemical Cycles 14*(281), 281–295.

Funtowicz, S. and J. Ravetz (1994). Emergent complex systems. *Futures 26*(6), 568–582.

Fursin, G. (2013). Collective mind: cleaning up the research and experimentation mess in computer engineering using crowdsourcing, big data and machine learning. *Computing Research Repository (CoRR) abs/1308.2410*, 1–15.

Gabbay, D., C. Hooker, P. Thagard, and J. Woods (2011). *Philosophy of Complex Systems* (1st ed.), Volume 10 of *Handbook of Science (Hooker, C., Ed.)*. Oxford, UK: North Holland (Elsevier B.V.).

Galla, T. and J. Farmer (2013). Complex dynamics in learning complicated games. *Proceedings of the National Academy of Sciences 110*(4), 1232–1236.

Gallagher, J., E. Hubal, L. Jackson, J. Inmon, E. Hudgens, A. Williams, D. Lobdell, J. Rogers, and T. Wade (2013). Sustainability, health and environmental metrics: impact on ranking and associations with socioeconomic measures for 50 US cities. *Sustainability 5*, 789–804.

Gao, J., Y. Cao, T. W.-W. Tang, and J. Hu (2007). *Multiscale Analysis of Complex Time Series: Integration of Chaos and Random Fractal Theory, and Beyond*. Hoboken, NJ, USA: Wiley-Interscience, John Wiley & Sons, Inc.

Gao, Z., D. Kong, C. Gao, and M.-C. Chen (2013). Modeling and control of complex dynamic systems 2013. *Journal of Applied Mathematics 2013*, 3.

García, M., C. Oyonarte, L. Villagarcía, S. Contreras, F. Domingo, and Y. Puigdefábregas (2008). Monitoring land degradation risk using ASTER data: The non-evaporative fraction as an indicator of ecosystem function. *Remote Sensing of Environment 112*, 3720–3736.

Gasparatos, A., P. Stromberg, and K. Takeuchi (2011). Biofuels, ecosystem services and human well-being: Putting biofuels in the ecosystem services narrative. *Agriculture, Ecosystems and Environment 142*, 111–128.

Gasser, R. and M. Huhns (1989). *Distributed Artificial Inteligence*. Research Notes in Artificial Intelligence. Pitman Publishing (Longman Group Ltd., UK) and Morgan Kaufmann Publishers, Inc. USA.

Geels, F. (2004). From sectoral systems of innovation to socio-technical systems: Insights about dynamics and change from sociology and institutional theory. *Research Policy 33*, 897–920.

Geels, F. (2007). Typology of sociotechnical transition pathways. *Research Policy 36*, 399–417.

Gelman, A., J. Carlin, H. Stern, D. Dudson, A. Vehtari, and D. Rubin (2013). *Bayesian Data Analysis* (3rd. ed.), Volume 106 of *Texts in Statistical Science*. Chapman and Hall/CRC Press.

Genton, M. and W. Kleiber (2015). Cross-covariance functions for multivariate geostatistics. *Statistical Science 30*(2), 147–163.

Gerbens-Leenes, W., A. Hoekstra, and T. van der Meer (2009). The water footprint of bioenergy. *Proceedings of the National Academy of Sciences (PNAS) 106*(25), 10219–10223.

Gerland, P., A. Raftery, H. Sevcikova, N. Li, D. Gu, T. Spoorenberg, L. Alkema, B. Fosdick, J. Chunn, N. Lalic, G. Bay, T. Buettner, G. Heilig, and J. Wilmoth (2014). World population stabilization unlikely this century. *Science 346*(6206), 234–237.

Gerten, D., J. Heinke, H. Hoff, H. Biemans, M. Fader, and K. Waha (2011). Global water availabilty and requirements for future food production. *Journal of Hydrometeorology 12*(5), 885–899.

Giannakis, D., A. Majda, and I. Horenko (2012). Information theory, model error, and predictive skill of stochastic models for complex nonlinear systems. *Physica D 241*, 1735–1752.

Gill, R., O. Ramos-Rodriguez, and N. Raine (2012). Combined pesticide exposure severely affects individual- and colony-level traits in bees. *Nature 491*, 105–109.

Gleeson, T., Y. Wada, M. Bierkens, and L. van Beek (2012). Water balance of global aquifers revealed by groundwater footprint. *Nature 488*, 197–200.

Glen, G. and K. Isaacs (2012). Estimating Sobol sensitivity indices using correlations. *Environmental Modelling and Software 37*, 157–166.

Gluckman, P. (2014). The art of science advice to government. *Nature 507*, 163–165.

Gneiting, T. (2002). Nonseparable, stationary covariance functions for space-time data. *Journal of the American Statistical Association 97*(458), 590–600.

Goldstein, E. and G. Coco (2015). Machine learning components in deterministic models: hybrid synergy in the age of data. *Frontiers in Environmental Science (Environmental Informatics) (Opinion Article), http://dx.doi.org/10.3389/fenvs.2015.00033 3*, 1–4.

Golev, A., M. Scott, P. Erskine, S. Ali, and G. Ballantyne (2014). Rare earths supply chains: Current status, constraints and opportunities. *Resources Policy 41*, 52–59.

Golinksi, M., C. Bauch, and M. Anand (2008). The effects of endogenous ecological memory on population stability and resilience in a variable environment. *Ecosystem Modelling 212*, 334–341.

Golińska, A. (2012). Detrended fluctuation analysis (DFA) in biomedical signal processing: selected examples. *Studies in Logic, Grammar and Rhetoric 29*(42), 107–115.

Gomes, C. (2009). Computational, sustainability: Computational methods for a sustainable environment, economy, and society. *The Bridge 39*(4), 5–13.

Gorbenko, A., V. Popov, and A. Sheka (2012). Robot self-awareness: Exploration of internal states. *Applied Mathematical Sciences 6*(14), 675–688.

Gordon, K. (2014). *Risky Business: The Economic Risks of Climate Change in the United States.* Risky Business, http://riskybusiness.org/site/assets/uploads/2015/09/RiskyBusiness_Report_WEB_09_08_14.pdf.

Graedel, T. and B. Allenby (2002). Hierarchical metrics for sustainability. *Environmental Quality Management 12*(2), 21–30.

Granger, C. (1996). Investigating causal relations by econometric models and cross-spectral methods. *Econometrics 37*, 424–438.

Grasman, J., O. van Herwaarden, and T. Hagenaars (2005). *Current Themes in Theoretical Biology (Resilience and Persistence in the Context of Stochastic Population Models)*, pp. 267–280. The Netherlands: Springer.

Grasso, M. and J. Roberts (2014). A compromise to break the climate impasse. *Nature Climate Change (Perspective) 4*, 543–549.

Graves, A., G. Wayne, and I. Danihelka (2014). Neural Turing Machines. *CoRR, dblp Computer Science Bibliography, http://arxiv.org/abs/1410.5401*, 1–26.

Green, D. (2014). *Encyclopedia of Life Support Systems (EOLSS) (Hierarchy, Complexity, and Agent-based Models)*, Volume Vol. III (Knowledge Management, Organizational Intelligence and Learning, Complexity), http://www.eolss.net/. Paris, France: UNESCO-EOLSS Joint Committee Secretariat.

Gregori, P., E. Porcu, J. Mateu, and Sasvári (2008). On potentially negative space time covariances obtained as sum of products of marginal ones. *Annals of the Institute of Statistical Mathematics 60*, 865–882.

Grimm, V., U. Berger, F. Bastiansen, S. Eliassen, V. Ginot, J. Giske, J. Goss-Custard, T. Grand, S. Heinz, G. Huse, A. Huth, J. Jepsen, C. Jørgensen, W. Mooij, B. Müller, G. Pe'er, C. Piou, S. Railsback, A. Robbins, M. Robbins, R. Vabø, U. Visser, and D. DeAngelis (2006). A standard protocol for describing individual-based and agent-based models. *Ecological Modelling 198*, 115–26.

Grimm, V., U. Berger, D. DeAngelis, J. Polhill, J. Giske, and S. Railsback (2010). The ODD Protocol: A review and first update. *Ecological Modelling 221*, 2760–2768.

Grimm, V. and S. Railsback (2005). *Individual-Based Modeling and Ecology*. Princeton, NJ, USA: Princeton University Press.

Groisman, P. and D. Easterling (1994). Variability and trends of total precipitation and snowfall over the United States and Canada. *Journal of Climate 7*, 184–205.

Gros, C. (2013). *Complex and Adaptive Dynamical Systems, a Primer* (3rd. ed.). Complexity. Berlin and Heidelberg: Springer-Verlag.

Grünwald, P. and P. Dawid (2004). Game Theory, Maximum Entropy, Minimum Discrepancy and Robust Bayesian Decision Theory. *The Annals of Statistics 32*(4), 1367–1433.

Grünwald, P. and J. Langford (2007). Suboptimal behaviour of Bayes and MDL in classification under misspecification. *Machine Learning 66*, 119–149.

Grush, R. (2004). The emulation theory of representation: Motor control, imagery, and perception. *Behavioural and Brain Sciences 27*, 377–442.

GSI-IISD (2007). *Biofuels - at what cost? Government support for ethanol and biodiesel in selected OECD countries*. A synthesis of reports addressing subsidies for biofuels in Australia, Canada, the European Union, Switzerland and the United States. Geneva, Switzerland: The Global Subsidies Initiative (GSI) of the International Institute for Sustainable Development (IISD), `https://www.iisd.org/gsi/sites/default/files/oecdbiofuels.pdf`.

Guan, J., B. Chang, and M. Aral (2013). A dynamic control system for global temperature change and sea level rise in response to CO_2 emissions. *Climate Research 58*, 55–66.

Guhaniyogi, R., A. Finley, S. Banerjee, and R. Kobe (2013). Modeling complex spatial dependencies: Low-rank spatially varying cross-covariances with application. *Journal of Agricultural and Applied Economics 18*(3), 274–298.

Gunderson, L. and C. Holling (2002). *Panarchy: Understanding Transformations in Human and Natural Systems*. Washington, DC, USA: Island Press.

Gunderson, L. and L. Pritchard (2002). *Resilience and the Behavior of Large-Scale Systems*. Washington, DC, USA: Island Press.

Gustavsson, J., C. Cederberg, U. Sonesson, R. van Otterdijk, and A. Meybeck (2011). *Global Food Losses and Food Waste: Extent, Causes, Prevention*. Rome, Italy: FAO, http://www.fao.org/docrep/014/mb060e/mb060e.pdf.

Gutenkunst, R., N. Newlands, M. Lutcavage, and L. Edelstein-Keshet (2007). Inferring resource distributions from Atlantic bluefin tuna movements: An analysis based on net displacement and length of track. *Theoretical Ecology 245*(2), 243–57.

Guzy, M., C. Smith, J. Bolte, D. Hulse, and S. Gregory (2008). Policy research using agent-based modeling to assess future impacts of urban expansion into farmlands and forests. *Ecology and Society 13*(1), 37.

Haas, T. (1990). Lognormal and moving window methods of estimating acid deposition. *Journal of the American Statistical Association 85*, 950–963.

Haase, D. and N. Schwarz (2009). Simulation models on human-nature interactions in urban landscapes: A review including spatial economics, system dynamics, cellular automata and agent-based approaches. *Living Reviews in Landscape Research 3*(2), 45.

Haddeland, I., J. Heinke, H. Biemans, S. Eisner, M. Flörkee, N. Hanasaki, M. Konzmann, F. Ludwig, Y. Masaki, J. Schewe, T. Stacke, Z. Tessler, Y. Wada, and D. Wisser (2014). Global water resources affected by human interventions and climate change. *Proceedings of the National Academy of Sciences 111*(9), 3251–3256.

Hadley, C. (2006). Science and Society, Analysis: Food allergies on the rise? *European Molecular Biology Organization (EMBO) Reports 7*(11), 1080–1083.

Hadzikadic, M., T. Carmichael, and C. Curtin (2010). Complex adaptive systems and game theory: An unlikely union. *Complexity 16*(1), 34–42.

Hahn, R. and M. Kühnen (2013). Determinants of sustainability reporting: a review of results, trends, theory, and opportunities in an expanding field of research. *Journal of Cleaner Production 59*, 5–21.

Haining, R. (1990). *Spatial Data Analysis in the Social and Environmental Sciences*. Cambridge, UK: Cambridge University Press.

Hall, R. (2009). *Inflation: Causes and Effects*. Chicago, IL, USA: University of Chicago Press.

Hall, R. (2010). *Compression: Meeting the Challenges of Sustainability Through Vigorous Learning Enterprises*. CRC Press (Taylor and Francis LCC).

Hammersley, J. and P. Clifford (1971). Markov fields on finite graphs and lattices. *(Unpublished)*, $http://www.statslab.cam.ac.uk/~grg/books/hammfest/hamm-cliff.pdf$, 1–36.

Hammon, L. and H. Hippner (2012). Crowdsourcing. *Business & Information Systems Engineering 3*, 163–166.

Haneveld, W. and M. Van der Vlerk (2006). Integrated chance constraints: Reduced forms and an algorithm. *Computational Management Science 3*(4), 245–269.

Hänggi, P. and P. Jung (1995). Coloured noise in dynamical systems. *Advances in Chemical Physics 89*, 239–326.

Hanington, I. and D. Suzuki (2012). *Everything Under the Sun: Toward a Brighter Future on a Small Blue Planet.* Vancouver, Canada: D&M Publishers, Inc. (Greystone Books).

Hanjra, M. and M. Qureshi (2010). Global water crisis and future food security in an era of climate change. *Food Policy 35*, 365–377.

Hansen, J., M. Sato, and R. Ruedy (2012). Perception of climate change. *Proceedings of the National Academy of Sciences (PNAS) 109*(37), E2415–E2423.

Harper, S. (2014). Economic and social implications of aging societies. *Science 346*(6209), 587–591.

Hasan, T., S. Ali, and M. Khan (2013). A comparative study of loss functions for Bayesian control in mixture models. *Electronic Journal of Applied Statistical Analysis (EJASA) 6*(2), 175–185.

Hasenauer, H., K. Merganicova, R. Petritsch, S. Pietsch, and P. Thornton (2003). Validating daily climate interpolations over complex terrain in Austria. *Agricultural and Forest Meteorology 119*, 87–107.

Hasselmann, K. (1976). Stochastic climate models. Part I. Theory. *Tellus 28*, 473–485.

Hastie, T., R. Tibshirani, and J. Friendman (2008). *The Elements of Statistical Learning: Data Mining, Inference, and Prediction* (2 ed.). Springer Series in Statistics. Springer, $http://www.springer.com/gp/book/9780387848570$.

Hawking, S, W. (2014). Information Preservation and Weather Forecasting for Black Holes. *arXiv:1401.5761 [hep-th] (High Energy Physics - Theory)*, $http://arxiv.org/abs/1401.5761$, 1–4.

Hayes, B. (2012). Computation and the Human Predicament: The Limits to Growth and the limits to computer modeling. *American Scientist 100*(3), 186–191.

Heckmann, I., T. Comes, and S. Nickel (2015). A critical review on supply chain risk: Definition, measure and modeling. *Omega 52*, 119–132.

Hedges, C. (2015). *Wages of Rebellion: The Moral Imperative of Revolt.* New York, NY, USA: Nation Books (Nation Institute and Perseus Books Group).

Hegmann, G., C. Cocklin, R. Creasey, S. Dupuis, A. Kennedy, L. Kingsley, W. Ross, S. H., and D. Stalker (1999). *Cumulative Effects Assessment Practitioners Guide.* Hull, QC, Canada: The Cumulative Effects Assessment Working Group (Canadian Environmental Assessment Agency (CEAA) and AXYS Environmental Consulting Ltd.

Heinsalu, E., M. Patriarca, I. Goychuk, G. Schmid, and P. Hänggi (2006). Fractional Fokker-Planck dynamics: Numerical algorithm and simulations. *Physical Review E 73*, 046133.

Heipke, C. (2010). Crowdsourcing geospatial data. *ISPRS Journal of Photogrammetry and Remote Sensing 65*, 550–557.

Helbing, D. (2012). *Social Self-Organization.* Understanding Complex Systems. Berlin: Springer-Verlag, http://www.springer.com/fr/book/9783642240034.

Hellmann, J., J. Byers, B. Bierwagen, and J. Dukes (2008). Five potential consequences of climate change for invasive species. *Conservation Biology 22*(3), 534–543.

Henderson, F. and A. Lewis (1998). *Manual of Remote Sensing* (3rd. ed.), Volume 2 of *Principles and Applications of Imaging Radar.* Hoboken, NJ, USA: John Wiley & Sons Inc.

Hermans, M., B. Schrauwen, P. Bienstman, and J. Dambre (2014). Automated design of complex dynamic systems. *PLoS ONE 9*(1), e86696.

Hickey, G., P. Craig, and A. Hart (2009). On the application of loss functions in determining assessment factors for ecological risk. *Ecotoxicology and Environmental Safety 72*, 293–300.

Hitz, S. and J. Smith (2004). Estimating global impacts from climate change. *Global Environmental Change 14*, 201–218.

Hoekstra, A. and P. Hung (2002). *Virtual Water Trade, A Quantification of virtual water flows between nations in relation to international crop trade.* Virtual Water Research Report Series. Delft, The Netherlands: IHE Delft, http://waterfootprint.org/media/downloads/Report11.pdf.

Hoff, H. (2011). *Understanding the Nexus: Background Paper for the Bonn2011 Conference: The Water, Energy and Food Security Nexus.* Stockholm, Sweden: Stockholm Environment Institute (SEI).

Hoffert, M., K. Caldeira, G. Benford, D. Criswell, G. Green, H. Herzog, A. Jain, H. S. Kheshgi, K. Lackner, J. Lewis, H. Lightfoot, W. Manheimer, J. Mankins, M. Mauel, L. Perkins, M. Schlesinger, T. Volk, and T. Wigley (2002). Advanced technology paths to global climate stability: Energy for a greenhouse planet. *Science 298*, 981–987.

Hoffman, A. and M. Hercus (2000). Environmental stress as an evolutionary force. *BioScience 50*(3), 217–226.

Hofkirchner, W. (2005). *Ludwig von Bertalanffy: Forerunner of Evolutionary Systems Theory.* The New Role of System Sciences for a Knowledge-Based Society. Proceedings of the First World Congress of the International Federation of Systems Research (November 14-17, Kobe, Japan), `http://www.jaist.ac.jp/library/jaist-press/catalogue/cd_ifsr.html`.

Holling, C. (1973). Resilience and stability of ecological systems. *Annual Review of Ecology and Systematics 4*, 1–24.

Holling, C. (2001). Understanding the complexity of economic, ecological, and social systems. *Ecosystems 4*(1), 390–405.

Holman, I., M. Rounsevell, G. Cojacaru, S. Shackley, C. McLachlan, E. Audsley, P. Berry, C. Fontaine, P. Harrison, C. Henriques, M. Mokrech, R. J. Nicholls, K. Pearn, and J. Richards (2008). The concepts and development of a participatory regional integrated assessment tool. *Climate Change 90*, 5–30.

Homer-Dixon, T. (2006). *The Upside of Down: Catastrophe, Creativity and the Renewal of Civilization.* Toronto, Canada: Random House.

Hoornweg, D., P. Bhada-Tata, and C. Kennedy (2013). Waste production must peak this century. *Nature 502*, 615–617.

Hooten, M. and C. Wilke (2010). Statistical agent-based models for discrete spatio-temporal systems. *Journal of the American Statistical Association (Theory and Methods) 105*(489), 236–248.

Hopkinson, R. and D. McKenney (2011). Impact of aligning climatological day on gridding daily maximum-minimum temperature and precipitation over Canada. *Journal of Applied Meteorology and Climatology 50*, 1654–1665.

Hosmer, D. and S. Lemeshow (2000). *Applied Logistic Regression.* New York, NY, USA: John Wiley & Sons Ltd.

Hosseini, R., N. Newlands, C. Dean, and A. Takemura (2015). Statistical modeling of soil moisture, integrating satellite, remote-sensing (SAR) and ground-based data. *Remote Sensing 7*, 2752–2780.

Host, G., H. Omre, and P. Switzer (1995). Spatial interpolation errors for monitoring data. *Journal of the American Statistical Association 90*(431), 853–861.

Hougaard, P. (1999). Multi-state models: A review. *Lifetime Data Analysis 5*, 239–264.

Hovmöller, M., A. Yahyaoi, E. Milus, and A.-M. F. Justesen (2008). Rapid global spread of two aggressive strains of a wheat rust fungus. *Molecular Ecology 17*, 3818–3826.

Hsu, A. e. a. (2016). *Global Metrics for the Environment: Environmental Performance Index (EPI), 2016 Report.* New Haven, CT, USA: Yale University, www.epi.yale.edu.

Huang, G., G.-B. Huang, S. Song, and K. You (2015). Trends in extreme learning machines: A review. *Neural Networks 61*, 32–48.

Huang, J., M. Tichit, M. Poulot, S. Darly, S. Li, C. Petit, and C. Aubry (2015). Comparative review of multifunctionality and ecosystem services in sustainable agriculture. *Journal of Environmental Management 149*, 138–147.

Huang, Z., J. Li, H. Su, G. Watts, and H. Chen (2007). Large-scale regulatory network analysis from microarray data: Modified Bayesian network learning and association rule mining. *Decision Support Systems 43*, 1207–1225.

Hubacek, K., K. Feng, J. Minx, S. Pfister, and N. Zhou (2014). Teleconnecting consumption to environmental impacts at multiple spatial scales. *Journal of Industrial Ecology 18*(1), 7–9.

Huber, V., H. Schellhuber, N. Arnell, K. Frieler, A. Friend, D. Gerten, I. Haddeland, P. Kabat, H. Lotze-Campen, W. Lucht, M. Parry, F. Piontek, C. Rosenzweig, J. Schewe, and L. Warszawski (2014). Climate impact research: Beyond patchwork. *Earth System Dynamics 5*, 399–408.

Hughes, G. and M. Morgan (1973). The structure of fish gills in relation to their respiratory function. *Biological Reviews 48*(4), 419–475.

Hughes, L. and J. Rudolph (2011). Future world oil production: Growth, plateau or peak? *Current Opinion in Environmental Sustainability 3*, 225–234.

Hunter, R. and R. Meentemeyer (2005). Climatologically aided mapping of daily precipitation and temperature. *Journal of Applied Meteorology 44*, 1501–1510.

Huntington, T. (2006). Evidence for intensification of the global water cycle: Review and synthesis. *Journal of Hydrology 319*, 83–95.

Hutchinson, M. (1995). Interpolating mean rainfall using thin plate smoothing splines. *International Journal of Geographical Information Systems 9*(4), 385–403.

Hutchinson, M. (1998a). Interpolation of rainfall data with thin plate smoothing splines - part I: Two dimensional smoothing of data with short-range correlation. *Journal of Geographic Information and Decision Analysis 2*(2), 139–151.

Hutchinson, M. (1998b). Interpolation of rainfall data with thin plate smoothing splines - part II: Analysis of topographic dependence. *Journal of Geographic Information and Decision Analysis 2*(2), 152–167.

IAASTD (2009a). *Agriculture at a Crossroads: Global Report.* Washington, DC, Covelo, CA, London, UK: Island Press, http://www.globalagriculture.org/.

IAASTD (2009b). *Agriculture at a Crossroads: Synthesis Report - A Synthesis of the Global and Sub-Global IAASTD Reports.* International Assessment of Agricultural Knowledge, Science and Technology for Development (IAASTD), http://www.unep.org/dewa/agassessment/.

IEA-Bioenergy (2007). *Potential Contribution of Bioenergy to the World's Future Energy Demand.* International Energy Agency (IEA), September, http://www.ieabioenergy.com/publications/.

IFC (2012). *IFC Sustainability Framework: Policy and Performance Standards on Environmental and Social Sustainability.* Washington, DC, USA: International Finance Corporation (IFC), www.ifc.org/sustainabilityframework2012.

IIRC (2013). *The International <IR> Framework, Summary of Significant Issues, Basis for Conclusions Online Reports.* Washington, DC, USA: International Integrated Reporting Council, http://integratedreporting.org/wp-content/uploads/2013/12/13-12-08-Summary-of-significant-issues-IR.pdf.

IISD (2014). *Food Security Indicator and Policy Analysis Tool (FIPAT) Guidebook.* Winnipeg, Canada: International Institute for Sustainable Development (IISD), https://www.iisd.org/publications/.

Iizumi, T., J. Luo, A. Challinor, G. Sakurai, M. Yokozawa, H. Sakuma, M. Brown, and T. Yamagata (2014). Impacts of El Niño Southern Oscillation on the global yields of major crops. *Nature Communications 5*, 3712.

Illari, P., F. Russo, and J. Williamson (2011). *Causality in the Sciences.* Oxford, UK: Oxford University Press.

Imkeller, P. and A. Monahan (2002). Conceptual stochastic climate models. *Stochastic Dynamics 2*(3), 16.

INFASA (2006). *From Common Principles to Common Practice.* Proceedings and outputs of the first Symposium of the International Forum on Assessing Sustainability in Agriculture (INFASA) (March 16, 2006, Bern, Switzerland). INFASA, https://www.iisd.org/pdf/2007/infasa_common_principles.pdf.

Innes, J. (1998). *Measuring Environmental Change*, pp. 429–457. Ecological Scale: Theory and Applications (Peterson, D.L. and Parker, V.T., Eds.). New York, NY, USA: Columbia University Press.

IPCC (2012). *IPCC Special Report on Renewable Energy Sources and Climate Change Mitigation.* United Kingdom and New York, NY, USA: Cambridge University Press, Cambridge, http://srren.ipcc-wg3.de/report.

IPCC (2013). *Climate Change 2013: The Physical Science Basis*, Volume Contribution of Working Group I to the Fifth Assessment Report. New York, NY, USA: Intergovernmental Panel on Climate Change (IPCC), http://www.climatechange2013.org/.

IPCC (2014a). *Climate Change 2014: Impacts, Adaptation, and Vulnerability.* Contribution of Working Group II to the Fifth Assessment Report (AR5). Cambridge, United Kingdom and New York, NY, USA: Cambridge University Press, http://www.ipcc.ch/report/ar5/wg2/.

IPCC (2014b). *Climate Change 2014: Synthesis Report (SYR)*, Volume Contribution of Working Group II to the Fifth Assessment Report (AR5). New York, NY, USA: Intergovernmental Panel on Climate Change (IPCC), http://www.ipcc.ch/report/ar5/syr/.

ISI (2014). *Statistics and Science: A Report of the London Workshop on the Future of the Statistical Sciences.* 2014 London Workshop on the Future of the Statistical Sciences, http://www.worldofstatistics.org/wos/pdfs/Statistics&Science-TheLondonWorkshopReport.pdf.

Jackson, T. (2009). *Prosperity without Growth: Economics for a Finite Planet.* London, UK and New York, NY, USA: Earthscan (Taylor & Francis).

Jain, R. (2005). Sustainability: Metrics, specific indicators and preference index. *Clean Technology and Environmental Policy 7*, 71–72.

Jakeman, A., R. Letcher, and J. Norton (2006). Ten iterative steps in development and evaluation of environmental models. *Environmental Modelling and Software 21*, 602–614.

James, A., J. Pitchford, and M. Plank (2012). Disentangling nestedness from models of ecological complexity. *Nature 487*, 227–230.

Jansen, R., H. Yu, D. Greenbaum, Y. Kluger, N. Krogan, S. Chung, A. Emili, M. Snyder, J. Greenblatt, and M. Gerstein (2003). A Bayesian networks approach for predicting protein-protein interactions from genomic data. *Science 302*(17 October), 449–453.

Janssen, M. and S. Carpenter (1999). Managing the resilience of lakes: A multi-agent modeling approach. *Ecology and Society 3*(2), 1–15.

Janssen, N., M. Gerlofs-Nijland, T. Lanki, R. Salonen, F. Cassee, G. Hoek, P. Fischer, B. Brunekreef, and M. Krzyzanowsji (2012). *Health effects of black carbon (BC).* Copenhagen, Denmark: World Health Organization (WHO), http://www.euro.who.int/__data/assets/pdf_file/0004/162535/e96541.pdf.

Jennings, B. (2007). On the Nature of Science. *Physics in Canada 63*(1), 7–20.

Jensen, P., L. Basson, and M. Leach (2011). Reinterpreting industrial ecology. *Journal of Industrial Ecology 15*(5), 680–692.

Jiang, L. (2014). Internal consistency of demographic assumptions in the shared socioeconomic pathways. *Population and Environment 35*, 261–285.

Jindal, A. and K. Psounis (2006). Modeling spatially correlated data in sensor networks. *ACM Transactions on Sensor Networks 2*(4), 466–499.

Johnson, G., C. Daly, and G. Taylor (2000). Spatial variability and interpolation of stochastic weather simulation model parameters. *Journal of Applied Meteorology 39*, 778–796.

Johnson, S., S. Low-Choy, and K. Mengersen (2011). Integrating Bayesian networks and geographical information systems: Good practice examples. *Integrated Environmental Assessment and Management 8*(3), 473–479.

Jolly, W., J. Graham, A. Michaelis, R. Nemani, and S. Running (2005). A flexible, integrated system for generating meteorological surfaces derived from point sources across multiple geographic scales. *Environmental Modelling and Software 20*, 873–882.

Jones, J., G. Hoogenboom, C. Porter, K. Boote, W. Batchelor, L. Hunt, P. Wilkens, U. Singh, A. Gijsman, and J. Ritchie (2003). DSSAT Cropping System Model. *European Journal of Agronomy 18*, 235–265.

Jorgensen, S. and S. Nielsen (2015). Hierarchical networks. *Ecological Modelling 295*, 59–65.

Juroszek, P. and A. von Tiedemann (2013). Climate change and potential future risks through wheat diseases: A review. *European Journal of Plant Pathology 136*, 21–23.

Kabat, P. and M. Contestabile (2013). Water at a crossroads. *Nature Climate Change 3*, 11–12.

Kahan, D. (2012). *Cultural cognition as a conception of the cultural theory of risk.* Handbook of Risk Theory: Epistemology, Decision Theory, Ethics and Social Implications of Risk (Roeser, S., Hillerbrand, R., Sandin, P and Peterson, M., Eds.). London, UK: Springer Science+Business Media B.V.

Kahan, D., E. Peters, M. Wittlin, P. Slovic, L. Ouellette, D. Braman, and G. Mandel (2012). The polarizing impact of science literacy and numeracy on perceived climate change risks. *Nature Climate Change (Letters) 2*, 732–735.

Kale, M. and F. Butar (2011). Fractal analysis of time series and distribution properties of Hurst exponent. *Journal of Mathematical Sciences & Mathematics Education 5*(1), 8–19.

Kander, A., M. Jiborn, D. Moran, and T. Wiedmann (2015). National greenhouse-gas accounting for effective climate policy on international trade. *Nature Climate Change 5*, 431–434.

Kang, Q., L. Wang, and Q. Wu (2008). A novel ecological particle swarm optimization algorithm and its population dynamics analysis. *Applied Mathematics and Computation 205*, 61–72.

Kapitza, S. (1996). The phenomenological theory of world population growth. *Phys. Usp. 39*(1), 57–71.

Karl, T., A. Arguez, B. Huang, J. Lawrimore, J. McMahon, M. Menne, T. Peterson, R. Vose, and H.-M. Zhang (2015). Possible artifacts of data biases in the recent global surface warming hiatus. *Science 348*(6242), 1469–1472.

Karp, L. (2005). Global warming and hyperbolic discounting. *Journal of Public Economics 89*, 261–282.

Katenka, N. (2009). *Statistical problems in wireless sensor networks.* PhD. diss., (University of Michigan).

Kavanagh, P., N. Newlands, V. Christensen, and D. Pauly (2004). Automated parameter optimization for Ecopath ecosystem models. *Ecological Modelling 172*, 141–149.

Kazianka, H. and J. Pilz (2010). Copula-based geostatistical modeling of continuous and discrete data including covariates. *Stochastic Environmental Research and Risk Assessment 24*, 661–673.

Keledjian, A., G. Brogan, B. Lowell, J. Warrenchuk, B. Enticknap, G. Shester, M. Hirshfield, and D. Cano-Stocco (2014). *Wasted Catch - Unsolved Problems in US Fisheries.* Washington, DC, USA: Oceana, http://oceana.org/reports/wasted-catch-unsolved-problems-us-fisheries.

Keller, C., A. Bostrom, M. Kuttschreuter, L. Savadori, A. Spence, and M. White (2012). Bringing appraisal theory to environmental risk perception: A review of conceptual approaches of the past 40 years and suggestions for future research. *Journal of Risk Research 15*(3), 237–256.

Kelly, R., A. Jakeman, O. Barreteau, M. Borsuk, S. El-Sawah, S. Hamilton, H. Henriksen, S. Kuikka, H. Maier, A. Rizzoli, H. van Delden, and A. Voinov (2013). Selecting among five common modelling approaches for integrated environmental assessment and management. *Environmental Modelling and Software 47*, 159–181.

Kenall, A., S. Edmunds, L. Goodman, L. Bal, L. Flintoft, D. Shanahan, and T. Shipley (2015). Better reporting for better research: A checklist for reproducibility. *GigaScience 4*, 32.

Kennedy, C. and D. Hoornweg (2012). Mainstreaming urban metabolism. *Journal of Industrial Ecology, Special Issue: Sustainable Urban Systems 16*(6), 780–782.

Keohane, R. and D. Victor (2010). *The Regime Complex for Climate Change.* Harvard Kennedy School.

Kerr, R. (2012). Technology is turning US oil around but not the World's. *Science 335*, 522–523.

Khan, J., S. Van Aelst, and R. Zamar (2010). Fast robust estimation of prediction error based on resampling. *Computational Statistics and Data Analysis 54*, 3121–3130.

Khan, J., S. Van Aelst, and R. H. Zamar (2007). Robust linear model selection based on least angle regression. *Journal of the American Statistical Association 102*, 1289–1299.

Kiiza, B. and G. Pederson (2012). ICT-based market information and adoption of agricultural seed technologies: Insights from Uganda. *Telecommunications Policy 36*, 253–259.

King, C. and H. Jaafar (2015). Rapid assessment of the water-energy-food-climate nexus in six selected basins of North Africa and West Asia undergoing transitions and scarcity threats. *International Journal of Water Resources Development 31*(3), 343–359.

King, R. (2012). A review of Bayesian state-space modelling of capture-recapture recovery data. *Interface Focus 2*, 190–204.

King, R. (2014). Statistical Ecology. *Annual Review of Statistics and Its Application 1*, 401–426.

King, R., S. Brooks, B. Morgan, and T. Coulson (2006). Factors influencing soay sheep survival: A Bayesian analysis. *Biometrics 62*(1), 211–220.

Kinney, P. (2008). Climate change, air quality, and human health. *American Journal of Preventive Medicine 35*(5), 459–467.

Kintisch, E. (2014). Is Atlantic holding Earth's missing heat? New leads in the hunt to explain the global warming hiatus. *Science 345*, 6199.

Klein, A.-M., B. Vaissiere, J. Cane, I. Steffan-Dewentre, S. Cunningham, C. Kremen, and T. Tscharntke (2007). Importance of pollinators in changing landscapes for world crops. *Proceedings of the Royal Society B: Biological Sciences 274*, 303–313.

Klein, N. (2014). *This Changes Everything: Capitalism vs. The Climate.* Simon & Schuster, http://thischangeseverything.org/.

Klir, G. (2001). *Facets of Systems Science* (2nd. ed.), Volume 15 of *IFSR International Series on Systems Science and Engineering.* New York, NY, USA: Kluwer Academic/Plenum Press.

Kloeden, P. and E. Platen (1999). *Numerical Solution of Stochastic Differential Equations*, Volume 23 of *Stochastic Modelling and Applied Probability*. Berlin and Heidelberg: Springer-Verlag.

Koller, M. and W. Stahel (2011). Sharpening wald-type inference in robust regression for small samples. *Computational Statistics and Data Analysis 55*, 2504–2515.

Kolovos, A., J. Angulo, K. Modis, G. Papantonopoulos, J. Wang, and G. Christakos (2013). Model-driven development of covariances for spatiotemporal environmental health assessment. *Environmental Monitoring and Assessment 185*, 815–831.

Kong, E., H. Tong, and Y. Xia (2010). Statistical modeling of nonlinear long-term cumulative effects. *Statistica Sinica 20*, 1097–1123.

Kossin, J., K. Emanuel, and G. Vecchi (2014). The poleward migration of the location of tropical cyclone maximum intensity. *Nature 509*, 349–352.

Kotsiantis, S., I. Zaharakis, and P. Pintelas (2006). Machine learning: A review of classification and combining techniques. *Artificial Intelligence Review 26*, 159–190.

Kouadio, L. and N. Newlands (2014). *Data hungry models in a food hungry world – an interdisciplinary challenge bridged by statistics*. Statistics in Action: A Canadian Outlook (Chapter 21)(Lawless, J.F., Ed.). London, UK: CRC Press (Taylor & Francis).

Kouadio, L. and N. Newlands (2015). Building capacity for spatial-based sustainability metrics in agriculture. *Decision Analytics 2*(2), 18.

Koutsourelakis, P. (2008). Design of complex systems in the presence of large uncertainties: a statistical approach. *Computer Methods in Applied Mechanics and Engineering 197*, 4092–4103.

Kriegler, E., J. Hall, H. Held, R. Dawson, and H. Schellnhuber (2009). Imprecise probability assessment of tipping points in the climate system. *Proceedings of the National Academy of Sciences (PNAS) 106*(13), 5041–5046.

Krokhmal, P., M. Zabarankin, and S. Uryasev (2011). Modeling and optimization of risk. *Surveys in Operations Research and Management Science 16*, 49–66.

Krupinsky, J., K. Bailey, M. McMullen, B. Gossen, and T. Turkington (2002). Managing plant disease risk in diversified cropping systems. *Agronomy Journal 94*, 198–209.

Krzyzanowski, M., J. Apte, S. Bonjour, M. Brauer, A. Cohen, and A. Prüss-Ustun (2014). Air pollution in the mega-cities. *Current Environmental Health Reports 1*, 185–191.

Kuehn, C. (2011). A mathematical framework for critical transitions: Bifurcations, fast–slow systems and stochastic dynamics. *Physica D: Nonlinear Phenomena 240*(12), 1020–1035.

Kuiper, H., G. Kleter, H. J. M. Noteborn, and E. Kok (2001). Assessment of the food safety issues related to genetically modified foods. *The Plant Journal 27*(6), 503–528.

Kurzweil, R. (2013). *How to Create a Mind: The Secret of Human Thought Revealed.* New York, NY, USA: Penguin Books.

Kwasniok, F. (2013). Predicting critical transitions in dynamical systems from time series using nonstationary probability density modeling. *Physical Review E 88*(052917), 10.

Kwuimy, C., M. Samadani, and C. Nataraj (2014). Bifurcation analysis of a nonlinear pendulum using recurrence and statistical methods: Applications to fault diagnostics. *Nonlinear Dynamics 76*, 1963–1975.

Labuschagne, C., A. Brent, and R. Van Erck (2005). Assessing the sustainability performances of industries. *Journal of Cleaner Production 13*(4), 373–385.

Lackey, R. (2006). Perspective, ecological policy: Axioms of ecological policy. *Fisheries 31*(6), 286–290.

Laird, D., R. Brown, J. Amonette, and J. Lehmann (2009). Review of the pyrolysis platform for coproducing bio-oil and biochar. *Biofuels, Bioproducts and Biorefining 3*(5), 547–562.

Laszlo, A. and S. Krippner (1998). *Systems Theories: Their Origins, Foundations, and Development.* Advances in Psychology, Volume 126, Systems Theories and A Priori Aspects of Perception (Chapter 3) (Jordan, J.S., Ed.). Amsterdam, The Netherlands: North-Holland/Elsevier Science B.V.

Lathers, C. (2013). Endocrine disruptors: A new scientific role for clinical pharmacologists? Impact on human health, wildlife and the environment. *The Journal of Clinical Pharmacology 42*(1), 7–23.

Law, K. and R. Thomson (2014). Microplastics in the seas. *Science 345*(6193), 144–145.

Lawless, J. (2002). *Statistical models and methods for lifetime data* (2nd. ed.). Wiley Series in Probability and Statistics. London, UK: John Wiley & Sons Inc.

Leggett, M., N. Newlands, D. Greenshields, L. West, S. Inman, and M. Koivunen (2014). Maize yield response to a phosphorus-solubilizing microbial inoculant in field trials. *The Journal of Agricultural Science 153*(8), 1464–1478.

Lehmann, B. (2014). Economic geology of rare-earth elements in 2014: A global perspective. *European Geologist 37*, 1–4.

Lele, S. and B. Dennis (2009). Bayesian methods for hierarchical models: Are ecologists making a Faustian bargain? *Ecological Applications 19*(3), 581–584.

Lenton, T. (2011). Early warning of climate tipping points. *Nature Climate Change (Review) 1*, 201–209.

Lenton, T., H. Held, E. Kreigler, J. Hall, W. Lucht, S. Rahmstorf, and H. Schellnhuber (2008). Tipping elements in the Earth's climate system. *Proceedings of the National Academy of Sciences 105*, 1786–1793.

Lenton, T., V. Livina, V. Dakos, E. van Nes, and M. Scheffer (2012). Early warning of climate tipping points from critical slowing down: Comparing methods to improve robustness. *Philosphical Transactions: Mathematical, Physical and Engineering 370*(1962), 1185–1204.

Levin, S. (1998). Ecosystems and the biosphere as complex adaptive systems. *Ecosystems 1*, 431–436.

Levin, S. (2000). *Fragile Dominion: Complexity and the Commons*. Cambridge, MA, USA: Perseus Books Group (Perseus Publishing).

Levin, S. (2002). Complex adaptive systems: Exploring the known, the unknown and the unknowable. *Bulletin of the American Mathematical Society 40*(1), 9–19.

Levin, S. (2012). The trouble of discounting tomorrow. *Solutions (for a sustainable and desirable future) 3*(4), 1–5.

Levina, V. and T. Lenton (2007). A modified method for detecting incipient bifurcations in a dynamical system. *Geophysical Research Letters 34*, L03712.

Lewis, P., M. Platzner, and X. Yao (2012). An outlook for self-awareness in computing systems. *Awareness Magazine: Self-Awareness in Autonomic Systems, http: // www. awareness-mag. eu/ view. php? source= 004093-2012-03-19* (26 March), 3.

Li, G., H. Yang, L. Sun, P. Ji, and L. Feng (2009). The evolutionary complexity of complex adaptive supply networks: A simulation and case study. *International Journal of Production Economics 124*, 310–330.

Li, M. (2010). Fractal time series - A tutorial review. *Mathematical Problems in Engineering 2010*, 1–26.

Li, N. D. and J. V. Zidek (2006). *Statistical Analysis of Environmental Spatio-Temporal Processes*. Springer Series in Statistics. New York, NY, USA: Springer Science & Business Media Inc.

Likens, G., F. Bormann, and N. Johnson (1981). *Interactions between major biogeochemical cycles in terrestrial ecosystems*, Volume 17 of *Scientific Committee on Problems of the Environment (SCOPE): Some Perspectives of the Major Biogeochemical Cycles*. Chichester, NY, USA: John Wiley & Sons Ltd.

Lin, T., J.-y. Lin, S.-h. Cui, and S. Cameron (2009). Using a network framework to quantitatively select ecological indicators. *Ecological Indicators 9*, 1114–1120.

Ling, Q., X. Jin, Y. Wang, H. Li, and Z. Huang (2013). Lyapunov function construction for nonlinear stochastic dynamical systems. *Nonlinear Dynamics 72*, 853–864.

Linkov, I., T. Bridges, F. Creutzig, J. Decker, C. Fox-Lent, W. Kröger, J. Lambert, A. Levermann, B. Montreuil, J. Nathwani, R. Nyer, O. Renn, B. Scharte, A. Scheffler, M. Schreurs, and T. Thiel-Clemen (2014). Changing the resilience paradigm. *Nature Climate Change (Commentary) 4*, 407–409.

Lipsey, R., K. Carlaw, and C. Bekar (2005). *Economic Transformations: General Purpose Technologies and Long-run Economic Growth*. Oxford, UK: Oxford University Press.

Lissner, T., D. Reusser, J. Schewe, T. Lakes, and J. Kropp (2014). Climate impacts on human livelihoods: where uncertainty matters in projections of water availability. *Earth System Dynamics 5*, 355–373.

Littell, J., D. McKenzie, B. Kerns, S. Cushman, and C. Shaw (2011). Managing uncertainty in climate-driven ecological models to inform adaptation to climate change. *Ecosphere 360*(9), 1–19.

Lobell, D. (2013). The use of satellite data for crop yield gap analysis. *Field Crops Research 143*(1), 56–64.

Lobell, D., G. Bala, and P. Duffy (2006). Biogeophysical impacts of cropland management changes on climate. *Geophysical Research Letters 33*, L06708.

Lobell, D., K. G. Cassman, and C. Field (2009). Crop yield gaps: their importance, magnitude and causes. *Annual Review of Environment and Resources 343*, 179–204.

Long, N. (2000). *Mathematical models of resource and energy economics*. Mathematical Models in Economics (Chapter 8). Encyclopedia of Life Support Systems (EOLSS), http://www.eolss.net/.

Longhurst, A. and D. Pauly (1987). *Ecology of Tropical Oceans*. San Diego, CA, USA: Academic Press.

Longwell, H. (2002). The future of the oil and gas industry: Past approaches, new challenges. *World Energy 5*(3), 100–104.

Lontzek, T., Y. Cai, K. Judd, and T. Lenton (2015). Stochastic integrated assessment of climate tipping points indicates the need for strict climate policy. *Nature Climate Change 5*(May), 441–444.

Lòpez-Ruiz, R., H. L. Mancini, and X. Calbet (1995). A statistical measure of complexity. *Physical Letters A 209*, 321–326.

Lorenz, E. (1995). *The Essense of Chaos.* Seattle, WA, USA: University of Washington Press.

Lorenzoni, I., S. Nicholson-Cole, and L. Whitmarsh (2007). Barriers perceived to engaging with climate change among the UK public and their policy implications. *Global Environmental Change 17*, 445–459.

Loveland, T. and A. Belward (1997). The IGBP-DIS global 1 km land cover data set, DIScover: First results. *International Journal of Remote Sensing 18*, 3289–3295.

Ma, C. (2008). Recent developments on the construction of spatio-temporal covariance models. *Stochastic Environmental Research and Risk Assessment 22 (Suppl 1)*, S39–S47.

Macal, C. and M. North (2010). Tutorial on agent-based modelling and simulation. *Journal of Simulation 4*, 151–162.

Mace, G., B. Reyers, R. Alkemade, R. Biggs, F. Chapin III, S. Cornell, S. Diaz, S. Jennings, P. Leadley, P. Mumby, A. Purvis, R. Scholes, A. Seddon, M. Solan, W. Steffen, and G. Woodward (2014). Approaches to defining a planetary boundary for biodiversity. *Global Environmental Change 28*, 289–297.

Madden, M. (2009). On the classification performance of TAN and general Bayesian networks,. *Knowledge-Based Systems 22*, 489–495.

Mahalanobis, P. (1936). On the generalised distance in statistics. *Proceedings of the National Institute of Sciences of India 2*(1), 49–55.

Majda, A. (2012). Challenges in climate science and contemporary applied mathematics. *Communications on Pure and Applied Mathematics 65*(7), 920–948.

Mandelbrot, B. (1982). *The Fractal Geometry of Nature.* New York, NY, USA: W.H. Freeman.

Mann, R. (2010). *An Introduction to Particle Physics and the Standard Model.* CRC Press (Taylor & Francis).

Manson, S. (2001). Simplifying complexity: A review of complexity theory. *Geoforum 32*(3), 405–414.

Manson, S. (2003). Epistemological possibilities and imperitives of complexity research: A reply to reitsma. *Geoforum 34*, 17–20.

Mao, X. (1997). *Stochastic Differential Equations and Their Applications.* Chichester, UK: Horwood Publishing.

Marchal, V., R. Dellink, D. van Vuuren, C. Clapp, J. Château, E. Lanzi, Magné, and J. van Vliet (2011). *OECD Environmental Outlook to 2050: The Consequences of Inaction.* OECD Environment Directorate (ENV) and the PBL Netherlands Environmental Assessment Agency (PBL), `www.oecd.org/environment/outlookto2050`.

Mariethoz, G. and J. Caers (2014). *Multiple-point Geostatistics: Stochastic Modeling with Training Images*. Hoboken, NJ, USA: Wiley-Blackwell.

Markose, S. (2004). Novelty in complex adaptive systems (CAS) dynamics: A computational theory of actor innovation. *Physica A. 344*, 41–49.

Marquet, P., R. Quiñones, S. Abades, F. Labra, M. Tognelli, M. Arim, and M. Rivadeneira (2005). Scaling and power-laws in ecological systems. *The Journal of Experimental Biology 208*, 1749–1769.

Marshall, J. (2014). Springtime for the artificial leaf: Researchers make headway in turning photons into fuel. *Nature 510*, 22–24.

Martin, S., G. Deffuant, and J. Calabrese (2011). *Defining Resilience Mathematically: From Attractors to Viability*. Viability and Resilience of Complex Systems: Concepts, Methods and Case Studies from Ecology and Society (Deffuant, G. and Gilbert, N., Eds.). Berlin and Heidelberg: Springer Science & Business Media.

Martins, A., T. Mata, C. Costa, and S. Sikdar (2007). Framework for sustainability metrics. *Industrial Engineering and Chemical Research 46*, 2962–2973.

May, R. (1972). Will a large complex system be stable? *Nature 238*, 413 – 414.

May, R. (2001). *Stability and Complexity in Model Ecosystems*. Princeton, NJ, USA: Princeton University Press.

Mayer, A., R. Donovan, and C. Pawlowski (2014). Information and entropy theory for the sustainability of coupled human and natural systems. *Ecology and Society 19*(3), 11.

McAllister, T., K. Beauchemin, X. Hao, and S. McGinn (2011). Greenhouse gases in animal agriculture - Finding a balance between food and emissions. *Animal Feed Science and Technology*, 166–167.

McBratney, A., B. Whclan, T. Ancev, and J. Bouma (2005). Future directions of precision agriculture. *Precision Agriculture 6*, 7–23.

McBride, A., V. H. Dale, L. Baskarana, M. E. Downing, L. Eaton, R. Efroymson, C. G. Jr., K. L. Kline, H. I. Jager, P. J. Mulholland, E. S. Parish, P. E. Schweize, and J. M. Storey (2011). Indicators to support environmental sustainability of bioenergy systems. *Ecological Indicators 11*, 1277–1289.

McCann, R., B. Marcot, and R. Ellis (2006). Bayesian belief networks: Applications in ecology and natural resource management. *Canadian Journal of Forestry Research 36*, 3053–3062.

McCloskey, J., R. Lilieholm, and C. Cronan (2011). Using Bayesian belief networks to identify potential compatibilities and conflicts between development and landscape conservation. *Landscape and Urban Planning 101*, 190–203.

McCullagh, P. and J. Nelder (1989). *Generalized Linear Models* (2nd. ed.), Volume 37 of *Monographs on Statistics and Applied Probability (Volume 37)*. London, UK: Chapman and Hall/CRC.

McDonald-Madden, E., P. Baxter, R. Fuller, T. Martin, E. Game, J. Montambault, and H. Possingham (2010). Monitoring does not always count. *Trends in Ecology and Evolution 25*(10), 547–550.

McDonough, W. and M. Braungart (2013). *The UpCycle: Beyond Sustainability - Designing for Abundance* (1st. ed.). New York, NY, USA: North Point Press.

McKenney, D., J. Pedlar, P. Papadopol, and M. Hutchinson (2006). The development of 1901-2000 historical monthly climate models for Canada and the United States. *Agricultural and Forest Meteorology 138*(1-4), 69–81.

MEA (2005). *Millennium Ecosystem Assessment. Ecosystems and Human Well-being: Synthesis*. Washington, DC, USA: World Resources Institute (WRI),`http://www.millenniumassessment.org/en/Synthesis.html`.

Meadows, D. (2008). *Thinking in Systems: A Primer*. White River Junction, VT, USA: Chelsea Green Publishing.

Meadows, D., D. Meadows, and J. Randers (1992). *Beyond the Limits*. London, UK: Earthscan.

Meadows, D., D. Meadows, J. Randers, and W. Behrens (1972). *The Limits to Growth*. New York, NY, USA: Universe Books.

Meadows, D., J. Randers, and D. Meadows (2004). *Limits to Growth: the 30-year update*. White River Junction, VT, Chelsea Green Publishing.

Meadows, D. L., W. Behrens III, D. Meadows, R. Naill, J. Randers, and E. Zahn (1974). *Dynamics of Growth in a Finite World*. London, UK: Productivity Press, Inc. (CRC Press).

Meinke, H., S. Howden, P. Struik, R. Nelson, D. Rodriguez, and S. Chapman (2009). Adaptation science for agriculture and natural resource management - Urgency and theoretical basis. *Current Opinion in Environmental Sustainability 1*(1), 69–76.

Mekis, E. and W. Hogg (1998). Rehabilitation and analysis of Canadian daily precipitation time series. *Atmosphere-Ocean 37*(1), 53–85.

Mekonnen, M. and A. Hoekstra (2011). *National Water Footprint Accounts: The Green, Blue and Grey Water Footprint of Production and Consumption*, Volume 1 and 2 of *Value of Water Research Report*. Delft, The Netherlands: IHE Delft, `http://waterfootprint.org/media/downloads/Report50-NationalWaterFootprints-Vol1.pdf`.

Menetrez, M. (2012). An overview of algae biofuel production and potential environmental impact. *Environmental Science and Technology 46*, 7073–7085.

Mercer (2011). *Climate Change Scenarios - Implications for Strategic Asset Allocation.* International Finance Corporation (IFC) Advisory Services in Sustainable Business. Washington, DC, USA: IFC (in partnership with Italy, Luxembourg, the Netherlands, and Norway), and Carbon Trust, `http://www.mercer.com/insights/point/2014/climate-change-scenarios-implications-for-strategic-asset-allocation.html`.

Merrick, K. and K. Shafi (2013). A game theoretic framework for incentive-based models of intrinsic motivation in artificial systems. *Frontiers in Psychology 4*, 1.

Mesle, R. (1993). *Process Theology: A Basic Introduction.* St. Louis, MO, USA: Chalice Press.

Meyer, M., M. Distelkamp, G. Ahlert, and B. Meyer (2013). *Macroeconomic Modelling of the Global Economy-Energy-Environment Nexus: An Overview of Recent Advancements of the Dynamic Simulation Model GINFORS.* Osnabrück: Gesellschaft für Wirtschaftliche Strukturforschung mbH (gws Discussion Paper 2013/5), `http://www.gws-os.com/discussionpapers/gws-paper13-5.pdf`.

Meyer, P. and J. Ausubel (1999). Carrying capacity: a model with logistically varying limits. *Technological Forecasting and Social Change 61*(3), 209–214.

MGI (2015). *Debt And (Not much) Deleveraging.* McKinsey Global Institute (MGI), `http://www.mckinsey.com/insights/economic_studies/debt_and_not_much_deleveraging`.

Milewska, E. and W. Hogg (2001). Spatial representativeness of a long-term climate network in Canada. *Atmosphere-Ocean 39*(2), 145–161.

Mills, M. (2013). *The Cloud begins with Coal: Big Data, Big Networks, Big Infrastructure, and Big Power: An Overview of Electricity used by the Global Digital Ecosystem.* Digital Power Group, `http://www.cepi.org/node/16427`.

Minsky, M. (2006). *The Emotion Machine: Commonsense Thinking, Artificial Intelligence, and the Future of the Human Mind.* New York, NY, USA: Simon & Schuster.

Mnih, V., K. Kavukcuoglu, D. Silver, A. Graves, I. Antonoglou, D. Wierstra, and M. Riedmiller (2013). Playing Atari with Deep Reinforcement Learning. *dblp Computer Science Bibliography (arXiv:1312.5602)*, `http://dblp.uni-trier.de/rec/bib/journals/corr/MnihKSGAWR13`, 1–9.

Mnih, V., K. Kavukcuoglu, D. Silver, A. Rusu, J. Veness, M. Bellemare, A. Graves, M. Riedmiller, A. Fidjeland, G. Ostrovski, S. Petersen, C. Beattie, A. Sadik, I. Antonoglou, H. King, D. Kumaran, D. Wierstra, S. Legg, and D. Hassabis

(2015). Human-level control through deep reinforcement learning. *Nature 518*, 529–533.

Molden, D. (2007). *Water for Food, Water for Life: A Comprehensive Assessment of Water Management in Agriculture*. London, UK and Columbo, USA: Earthscan and International Water Management Institute, `http://www.iwmi.cgiar.org/assessment/Publications/books.htm`.

Moon, W. and J. Wettlaufer (2013). A stochastic perturbation theory for non-autonomous systems. *Journal of Mathematical Physics 54*(123303), 1–31.

Mooney, H., A. Duraiappah, and A. Largauderie (2013). Evolution of natural and social science interactions in global change research programs. *Proceedings of the National Academy of Sciences 110*(1), 3665–3672.

Moran, A. (2006). Levels of consciousness and self-awareness: A comparison and integration of various neurocognitive views. *Consciousness and Cognition 15*(2), 358–371.

Morgan, M. and H. Dowlatabadi (1996). Learning from integrated assessment of climate change. *Climate Change 34*(3/4), 337–368.

Morozov, A. and J.-C. Poggiale (2012). From spatially explicit ecological models to mean-field dynamics: The state of the art and perspectives. *Ecological Complexity 10*, 1–11.

Morrison, J., M. Morikawa, M. Murphy, and P. Schulte (2009). *Water Scarcity & Climate Change: Growing Risks for Businesses & Investors*. Oakland, CA and Boston, MA, USA: Pacific Institute, `http://www.ceres.org/resources/reports/water-scarcity-climate-change-risks-for-investors-2009`.

Morse, S., N. McNamara, M. Acholo, and B. Okwoli (2001). Sustainability indicators: the problem of integration. *Sustainable Development 9*, 1–15.

Moss, R., J. Edmonds, K. Hibbard, M. Manning, S. Rose, D. van Vuuren, T. Carter, S. Emori, M. Kainuma, T. Kram, G. Meehl, J. Mitchell, N. Nakicenovic, K. Riahi, S. Smith, R. Stouffer, A. Thomson, J. Weyant, and T. Wilbanks (2010). The next generation of scenarios for climate change research and assessment. *Nature 463*, 747–756.

Mowery, D., R. Nelson, and B. Martin (2010). Technology policy and global warming: Why new policy models are needed (or why putting new wine in old bottles won't work). *Research Policy 39*, 1011–1023.

Muller, C., L. Chapman, S. Grimmond, D. Young, and X. Cai (2013). Sensors and the city: A review of urban meteorological networks. *International Journal of Climatology 33*(7), 1585–1600.

Müller, C. and T. Clough (2013). Advances in understanding nitrogen flows and transformations: gaps and research pathways. *Journal of Agricultural Science 152*(S1), 34–44.

Murray, J. and D. King (2012). Oil's tipping point has passed. *Nature 481*, 433–435.

Nagel, E. (1979). *The Structure of Science: Problems in the Logic of Scientific Explanation.* Indianapolis, IN, USA: Hackett Publishing Company.

Nagel, N. (1996). *Learning through Real-World Problem Solving: The Power of Integrated Teaching.* USA: SAGE Publications, Inc.

Nakau, K. (2004). A model for evaluating extreme risks with stochastic sustainability criteria: a case study of soil remediation on landfill sites. *Journal of Risk Research 7*(7-8), 689–704.

NAP (2010). *New Research Directions for the National Geospatial-Intelligence Agency Workshop Report.* Washington, DC, USA: Steering Committee on New Research Directions for the National Geospatial-Intelligence Agency (NGA); Mapping Science Committee; National Research Council, `http://www.nap.edu/catalog/12964/`.

NAP (2013). *Future US Workforce for Geospatial Intelligence.* Washington, DC, USA: Committee on the Future U.S. Workforce for Geospatial Intelligence; Board on Earth Sciences and Resources; Board on Higher Education and Workforce; Division on Earth and Life Studies; National Research Council, `http://www.nap.edu/catalog/18265/future-us-workforce-for-geospatial-intelligence`.

NAP (2015). *Sea Change: 2015-2025 Decadal Survey of Ocean Sciences.* Washington, D.C., USA: Committee on Guidance for NSF on National Ocean Science Research Priorities: Decadal Survey of Ocean Sciences; Ocean Studies Board; Division on Earth and Life Studies; National Research Council, `http://www.nap.edu/catalog/21655/sea-change-2015-2025-decadal-survey-of-ocean-sciences`.

Nariai, N., Y. Tamada, S. Imoto, and S. Miyano (2005). Estimating gene regulatory networks and protein–protein interactions of *saccharomyces cerevisiae* from multiple genome-wide data. *Bioinformatics 21*(2), 206–212.

NASA (1988). *Earth System Science: A Closer View.* Washington, D.C., USA: NASA Advisory Council. Earth System Sciences Committee, United States. National Aeronautics and Space Administration.

Nash, J. (1951). Non-cooperative games. *Annals of Mathematics 54*, 286–295.

Nature (2013). Fuelling the future. *Nature (Video Debate) 502*, S60–S61.

Nature (2015a). Rethinking the brain. *Nature 519*(7544), 389.

Nature (2015b). Time for the social sciences. *Nature 517*(5), 1.

Neilsen, D., G. Duke, B. Taylor, J. Byrne, S. Kienzle, and T. Van der Gulik (2013). Development and verification of daily gridded climate surfaces in the Okanagan Basin of British Columbia. *Canadian Water Resources Journal 35*(2), 131–154.

Nelson, G., M. Rosegrant, A. Palazzo, I. Gray, C. Ingersoll, R. Robertson, S. Tokgoz, T. Zhu, T. B. Sulser, C. Ringler, S. Msangi, and L. You (2010). *Food Security, Farming, and Climate Change to 2050: Scenarios, Results, Policy Options.* Washington, DC, USA: International Food Policy Research Institute.

Nelson, G., H. Valin, R. Sands, P. Havlík, H. Ahammadd, D. Derynge, J. Elliott, S. Fujimori, T. Hasegawah, E. Heyhoe, P. Kyle, M. Von Lampe, H. Lotze-Campen, D. Mason d'Croza, H. van Meijl, D. van der Mensbrugghe, C. Müller, A. Popp, R. Robertson, S. Robinson, E. Schmid, C. Schmitz, A. Tabeau, and D. Willenbockel (2013). Climate change effects on agriculture: Economic responses to biophysical shocks. *Proceedings of the National Academy of Sciences (PNAS) 111*(9), 3274–3279.

Newlands, N. (2002). *Shoaling dynamics and abundance estimation: Atlantic bluefin tuna (Thunnus thynnus).* PhD., diss. (University of British Columbia).

Newlands, N. (2006). *Modeling agroecosystems as complex adaptive systems.* Proceedings of the Annual Conference of the Canadian Society for Engineering in Agricultural, Food and Biological systems (CSBE/SCGAB) (July 16–19, Edmonton, AB, Canada) (Paper No. 06-300). Edmonton, AB, Canada: The Canadian Society for Bioengineering (CSBE), http://www.csbe-scgab.ca/docs/meetings/2006/CSBE06300.pdf.

Newlands, N. (2007). *GHGFarm: a software tool to estimate and reduce net greenhouse gas emissions from farms in Canada.* Proceedings of the Third IASTED International Conference in Environmental Modelling and Simulation (EMS). Honolulu, HI, USA: ACTA Press (Anaheim, CA, USA), ACM Digital Library, http://dl.acm.org/citation.cfm?id=1659821.1659824.

Newlands, N. (2008). *The Promise of Biofuels: An Opportunity and a Challenge.* Better Farming, Better Air: A scientific analysis of farming practice and greenhouse gases in Canada (Janzen, H.H., Desjardins, R.L., Rochette, P., Boehm, M., Worth, D., Eds.). Ottawa, ON, Canada: Agriculture and Agri-Food Canada (AAFC) (Government of Canada).

Newlands, N. (2010). *Predicting energy crop yield using Bayesian networks.* Proceedings of the 5th IASTED International Conference in Computational Intelligence (CI 2010). Maui, HI, USA: International Association of Science and Technology for Development (IASTED), ACTA Press (Anaheim, CA, USA), https://www.actapress.com/Abstract.aspx?paperId=43006.

Newlands, N., A. Davidson, A. Howard, and H. Hill (2011). Validation and intercomparison of three methodologies for interpolating daily precipitation and temperature across Canada. *Environmetrics 22*(2), 205–223.

Newlands, N., G. Espino-Hernández, and R. Erickson (2012). Understanding crop response to climate variability with complex agroecosystem models. *International Journal of Ecology 2012*, 13.

Newlands, N. and M. Lutcavage (2001). *From individuals to local population densities: Movements of North Atlantic bluefin tuna (Thunnus thynnus) in the Gulf of Maine/Northwestern Atlantic.* Electronic Tagging and Tracking in Marine Fisheries (Sibert, J.R., Nielsen, J.L., Eds.). The Netherlands: Kluwer Academic Press.

Newlands, N., M. Lutcavage, and T. Pitcher (2004). Analysis of foraging movements of atlantic bluefin tuna (*thunnus thynnus*): Individuals switch between two modes of search behavior. *Population Ecology 46*, 39–53.

Newlands, N., M. Lutcavage, and T. Pitcher (2006). Atlantic bluefin tuna in the Gulf of Maine, I: Estimation of seasonal abundance accounting for movement, school and school-aggregation behaviour. *Environmental Biology of Fishes 77*, 177–195.

Newlands, N., M. Lutcavage, and T. Pitcher (2007). Atlantic bluefin tuna in the Gulf of Maine, II: Efficiency of alternative sampling designs in estimating seasonal abundance accounting for changes in tuna behaviour. *Environmental Biology of Fishes 80*(4), 405–420.

Newlands, N. and T. Porcelli (2008). Measurement of the size, shape and structure of Atlantic bluefin tuna schools in the open ocean. *Fisheries Research 91*(1), 42–55.

Newlands, N. and T. Porcelli (2015). *Downscaling of regional climate scenarios within agricultural areas in Canada with a multi-variate, multi-site model*, Volume 117 of *Interdisciplinary Topics in Applied Mathematics, Modeling and Computational Science (Cojocaru, M., Kotsireas, I. S., Makarov, R. N., Melnik, R., Shodiev, H., Eds.)*. Berlin and Heidelberg: Springer-Verlag.

Newlands, N. and D. Stephens (2015). *Increasing confidence in agricultural crop forecasts and climate adaptation decisions with causality analysis.* Technical Report No. 275. Vancouver, BC, Canada: University of British Columbia (UBC), https://www.stat.ubc.ca/Research/TechReports/tr/275.pdf.

Newlands, N. and L. Townley-Smith (2012). Biodiesel from oilseeds in the Canadian Prairies and supply-chain models for exploring production cost scenarios - A review. *International Scholarly Research Network (ISRN Agronomy)*, *http://www.hindawi.com/journals/isrn/2012/980621/ 2012*(980621), 11.

Newlands, N., L. Townley-Smith, and T. Porcelli (2012). *A renewable source of jetfuel from alternative oilseeds? Predicting crop response under environmental uncertainty*, Volume 2012 of *Proceedings of the 2012 Modeling and Simulation, International Association of Science and Technology for Development (IASTED)*. Banff, AB, Canada: Acta Press (Paper No. 783-063).

Newlands, N., D. Zamar, O. Clark, Y. Zhang, and B. McConkey (2013). An integrated assessment model for exploring potential impacts of global change scenarios on the Canadian agricultural system. *46th Hawaii International Conference on System Sciences (HICSS-46) (Maui, HI, USA, 7–10 January) 13385069*, 915–924.

Newlands, N., D. Zamar, L. Kouadio, Y. Zhang, A. Chipanshi, A. Potgieter, S. Toure, and H. S. Hill (2014). An integrated, probabilistic model for improved seasonal forecasting of agricultural crop yield under environmental uncertainty. *Frontiers in Environmental Science (Interdisciplinary Climate Studies), http: // dx. doi. org/ 10. 3389/ fenvs. 2014. 00017 2*(17), 1–21.

Newlands, N. K. and D. Zamar (2012). *In-season probabilistic crop yield forecasting - integrating agro-climate, remote-sensing and crop phenology data.* Proceedings of the Joint Statistical Meetings (JSM), Statistics - Growing to Serve a Data Dependent Society, Section on Statistics and the Environment (San Diego, CA, USA). Alexandria, VA, USA: (CD-ROM Digital/Online Library) e-Paper, American Statistical Association (AMS) , 15 pages. (July 30, Abstract 304705), http://www.amstat.org/meetings/jsm/2012/onlineprogram/AbstractDetails.cfm?abstractid=304705.

Newman, M. (2005). Power laws, Pareto distributions and Zipf's law. *Contemporary Physics 46*, 323–351.

Newman, M. (2011). Complex Systems: A Survey. *American Journal of Physics 79*, 800–810.

Ng, T., J. Eheart, X. Cai, and J. Braden (2011). An agent-based model of farmer decision-making and water quality impacts at the watershed scale under markets for carbon allowances and a second-generation biofuel crop. *Water Resources Research 47*(W09519), 17.

Niccolucci, V., E. Tiezzi, F. Pulselli, and C. Capinerib (2012). Biocapacity vs Ecological Footprint of world regions: A geopolitical interpretation. *Ecological Indicators 16*, 23–30.

Niemeijer, D. and R. de Groot (2008). Framing environmental indicators: Moving from causal chains to causal networks. *Environment, Development and Sustainability 10*, 89–106.

Ninyerloa, M., X. Pons, and J. Roure (2000). A methodological approach of climatological modelling of air temperature and precipitation through GIS techniques. *International Journal of Climatology 20*(14), 1823–1841.

Nisan, N., T. Roughgarden, E. Tardos, and W. Vazirani (2007). *Algorithmic Game Theory.* Cambridge, UK: Cambridge University Press.

Norberg, J. and G. Cumming (2008). *Complexity Theory for a Sustainable Future.* New York, NY, USA: Columbia University Press.

Nordhaus, W. (1973). World Dynamics: Measurement Without Data. *Economic Journal 83*(332), 1156–1183.

Norstram, J. (1996). The use of Precautionary loss functions in Risk Analysis. *IEEE Transactions on Reliability 45*(3), 400–403.

North, M. (2014). A theoretical formalism for analyzing agent-based models. *Complex Adaptive Systems Modeling 2*(3), 34.

Northrop, R. (2011). *Introduction to Complexity and Complex Systems.* Boca Raton, FL, USA: CRC Press (Taylor & Francis Group).

Noyes, P., M. McElwee, H. Miller, B. Clark, L. Van Tiem, K. Walcott, K. Erwin, and E. Levin (2009). The toxicology of climate change: Environmental contaminants in a warming world. *Environment International 35*, 971–986.

NRC (1997a). *Preparing for the 21st Century: The Education Imperative.* Washington, DC, USA: National Academies Press, http://www.nap.edu/read/9537/chapter/1.

NRC (1997b). *Preparing for the 21st Century: The Environment and the Human Future.* Washington, DC, USA: National Academies Press, http://www.nap.edu/read/9536/chapter/1.

NRC (2010). *New Research Directions for the National Geospatial-Intelligence Agency.* Steering Committee on New Research Directions for the National Geospatial-Intelligence Agency; Mapping Science Committee. National Academies Press, http://www.nap.edu/read/12964/chapter/1.

Nuzzo, R. (2014). Scientific method: Statistical errors. *Nature (News Feature) 506*(7487), 150–152.

Nychka, D., C. Wilke, and J. Royle (2002). Multiresolution models for nonstationary spatial covariance functions. *Statistical Modelling 2*, 315–331.

Oakley, J. and A. O-Hagan (2004). Probabilistic sensitivity analysis of complex models: A Bayesian approach. *Journal of the Royal Society Series B (Statistical Methodology) 66*(3), 751–769.

O'Conner, F., O. Boucher, N. Gedney, C. Jones, G. Folberth, R. Coppell, P. Friedlingstein, W. Collins, J. Chappellaz, J. Ridley, and C. Johnson (2010). Possible role of wetlands, permafrost, and methane hydrates in the methane cycle under future climate change: A review. *Reviews of Geophysics 48*, RG4005.

OECD (2009). *Sustainable Manufacturing and Eco-Innovation: Framework, Practices and Measurement (Synthesis Report).* OECD, http://www.oecd.org/innovation/inno/43423689.pdf.

OECD (2010a). *OECD Information Technology Outlook.* OECD, http://www.oecd.org/sti/ieconomy/information-technology-outlook-19991444.htm.

OECD (2010b). *The Scope of Fossil-Fuel Subsidies in 2009 and a Roadmap for Phasing Out Fossil-Fuel Subsidies: An IEA, OECD and World Bank Joint Report, Prepared for the G-20 Summit, Seoul (Republic of Korea 11-12 November)*. International Energy Agency (IEA), OECD and The World Bank,`http://www.oecd.org/env/cc/46575783.pdf`.

OECD (2013). *Putting Green Growth at the Heart of Development.* OECD Green Growth Studies. OECD, `http://www.oecd.org/dac/environment-development/Putting%20Green%20Growth%20at%20the%20Heart%20of%20Development_Summary%20For%20Policymakers.pdf`.

OECD/IEA (2013). *World Energy Outlook 2013.* IEA Publishing, `http://www.worldenergyoutlook.org/weo2013/`.

OECD/IEA (2015). *World Energy Outlook Special Report 2015: Energy and Climate Change.* IEA Publishing, `http://www.iea.org/publications/freepublications/publication/weo-2015-special-report-energy-climate-change.html`.

Oerter, R. (2006). *The Theory of Almost Everything: The Standard Model, the Unsung Triumph of Modern Physics*. New York, NY, USA: Penguin Group.

Oki, T., D. Entekhabi, and T. Harrold (2004). *The Global Water Cycle*, Volume Geophysical Monograph 150, pp. 225–237. International Union of Geodesy and Geophysics (IUGG) and the American Geophysical Union (AGU).

Oksendal, B. (2010). *Stochastic Differential Equations: An Introduction with Applications* (6th ed.). Universitext. Berlin and Heidelberg: Springer.

Oliveira, M., C. Bastos-Filho, and R. Menezes (2015). Using network science to assess particle swarm optimizers. *Social Network Analysis and Mining 5*, 3–13.

O'Malley, A. and A. Zaslavsky (2008). Domain-level covariance analysis for survey data with structured nonresponse. *Journal of the American Statistical Association 103*(484), 1405–1418.

Omohundro, S. (2012). *Rational artificial intelligence for the greater good.* The Singularity Hypothesis: A Scientific and Philosophical Assessment (Eden, A.H., Moor, J.H., Soraker, J., Steinhart, E., Eds.). Berlin: Springer-Verlag.

Oremland, M. (2011). *Optimization and Optimal Control of Agent-Based Models.* MSc., diss. (Virginia Polytechnic Institute and State University).

Oremland, M. and R. Laubenbacher (2014). Optimization of agent-based models: scaling methods and heuristic algorithms. *Journal of Artificial Societies and Social Simulation 17*(2), 6.

Ortiz, M. and M. Wolff (2002). Dynamical simulation of mass-balace trophic models for benthic communities of north-central Chile: assessment of resilience time under alternative management scenarios. *Ecological Modelling 148*, 277–291.

Osborne, B., M. Saunders, D. Walmsley, M. Jones, and P. Smith (2010). Key questions and uncertainties associated with the assessment of the cropland greenhouse gas balance. *Agriculture, Ecosystems and Environment 139*, 293–301.

O'Sullivan, D. (2004). Complexity science and human geography. *Transactions of the Institute of British Geographers 29*(3), 282–295.

Otto, S. (2011). *Fool Me Twice: Fighting the Assault on Science.* New York, NY, USA: Rodale Books.

Paciorek, C. and M. Schervish (2006). Spatial modelling using a new class of nonstationary covariance functions. *Environmetrics 17*, 483–506.

Paegelow, M. and M. Olmedo (2008). *Modelling Environmental Dynamics: Advances in Geomatic Solutions.* Environmental Science and Engineering, Subseries: Environmental Science (Allan, R., Förstner, U., Salomons, W., Eds.). Berlin and Heidelberg: Springer-Verlag.

Pahl-Wostl, C., C. Vörösmarty, A. Bhaduri, J. Bogardi, J. Rockström, and J. Alcamo (2013). Towards a sustainable water future: Shaping the next decade of global water research. *Current Opinions in Environmental Sustainability 5*, 708–714.

Panigrahi, B., Y. Shi, and M.-H. Lim (2011). *Handbook of Swarm Intelligence: Concepts, Principles and Applications*, Volume 8 of *Adaptation, Learning and Optimization (Lim, M.-H., Ong, Y.-S., Eds.).* Berlin: Springer-Verlag.

Paperin, G., D. Green, and S. Sadedin (2011). Dual-phase evolution in complex adaptive systems. *Journal of the Royal Society Interface 8*, 609–629.

Paracchini, M., C. Pacini, M. Jones, and M. Pérez-Soba (2011). An aggregation framework to link indicators associated with multifunctional land use to the stakeholder evaluation of policy options. *Ecological Indicators 11*, 71–80.

Parrish, J. and W. Hamner (1997). *Animal Groups in Three Dimensions.* Cambridge, UK: Cambridge University Press.

Parry, M., C. Rosenzweig, A. Iglesias, M. Livermore, and G. Fischer (2004). Effects of climate change on global food production under SRES emissions and socioeconomic scenarios. *Global Environmental Change 14*, 53–67.

Parson, E. and K. Fisher-Vanden (1997). Integrated assessment models of global climate change. *Annual Reviews in Energy and the Environment, 22*, 589–662.

Patz, J., D. Campbell-Lendrum, T. Holloway, and J. Foley (2005). Reviews: Impact of regional climate change on human health. *Nature 438*, 310–317.

Patz, J., H. Gibbs, J. Foley, J. Rogers, and K. Smith (2007). Climate change and global health: Quantifying a growing ethical crisis. *EcoHealth 4*, 397–405.

Pauly, D. (1981). The relationship between gill surface area and growth performance in fish: A generalization of von Bertalanffy's theory of growth. *Reports
on Marine Research (Berichte der Deutschen Wissenchaftlichen Kommission für
Meeresforschung) 28*(4), 251–282.

Pauly, D. and R. Watson (2005). Background and interpretation of the 'Marine
Trophic Index' as a measure of biodiversity. *Philosophical Transactions of the
Royal Society B: Biological Sciences 360*, 415–423.

Pauly, P., V. Christensen, S. Guénette, T. Pitcher, U. Sumaila, C. Walters, R. Watson, and D. Zeller (2002). Towards sustainability in world fisheries. *Nature 418*,
689–695.

Pavlou, M., G. Ambler, S. Seaman, O. Guttmann, P. Elliott, M. King, and R. Omar
(2015). How to develop a more accurate risk prediction model when there are few
events. *theBMJ (Research Methods and Reporting) 351*, 3868.

Peñuelas, J., J. Sardans, A. Rivas-Ubach, and I. Janssens (2012). The human-
induced imbalance between C, N and P in Earth's life system. *Global Change
Biology 18*, 3–6.

Pearl, J. (1982). The solution for the branching factor of the alpha–beta pruning
algorithm and its optimality. *Communications of the ACM 25*(8), 559–564.

Pearl, J. (2004). *Graphical models for probabilistic and causal reasoning* (2nd. ed.).
Computer Science Handbook (Tucker, A.B., Ed.). London, UK: CRC Press.

Pearl, J. (2009). Causal inference in statistics: An overview. *Statistics Surveys 3*,
96–146.

Peavoy, D., C. Franzke, and G. Roberts (2015). Systematic physics constrained
parameter estimation of stochastic differential equations. *Computational Statistics
and Data Analysis 83*, 182–199.

Peden, E., M. Boehm, D. Mulder, R. Davis, W. Old, P. King, M. Ghirardi, and
A. Dubini (2014). Identification of global ferredoxin interaction networks in
Chlamydomonas reinhardtii. *The Journal of Biological Chemistry 288*, 35192–
35209.

Peel, M., B. Finlayson, and T. Mcmahon (2007). Updated world map of the Köppen-
Geiger climate classification. *Hydrology and Earth System Sciences 4*(2), 439–473.

Peng, C. K., S. Havlin, H. E. Stanley, and A. L. Goldberger (1995). Quantification
of scaling exponents and crossover phenomena in nonstationary heartbeat time
series. *Chaos 5*, 82–87.

Perera, O. (2012). *Basel III: To what extent will it promote sustainable development?* Winnipeg, Manitoba, Canada: International Institute for Sustainable
Development (IISD), http://www.iisd.org/pdf/2012/basel13.pdf.

Pernkopf, F. (2005). Bayesian network classifiers versus selective k-NN classifier. *Pattern Recognition 38*, 1–10.

Perold, A. (2004). The Capital Asset Pricing Model. *Journal of Economic Perspectives 18*(3), 3–24.

Peters, D. (2010). Accessible ecology: synthesis of the long, deep, and broad. *Trends in Ecology and Evolution 25*(10), 592–601.

Peters, G., S. Davis, and R. Andrew (2012). A synthesis of carbon in international trade. *Biogeosciences 9*, 3247–3276.

PEW (2014). *AI, Robotics, and the Future of Jobs.* Pew Research Center, (August 6) (Smith, A. and Anderson, J.) , `http://www.pewinternet.org/2014/08/06/future-of-jobs/`.

Phillips, A., N. Newlands, S. Liang, and B. Ellert (2014). Integrated sensing of soil moisture at the field-scale: Sampling, modelling and sharing for improved agricultural decision-support. *Computers and Electronics in Agriculture 107*, 73–88.

Picheny, V. (2014). Multiobjective optimization using Gaussian process emulators via stepwise uncertainty reduction. *Statistics and Computing 25*(6), 1265–1280.

Pidgeon, N. and B. Fischhoff (2011). The role of social and decision sciences in communicating uncertain risks. *Nature Climate Change 1*, 35–40.

Pinter, L., P. Hardi, and P. Bartelmus (2005). *Sustainable Development Indicators: Proposals for a Way Forward.* Discussion Paper Prepared under a Consulting Agreement on behalf of the UN Division for Sustainable Development (UNSD). Winnipeg, Canada: International Institute for Sustainable Development (IISD), `https://www.iisd.org/pdf/2005/measure_indicators_sd_way_forward.pdf`.

Pionteka, F., C. Müller, T. Pugh, D. Clark, D. Deryng, J. Elliott, F. de Jesus Colón González, M. Flörkeg, C. Folberth, K. Frielera, A. Friend, S. Gosling, D. Hemming, N. Khaborov, H. Kim, M. Lomas, Y. Masaki, M. Mengel, A. Morse, K. Neumann, K. Nishina, S. Ostberg, R. Pavlick, A. Ruane, J. Schewe, E. Schmid, T. Stacke, Q. Tang, Z. Tessler, A. Tompkins, L. Warszawskia, D. Wisser, and H. Schellnhuber (2014). Multisectoral climate impact hotspots in a warming world. *Proceedings of the National Academy of Sciences 111*(9), 3233–3238.

Piou, C., U. Berger, and V. Grimm (2009). Proposing an information criterion for individual-based models developed in a pattern-oriented modelling framework. *Ecological Modelling 220*(17), 1957–1967.

Pongratz, J., C. Reick, T. Raddatz, and M. Claussen (2008). A reconstruction of global agricultural areas and land cover for the last millennium. *Global Biogeochemical Cycles 22*(3), GB3018.

Porcelli, T. (2008). *Structural, trend and extreme-value analysis of daily temperature and precipitation within southern British Columbia.* British Columbia Agriculture Research and Development Corporation (ARDCorp) (Unpublished, Technical Report).

Porcu, E., P. Gregori, and J. Mateu (2006). Nonseparable stationary anisotropic space-time covariance functions. *Stochastic Environmental Research and Risk Assessment 21*, 113–122.

Porporato, A., P. D'Odorico, F. Laio, and I. Rodriquez-Iturbe (2003). Hydrological controls on soil carbon and nitrogen cycles, I: Modeling scheme. *Advances in Water Resources 26*, 45–58.

Porritt, J. (2005). *Capitalism as if the World Matters.* London, UK: Earthscan.

Power, A. (2010). Ecosystem services and agriculture: tradeoffs and synergies. *Philosophical Transactions of the Royal Society B 365*, 2959–2971.

Powers, D. (2011). Evaluation: From Precision, Recall and F-Measure to ROC, Informedness, Markedness & Correlation. *Journal of Machine Learning Technologies 2*(1), 37–63.

Powers, D. and A. Atyabi (2012). *The problem of cross-validation: averaging and bias, repetition and significance.* Proceedings of the Spring World Congress on Engineering and Technology (SCET) (May 27-30, Xi'an, China). Xi'an, China: Institute of Electrical and Electronics Engineers (IEEE).

Prather, M. J., C. D. Holmes, and J. Hsu (2012). Reactive greenhouse gas scenarios: Systematic exploration of uncertainties and the role of atmospheric chemistry. *Geophysical Research Letters 39*, L09803.

Prescott, T., J. Bryson, and A. Seth (2007). Introduction. Modelling natural action selection. *Philosophical Transactions of the Royal Society B (Biological Science) 362*(1485), 1521–1529.

Puetz, S., A. Prokoph, G. Borchardt, and E. Mason (2014). Evidence of synchronous, decadal to billion year cycles in geological, genetic, and astronomical events. *Chaos, Solitons and Fractals 62-63*, 55–75.

Pukdeboon, C. (2011). A review of fundamental Lyapunov theory. *The Journal of Applied Science 10*(2), 55–61.

Raatikainen, P. (2015). *Gödel's Incompleteness Theorems.* The Stanford Encyclopedia of Philosophy (Zalta, E.N., Ed.). Stanford, CA, USA: Stanford University (The Metaphysics Research Lab), Center for the Study of Language and Information (CSLI), http://plato.stanford.edu/.

Rabitz, H., d. Alǐ, J. Shorter, and K. Shim (1999). Efficient input-output model representations. *Computer Physics Communications 117*, 11–20.

Raftery, A., N. Li, H. Sevcikova, P. Gerland, and G. Heilig (2012). Bayesian probabilistic population projections for all countries. *Proceedings of the National Academy of Sciences (PNAS) 109*(35), 13915–13921.

Rakotoarisoa, M. (2011). The impact of agricultural policy distortions on the productivity gap: Evidence from rice production. *Food Policy 36*, 147–157.

Ramakutty, N. and J. Foley (1998). Characterizing patterns of global land use: An analysis of global croplands area. *Global Biogeochemical Cycles 12*, 667–685.

Rammel, C., S. Stagl, and H. Wilfing (2007). Managing complex adaptive systems - A co-evolutionary perspective on natural resource management. *Ecological Economics 63*, 9–23.

Randolph, J., K. Falbe, A. Manuel, and J. Balloun (2014). A step-by-step guide to propensity score matching in R. *Practical Assessment, Research & Evaluation 19*(18), 1–6.

Ratti, R. and J. Vespignani (2015). OPEC and non-OPEC oil production and the global economy. *Energy Economics 50*, 364–378.

Ratzé, C., F. Gillet, J.-P. Müller, and K. Stoffel (2007). Simulation modelling of ecological hierarchies in constructive dynamical systems. *Ecological Complexity 4*, 13–25.

Ray, D., N. Muelleer, P. West, and J. Foley (2013). Yield trends are insufficient to double global crop production by 2050. *PLoS ONE 8*(6), e66428.

REAP (2008). *Analyzing Biofuel Options: Greenhouse Gas Mitigation Efficiency and Costs*. Ste. Anne de Bellevue, Quebec, Canada: Resource Efficient Agricultural Production (R.E.A.P.) - Canada, Brief to the House of Commons, Ottawa, Canada, 39th Parliament (2nd. Session) Standing Committee on Agriculture and Agri-Food, Study: Bill C-33, Act to amend the Canadian Environmental Protection Act, 1999.

Reggia, J. (2014). *Conscious Machines: The AI Perspective*. 2014 Association for the Advancement of Artificial Intelligence (AAAI) Symposium on The Nature of Humans and Machines. Québec, Canada: AAAI.

Rehman, A., A. Abbassi, N. Islam, and Z. Shaikh (2014). A review of wireless sensors and networks' applications in agriculture. *Computer Standards & Interfaces 36*(2), 263–270.

Reitsma, F. (2003). A response to simplifying complexity. *Geoforum 34*(1), 13–16.

Resch, B., M. Mittleboeck, S. Lipson, M. Welsh, J. Bers, R. Britter, C. Ratti, and T. Blaschke (2015). Integrated urban sensing: A geo-sensor network for public health monitoring and beyond. *International Journal of Geographical Information Science, http://dspace.mit.edu/handle/1721.1/64636 268*, 1–21.

Reusch, T. (2013). Climate change in the oceans: Evolutionary versus phenotypically plastic responses of marine animals and plants. *Evolutionary Applications (Special issue in Climate change, adaptation and phenotypic plasticity 7*(1), 104–122.

Richards, G. (2004). A fractal forecasting model for financial time series. *Journal of Forecasting 23*, 587–602.

Richardson, A. and R. Simpson (2011). Soil microorganisms mediating phosphorus availability update on microbial phosphorus. *Plant Physiology 156*, 989–996.

Roca, L. and C. Searcy (2012). An analysis of indicators disclosed in corporate sustainability reports. *Journal of Cleaner Production 20*, 103–118.

Rockström, J. (2003). Water for food and nature in drought–prone tropics: Vapour shift in rain–fed agriculture. *Philosophical Transactions of the Royal Society B 358*(1440), 1997–2009.

Rockström, J., M. Lannerstad, and M. Falkenmark (2007). Assessing the water challenge of a new green revolution in developing countries. *Proceedings of the National Academy of Sciences 104*(15), 6253–6260.

Rockström, J., W. Steffen, K. Noone, A. Persson, F. Chapin, E. Lambin, T. Lenton, M. Scheffer, C. Folke, H. Schellnhuber, B. Nykvist, C. de Wit, T. Hughes, S. van der Leeuw, H. Rodhe, S. Sorlin, P. Snyder, R. Costanza, U. Svedin, M. Falkenmark, L. Karlberg, R. Corell, V. Fabry, J. Hansen, B. Walker, D. Liverman, K. Richardson, P. Crutzen, and J. Foley (2009a). Planetary boundaries: Exploring the safe operating space for humanity. *Ecology and Society 14*(2), 32.

Rockström, J., W. Steffen, K. Noone, A. Persson, F. Chapin, E. Lambin, T. Lenton, M. Scheffer, C. Folke, H. Schellnhuber, B. Nykvist, C. de Wit, T. Hughes, S. van der Leeuw, H. Rodhe, S. Sorlin, P. Snyder, R. Costanza, U. Svedin, M. Falkenmark, L. Karlberg, R. Corell, V. Fabry, J. Hansen, B. Walker, D. Liverman, K. Richardson, P. Crutzen, and J. Foley (2009b). A safe operating space for humanity. *Nature 461*(282), 472–475.

Romps, D., J. Seeley, D. Vollaro, and J. Molinari (2014). Projected increase in lightning strikes in the United States due to global warming. *Science 346*(6211), 851–854.

Rosegrant, M. (2012). *International Model for Policy Analysis of Agricultural Commodities and Trade (IMPACT): Model Description.* Washington, DC, USA: International Food Policy Research Institute (IFPRI), `http://ebrary.ifpri.org/cdm/ref/collection/p15738coll2/id/12735`.

Rosegrant, M., K. Wiebe, S. Robinson, D. Mason-D'Croz, S. Islam, and N. Perez (2014). *IMPACT Model, Baseline, and Scenarios: New Developments.* Proceedings of the Global Futures and Strategic Foresight Conference (Washington, DC, USA), `http://globalfutures.cgiar.org/`. Washington, D.C., USA: International Food Policy Research Institute (IFPRI).

Rosenbaum, P. (2010). *Design of Observational Studies*. New York, NY, USA: Springer Science+Business Media, LLC.

Rosenbaum, P. and D. Rubin (1983). The central role of the propensity score in observational studies for causal effects. *Biometrika 70*, 41–55.

Rosenthal, S., C. Twomey, A. Hartnett, H. Wu, and I. Couzin (2015). Revealing the hidden networks of interaction in mobile animal groups allows prediction of complex behavioral contagion. *Proceedings of the National Academy of Sciences 112*(15), 4690–4695.

Rosenzweig, C., J. Elliott, D. Deryng, A. Ruane, C. Müller, A. Arneth, K. Boote, C. Folberth, M. Glotter, N. Khaborov, K. Neumann, F. Piontek, T. Pugh, E. Schmid, E. Stehfest, H. Yang, and J. Jones (2014). Assessing agricultural risks of climate change in the 21st Century in a global gridded crop model inter-comparison. *Proceedings of the National Academy of Sciences 111*(9), 3268–3273.

Rosillo-Calle, F. (2012). Food versus fuel: Toward a new paradigm—the need for a holistic approach. *ISRN Renewable Energy 2012*, 15.

Rossi, G. (2007). Measurability. *Measurement 40*, 545–562.

Rost, S., D. Gerten, A. Bondeau, W. Lucht, J. Rohwer, and S. Schaphoff (2008). Agricultural green and blue water consumption and its influence on the global water system. *Water Resources Research 44*(W09405), 12.

Rothman, D., G. Fournier, K. French, E. Almc, E. Boyle, C. Caod, and R. Summons (2014). Methanogenic burst in the end-Permian carbon cycle. *Proceedings of the National Academy of Sciences (PNAS) 111*(15), 5462–5467.

Roy, S., S. Negrão, and M. Tester (2014). Salt resistant crop plants. *Current Opinion in Biotechnology 26*, 115–124.

Rubel, F. and M. Kottek (2010). Observed and projected climate shifts 1901-2100 depicted by world maps of the köppen-geiger climate classification. *Meteorol. Zeitschrift 19*, 135–141.

Rubin, J. (2012). *The End of Growth*. Toronto, ON, Canada: Random House.

Rue, H., S. Martino, and N. Chopin (2009). Approximate Bayesian inference for latent Gaussian models by using integrated nested Laplace approximations. *Journal of the Royal Statistical Society 71*(2), 319–392.

Running, S., D. Baldocchi, D. Turner, S. Gower, P. Bakwin, and K. Hibbard (1999). A global terrestrial monitoring network integrating tower fluxes, flask sampling, ecosystem modeling and eos satellite data. *Remote Sensing of Environment 70*, 108–127.

Russell, S. and P. Norvig (2010). *Artificial Intelligence - A Modern Approach* (3rd. ed.). NJ, USA: Pearson Education, Inc.

Russo, F. (2010). Are causal analysis and systems analysis compatible approaches? *International Studies in Philosophy of Science 24*(1), 1–24.

Saghaian, S. (2010). The impact of the oil sector on commodity prices: Correlation or causation? *Journal of Agricultural and Applied Economics 42*(3), 477–485.

Sagl, G. and T. Blaschke (2014). *Integrated Urban Sensing in the Twenty-First Century*, Book section 14, pp. 440. Boca Raton, FL, USA: CRC Press (Taylor and Francis Group LLC).

Salamí, E., C. Barrado, and E. Pastor (2014). UAV flight experiments applied to the remote sensing of vegetated areas. *Remote Sensing 6*, 11051–11081.

Salcido-Guevara, L., F. Arreguin-Sánchez, L. Palmeri, and A. Barausse (2012). Metabolic scaling regularity in aquatic ecosystems. *CICIMAR Oceaáides 27*(2), 1–9.

Saleur, H., C. Sammis, and D. Sornette (2012). Discrete scale invariance: Complex fractal dimensions, and log-periodic fluctuations in seismicity. *Journal of Geophysical Research: Solid Earth 101*(B8), 1978–2012.

Saltelli, A. (2002). Making best use of model evaluations to compute sensitivity indices. *Computer Physics Communications 145*, 280–297.

Saltelli, A., S. Tarantola, and F. Campolongo (2000). Sensitivity analysis as an ingredient in modeling. *Statistical Science 15*(4), 377–395.

Sampson, P., D. Damian, and P. Guttorp (2001). *Advances in modeling and inference for environmental processes with nonstationary spatial covariance*. National Research Center for Statistics and the Environment (NRCSE), Technical Report Series (NRCSE-TRS 061).

Sandsmark, M. and H. Vennemo (2007). A portfolio approach to climate investments: CAPM and endogeneous risk. *Environmental Resource Economics 37*, 681–695.

Santoro, M., F. Hassan, M. Wahab, R. Cerveny, and R. Balling (2015). An aggregated climate teleconnection index linked to historical Egyptian famines of the last thousand years. *The Holocene 25*(5), 872–879.

Santos, F. and G. Pacheco (2011). Risk of collective failure provides an escape from the Tragedy of the Commons. *Proceedings of the National Academy of Sciences 108*(26), 10421–10425.

Savage, V., J. Gillooly, W. Woodruff, G. West, A. Allen, B. Enquist, and J. Brown (2004). The predominance of quarter-power scaling in biology. *Functional Ecology 18*, 257–282.

Sayama, H., I. Pestov, J. Schmidt, B. Bush, C. Wong, J. Yamanoi, and T. Gross (2013). Modeling complex systems with adaptive networks. *Computers and Mathematics with Applications 65*, 1645–1664.

Sayer, J., T. Sunderland, J. Ghazoul, J.-L. Pfund, D. Sheil, E. Meijaard, V. Ventera, A. Boedhihartono, M. Dayb, C. Garcia, C. van Oosten, and L. Buck (2013). Ten principles for a landscape approach to reconciling agriculture, conservation, and other competing land uses. *Proceedings of the National Academy of Sciences 110*(21), 8349–8356.

SCBD (2010). *Global Biodiversity Outlook 3*. Montreal, QC, Canada: Secretariat of the Convetion on Biological Diversity, `https://www.cbd.int/gbo3/`.

Schader, C., J. Grenz, M. Meier, and M. Stolze (2014). Scope and precision of sustainability assessment approaches to food systems. *Ecology and Society 19*(3), 42.

Scheffer, M., J. Bascompte, W. Brock, V. Brovkin, S. Carpenter, V. Dakos, H. Held, E. van Nes, M. Rietkerk, and G. Sugihara (2009). Early-warning signals for critical transitions. *Nature 461*, 53–59.

Scheffer, M., S. Carpenter, J. Foley, C. Folke, and B. Walkerk (2001). Catastrophic shifts in ecosystems. *Nature 413*, 591–596.

Scheffer, M., S. Carpenter, T. Lenton, J. Bascompte, W. Brock, V. Dakos, J. de Koppel, I. van de Leemput, S. Levin, E. van Nes, M. Pascual, and J. Vandermeer (2012). Anticipating critical transitions. *Science 338*(19 October), 344–348.

Schellnhuber, H., K. Frieler, and P. Kabat (2014). The elephant, the blind, and the intersectoral intercomparison of climate impacts. *Proceedings of the National Academy of Sciences 111*(9), 3225–3227.

Schiedek, D., B. Sundelin, J. Readman, and R. Macdonald (2007). Interactions between climate change and contaminants. *Marine Pollution Bulletin 54*, 1845–1856.

Schimel, J. (2004). Playing scales in the methane cycle: From microbial ecology to the globe. *Proceedings of the National Academy of Sciences (PNAS) 101*(34), 12400–12401.

Schlickenrieder, J., S. Quiroga, A. Diz, and A. Iglesias (2011). Impacts and adaptive capacity as drivers for prioritising agricultural adaptation to climate change in europe. *Economía Agraria y Recursos Naturales 11*(1), 59–82.

Schmidhuber, J. (2007). *Gödel Machines: Fully Self-referential Optimal Universal Self-improvers*. Artificial General Intelligence. Berlin: Springer.

Schmidhuber, J. (2014). Deep learning in neural networks: An overview. *Neural Networks 61*, 85–117.

Schneider, S. (1997). Integrated assessment modeling of global climate change: Transparent rational tool for policy making or opaque screen hiding value-laden assumptions? *Environmental Modeling and Assessment 2*, 229–249.

Schneider, S. (2008). Geoengineering: Could we or should we make it work? *Philosophical Transactions of the Royal Society A 366*, 3843–3862.

Schultz, A., R. Wieland, and G. Lutze (2000). Neural networks in agroecological modelling - stylish application or helpful tool? *Computers and Electronics in Agriculture 20*, 73–97.

Schumacher, E. (1999). *Small is Beautiful: Economics as If People Mattered, 25 Years Later...with Commentaries*. Vancouver, BC, Canada: Hartley & Marks Publishers Inc.

Schuur, E., A. McGuire, C. Schädel, G. Grosse, J. Harden, D. Hayes, G. Hugelius, C. Koven, P. Kuhry, D. Lawrence, S. Natali, D. Olefeldt, V. Romanovsky, K. Schaefer, M. Turetsky, C. Treat, and J. Vonk (2015). Climate change and the permafrost carbon feedback. *Nature 520*, 171–179.

Schwager, M., K. Johst, and F. Jeltsch (2006). Does red noise increase or decrease extinction risk? Single extreme events versus series of unfavourable conditions. *The American Naturalist 167*, 879–888.

Schwarz, J., B. Beloff, and E. Beaver (2002). Use sustainability metrics to guide decision-making. *Environmental Protection July*, 58–63.

Scrucca, L., A. Santucci, and F. Aversa (2010). Regression modeling of competing risk using R: an in depth guide for clinicians. *Bone Marrow Transplantation 45*, 1388–1395.

Seddon, A., M. Macias-Fauria, P. Long, D. Benz, and K. Willis (2016). Sensitivity of global terrestrial ecosystems to climate variability. *Nature (Letter)* (17 February), 1–15.

Selin, N. (2009). Global biogeochemical cycling of mercury: A review. *Annual Review of Environment and Resources 34*, 43–63.

Seth, A., T. Prescott, and J. Bryson (2011). *Optimised agent-based modelling of action selection*. Modelling Natural Action Selection. Cambridge, UK: Cambridge University Press.

Shabbar, A. and W. Skinner (2004). Summer drought patterns in Canada and the relationship to global sea surface temperature. *Journal of Climate 17*, 2866–2880.

Sharpe, W. (1964). Capital asset prices. A theory of market equilibrium under conditions of risk. *J. Finance 19*(3), 425–442.

Shaw, M. (2009). Preparing for changes in plant disease due to climate change. *Plant Protection Science 45*(Special Issue), 3–10.

Shea, K., R. Truckner, R. Weber, and D. Peden (2008). Climate change and allergic disease. *Clinical Reviews in Allergy and Immunology 122*, 443–453.

Shen, S. (2015). *Climate Mathematics: A Suite of Basic Tools of Modern Mathematics, Statistics, and R Programming for Climate Science.* San Diego, CA, USA: Scripps Institution of Oceanography, UCSD (Lecture notes of SIOC 290: Climate Mathematics).

Shen, S., H. Yin, K. Cannon, A. Howard, S. Chetner, and T. Karl (2005). Temporal and spatial changes in the agroclimate in Alberta, Canada from 1901 to 2002. *Journal of Applied Meteorology 44*, 1090–1105.

Sherman, M. (2011). *Spatial Statistics and Spatio-Temporal Data: Covariance Functions and Directional Properties.* Wiley Series in Probability and Statistics (Balding, D.J., Cressie, N.A.C., Fitzmaurice, G.M., Goldstein, H., Johnsone, I.M., Molenberghs, G., Scott, D.W., Smith, A.F.M., Tsay, R.S., Weisberg, S., Eds.). Hoboken, NJ, USA: John Wiley & Sons Ltd.

Shields, G. (1996). Introduction: On the interface of analytic and process philosophy. *Process Studies 25*, 34–54.

Sideris, L. and K. Moore (2008). *Rachel Carson: Legacy and Challenge.* SUNY series in Environmental Philosophy and Ethics. New York, NY, USA: State University of New York (SUNY) Press.

Sidle, R., W. Benson, J. Carriger, and T. Kamai (2013). Broader perspective on ecosystem sustainability: Consequences for decision making. *Proceedings of the National Academy of Sciences 110*(23), 9201–9208.

Siebert, S., V. Henrich, K. Frenken, and J. Burke (2013). *Update of the Digital Global Map of Irrigation Areas to Version 5.* Bonn: Food and Agricultural Organization of the United Nations (FAO), `http://www.fao.org/nr/water/aquastat/irrigationmap/gmia_v5_lowres.pdf`.

Siebert, S., F. Portmann, and P. Dʼoll (2010). Global patterns of cropland use intensity. *Remote Sensing 2*(7), 1625–1643.

Siegelmann, H. (1995). Computation beyond the Turing limit. *Science 268*, 545–548.

Siegelmann, H. (1999). *Neural Networks and Analog Computation Beyond the Turing Limit.* Progress in Theoretical Computer Science (Book, R.V.). Berlin and Heidelberg: Springer Science+Business Media, LLC.

Siegelmann, H. (2013). Turing on Super-Turing and adaptivity. *Progress in Biophysics and Molecular Biology 113*, 117–126.

Siegelmann, H. and E. Sontag (1991). Turing Computability with Neural Nets. *Applied Mathematics Letters 4*(6), 77–80.

Siirola, J., S. Hauan, and A. Westerberg (2004a). Computing Pareto fronts using distributed agents. *Computers and Chemical Engineering 29*, 113–126.

Siirola, J., S. Hauan, and A. Westerberg (2004b). Computing Pareto fronts using distributed agents. *Computers and Chemical Engineering 29*, 113–126.

Sikdar, S. (2003). Sustainable Development and Sustainability Metrics. *AIChE Journal 49*(8), 1928–1932.

Sillmann, J., V. Kharin, F. Zwiers, X. Zhang, and D. Bronaugh (2013a). Climate extremes indices in the cmip5 multimodel ensemble: Part 1. model evaluation in the present climate. *Journal of Geophysical Research: Atmospheres 118*, 1716–1733.

Sillmann, J., V. Kharin, F. Zwiers, X. Zhang, and D. Bronaugh (2013b). Climate extremes indices in the cmip5 multimodel ensemble: Part 2. future climate projections. *Journal of Geophysical Research: Atmospheres 118*, 2473–2493.

Silvertown, J., J. Tallowin, C. Stevens, S. Power, V. Morgan, B. Emmett, A. Hester, P. Grime, M. Morecroft, R. Buxton, P. Poulton, R. Jinks, and R. Bardgett (2010). Environmental myopia: A diagnosis and a remedy. *Trends in Ecology and Evolution 25*(10), 556–561.

Singh, R., H. Murty, S. Gupta, and A. Dikshit (2008). An overview of sustainability assessment methodologies. *Ecological Indicators 15*, 281–299.

Sinha, S. (2005). Complexity vs. stability in small-world networks. *Physica A. 346*, 147–153.

Sinha, S. and S. Sinha (2006). Robust emergent activity in dynamical networks. *Physics Review E: Statistical Nonlinear Soft Matter Physics 74*(6), 066117.

Skowronska, M. and T. Filipek (2013). Life cycle assessment of fertilizers: A review. *International Agrophysics 28*, 101–110.

Slingo, J. and T. Palmer (2011). Uncertainty in weather and climate prediction. *Philosophical Transactions of the Royal Society, A 369*, 4751–4767.

Slotine, J.-J. and W. Li (1991). *Fundamentals of Lyapunov Theory (Chapter 3)*. Applied Nonlinear Control. Englewood Cliffs, NJ, USA: Prentice Hall.

Smith, E., H. Janzen, and N. Newlands (2007). Energy balances of biodiesel production from soybean and canola oil under Canadian conditions. *Canadian Journal of Plant Science 87*(4), 793–801.

Smith, P., D. Hutchinson, J. Sterbenz, Schöller, A. Fessi, M. Karaliopoulos, C. Lac, and B. Plattner (2011). Network resilience: A systematic approach. *IEEE Communications Magazine 49*(7), 88–97.

Sobol, I. (1993). Sensitivity estimates for nonlinear mathematical models. *Mathematics and Computers in Simulation 1*, 407–414.

Sobol, I. (2001). Global sensitivity indices for nonlinear mathematical models and their Monte Carlo estimates. *Mathematics and Computers in Simulation 55*(1-3), 271–280.

Solé, R. and J. Bascompte (2006). *Self-organization in complex ecosystems*. Monographs in population biology. Princeton, N.J., USA: Princeton University Press.

Solovyev, A., M. Mikheev, L. Zhou, J. Dutta-Moscato, C. Ziraldo, G. An, Y. Vodovotz, and Q. Mi (2010). SPARK: A Framework for Multi-Scale Agent-Based Biomedical Modeling. *International Journal of Agent Technologies and Systems 2*(3), 18–30.

Soubeyrand, S., G. Thébaud, and J. Chadoeuf (2007). Accounting for biological variability and sampling scale: A multi-scale approach to building epidemic models. *Journal of the Royal Society Interface 4*, 985–997.

Spaling, H. (1995). Cumulative effects assessment. *Impact Assessment 12*, 231–251.

St. Louis, M. and J. Hess (2008). Climate change: Impacts on and implications for global health. *American Journal of Preventive Medicine 35*(5), 527–538.

Staehelin, J., N. Harris, C. Appenzeller, and J. Eberhard (2001). Ozone trends: A review. *Reviews of Geophysics 39*(2), 231–290.

Stanton, E., F. Ackerman, and S. Kartha (2008). *Inside the Integrated Assessment Models: Four Issues in Climate Economics*. Stockholm, Sweden: Stockholm Environment Institute (SEI) (WP-US-0801).

Starzyk, J. and D. Prasad (2011). A Computational Model of Machine Consciousness. *International Journal of Machine Consciousness 3*(2), 255–281.

Stein, M. (2005). Space-time covariance functions. *Journal of the American Statistical Association 100*(469), 310–321.

Steinhauer, J. (2014). Observation of self-amplifying Hawking radiation in an analogue black-hole laser. *Nature Physics 10*, 864–869.

Stephenson, A. and E. Gilleland (2006). Software for the analysis of extreme events: The current state and future directions. *Extremes 8*, 87–109.

Stern, G., R. Macdonald, P. Outridge, S. Wilson, J. Chételat, A. cole, H. Hintelmann, L. Loseto, A. Steffen, F. Wang, and C. Zdanowicz (2012). How does climate change influence arctic mercury. *Science of the Total Environment 414*, 22–42.

Stern, N. (2007). *The Economics of Climate Change : The Stern Review*. Cambridge, UK: Cambridge University Press.

Sternburg, R. (1982). *Handbook of Human Intelligence.* Cambridge, UK: Cambridge University Press.

Still, C., J. Berry, G. Collatz, and R. DeFries (2003). Global distribution of C3 and C4 vegetation: Carbon cycle implications. *Global Biogeochemical Cycles 17*(1), 1006.

Stocker, T. (2011). *Introduction to Climate Modelling.* Advances in Geophysical and Environmental Mechanics and Mathematics (Hutter, K., Ed.). Berlin: Springer-Verlag.

Stocker, T. and S. Johnsen (2003). A minimum thermodynamic model for the biopolar seasaw. *Paleoceanography 18*(4), 9.

Stolarski, R., A. Krueger, M. Schoeberl, R. McPeters, P. Newman, and J. Alpert (1986). Nimbus 7 satellite measurements of the springtime Antarctic ozone decrease. *Nature 322*, 808–811.

Straussfogel, D. and C. von Schilling (2009). *Systems Theory*, Volume 11 of *International Encyclopedia of Human Geography (Kitchin, R., Thrift, N., Eds.).* Oxford, UK: Elsevier.

Sturtevant, N. (2003). *Multi-Player Games: Algorithms and Approaches.* PhD., diss. (University of California).

Sugihara, G. and R. May (1990). Nonlinear forecasting as a way of distinguishing chaos from measurement error in time series. *Nature 344*(6268), 734–741.

Sun, Y., L. Gu, R. Dickinson, R. Norby, S. Pallardy, and F. Hoffman (2014). Impact of mesophyll diffusion on estimated global land CO_2 fertilization. *Proceedings of the National Academy of Sciences 111*(44), 15774–15779.

Sundaresan, J., K. Santosh, A. Deri, R. Roggema, and R. Singh (2014). *Geospatial Technologies and Climate Change.* Geotechnologies and the Environment (Volume 19). Berlin and Heidelberg: Springer-Verlag.

Suppes, P. (1994). *Ernest Nagel, 1901-1985: A Biographical Memoir.* Washington, D.C., USA: National Academy of Sciences.

Suryawanshi, A. and D. Ghosh (2015). Wind speed prediction using spatio-temporal covariance. *Natural Hazards 75*, 1435–1449.

Sutton, M., A. Bleeker, C. Howard, M. Bekunda, B. Grizzetti, W. de Vries, H. van Grinsven, Y. Abrol, T. Adhya, G. Billen, E. Davidson, A. Datta, R. Diaz, J. Erisman, X. Liu, O. Oenema, C. Palm, N. Raghuram, S. Reis, R. Scholz, T. Sims, H. Westhoek, F. Zhang, S. Ayyappan, A. Bouwman, M. Bustamante, D. Fowler, J. Galloway, M. Gavito, J. Garnier, S. Greenwood, D. Hellums, M. Holland, C. Hoysall, V. Jaramillo, Z. Klimont, J. Ometto, H. Pathak, V. Plocq Fichelet, D. Powlson, K. Ramakrishna, A. Roy, K. Sanders, C. Sharma, B. Singh, U. Singh,

X. Yan, and Y. Zhang (2013). *Our Nutrient World: The challenge to produce more food and energy with less pollution.* Edinburgh, UK: NERC/Centre for Ecology and Hydrology (CEH),`http://nora.nerc.ac.uk/500700/`.

Swain, D., M. Hutchings, and G. Marion (2007). Using a spatially explicit model to understand the impact of search rate and search distance on spatial heterogeneity within an herbivore grazing system. *Ecological Modelling 203*(3-4), 319–326.

Swart, R., R. Biesbroek, and T. Loureno (2014). Science of adaptation to climate change and science for adaptation. *Frontiers in Environmental Science 2*(29), 1–8.

Synder, C., T. Bruulsema, T. Jensen, and P. Fixen (2009). Review of greenhouse gas emissions from crop production systems and fertilizer management effects. *Agriculture, Ecosystems and Environment 133*, 247–266.

Tadic, J., X. Qiu, V. Yadav, and A. Michalak (2014). Mapping of satellite Earth observations using moving window block kriging. *Geoscience Model Development 7*, 5381–5405.

Takens, F. (1981). *Detecting strange attractors in turbulence*, Volume 898. Berlin: Springer-Verlag.

Tanner, M. and W. Wong (1987). The calculation of posterior distribution by data augmentation. *Journal of the American Statistical Society 82*, 528–540.

Tavoni, A. (2013). Game theory: building up cooperation. *Nature Climate Change 3*(September), 782–783.

Tavoni, A., A. Dannenberg, G. Kallis, and A. Löschel (2011). Inequality, communication and the avoidance of disastrous climate change in a public goods game. *Proceedings of the National Academy of Sciences 108*, 11825–11829.

Teixeira, E., G. Fischer, H. van Velthuizen, C. Walter, and F. Ewert (2013). Global hot-spots of heat stress on agricultural crops due to climate change. *Agricultural and Forest Meteorology 170*, 206–215.

Thenkabail, P. (2010). Global croplands and their importance for water and food security in the twenty-first century: Towards an Ever Green Revolution that Combines a Second Green Revolution with a Blue Revolution. *Remote Sensing 2*(9), 2305–2312.

Thenkabail, P., M. Hanjra, V. Dheeravath, and M. Gumma (2010). A holistic view of global croplands and their water use for ensuring global food security in the 21st century through advanced remote sensing and non-remote sensing approaches. *Remote Sensing 2*(1), 211–261.

Therond, O., H. Belhouchette, S. Janssen, K. Louhichi, F. Ewert, J.-E. Bergez, J. Wery, T. Heckelei, J. Olsson, D. Leenhardt, and M. Van Ittersum (2009).

Methodology to translate policy assessment problems into scenarios: The example of the SEAMLESS integrated framework. *Environmental Science and Policy 12*, 619–630.

Thomas, C., A. Cameron, R. Green, M. Bakkenes, L. Beaumont, Y. Collingham, B. Erasmus, M. Ferreira de Siqueira, A. Grainger, L. Hannah, L. Hughes, B. Huntley, A. van Jaarsveld, G. Midgley, L. Miles, M. Ortega-Huerta, A. Peterson, O. Phillips, and S. Williams (2004). Extinction risk from climate change. *Nature 427*, 145–148.

Thompson, J. and J. Sieber (2011). Climate tipping as a noisy bifurcation: A predictive technique. *IMA Journal of Applied Mathematics 76*, 27–46.

Thompson, P. (2012). The agricultural ethics of biofuels: The food versus fuel debate. *Agriculture 2*, 339–358.

Thornton, P. (2010). Livestock production: Recent trends, future prospects. *Philosophical Transactions of the Royal Society B 365*, 2853–2867.

Thornton, P., H. Hasenauer, and M. White (2000). Simultaneous estimation of daily solar radiation and humidity from observed temperature and precipitation: An application over complex terrain in Austria. *Agricultural and Forest Meteorology 104*, 255–271.

Thrush, S., J. Hewitt, P. Dayton, G. Coco, A. Lohrer, A. Norkko, J. Norkko, and M. Chiantore (2009). Forecasting the limits of resilience: Integrating empirical research with theory. *Proceedings of the Royal Society B: Biological Sciences 276*(1671), 3209–3217.

Tibshirani, R. (1996). Regression shrinkage and selection via the Lasso. *Journal of the Royal Statistical Society: Series B (Methodology) 58*(1), 267–288.

Tilman, D., K. Cassman, P. Matson, R. Naylor, and S. Polasky (2002). Agricultural sustainability and intensive production practices. *Nature 418*(6898), 671–677.

Topsoe, F. (2002). *Maximum entropy versus minimum risk and applications to some classical discrete distributions*. Proceedings of the 2002 IEEE Information Theory Workshop (25-25 October, Bangalore, India). IEEE.

Torrens, P. (2010). Agent-based models and the spatial sciences. *Geography Compass 4/5*, 428–448.

Tothova, M. (2011). *Main Challenges of Price Volatility in Agricultural Commodity Markets*. Methods to Analyze Agricultural Commodity Price Volatility (Piot-Lepetit, I., M'Barek, R., Eds.). Berlin and Heidelberg: Springer Science+Business Media, LCC.

Traub, J. (2007). Do negative results from formal systems limit scientific knowledge? *Complexity 3*(1), 29–31.

Travis, W. (2014). Weather and climate extremes: Pacemakers of adaptation. *Weather and Climate Extremes 5-6*, 29–39.

Trenberth, K., L. Smith, T. Qian, A. Dai, and J. Fasullo (2006). Estimates of the Global Water Budget and its annual cycle using observational and model data. *Journal of Hydrometeorology - Special Section 8*(4), 758–769.

Tripathi, P., S. Bandyopadhyay, and S. Pal (2007). Multi-Objective Particle Swarm Optimization with time variant inertia and acceleration coefficients. *Information Sciences 177*, 5033–3049.

Troncale, L. (2009). The future of general systems research: Obstacles, potentials, case studies. *Systems Research and Behavioral Science 26*, 511–552.

Trostle, R. (2008). *Global Agricultural Supply and Demand: Factors Contributing to the Recent Increase in Food Commodity Prices*. USDA, A Report from the Economic Research Service (WRS-0801), http://www.ers.usda.gov/media/218027/wrs0801_1_.pdf.

Tuazon, D., G. Corder, and B. McLellan (2013). Sustainable development: A review of theoretical contributions. *Sustainable Future for Human Security 1*(1), 40–48.

Tubiello, F. and C. Rosenzweig (2008). Developing climate change impact metrics for agriculture. *The Integrated Assessment Journal Bridging Science and Policy (IAJ) 8*(1), 165–184.

Turcotte, D. (1999). Self-organized criticality. *Reports on Progress in Physics 62*, 1377–1429.

Turing, A. (1936). On computable numbers, with an application to The Entscheidungsproblem. *Proceedings of the London Mathematical Society 2*(42), 230–265.

Turner, G. (2008). A comparison of *The Limits to Growth* with 30 years of reality. *Global Environmental Change 18*, 397–411.

Turnhout, E., M. Hisschemöller, and H. Eijsackers (2007). Ecological indicators: Between the two fires of science and politics. *Ecological Indicators 7*, 215–228.

UN (2011). *World Population Prospects: The 2010 Revision, Highlights and Advance Tables*. New York, NY, USA: UNDESA (Population Division), http://esa.un.org/unpd/wpp/.

UN (2013). *Water Security and the Global Water Agenda: A UN-Water Analytical Brief*. Hamilton, ON, Canada: UN-Water Task Force on Water Security, http://www.unwater.org/downloads/watersecurity_analyticalbrief.pdf.

UNCCD (1996). *United Nations Convention to Combat Desertification (UNCCD) in Those Countries Experiencing Serious Drought and/or Desertification, Particularly in Africa*. Washington, DC, USA: UN, http://www.unccd.int/.

UNDESA (2013). *Global Sustainable Development Report: Building the Common Future We Want.* New York, NY, USA: UNDESA, `https://sustainabledevelopment.un.org/globalsdreport/`.

UNDESA (2014). *Compendium of Technical Support Team (TST) Issues Briefs.* New York, NY, USA: UNDESA, Open Working Group on Sustainable Development Goals, `https://sustainabledevelopment.un.org/index.php?page=view&type=400&nr=1554&menu=35`.

UNEP (2009). *The Environmental Food Crisis - The Environment's Role in Averting Future Food Crises: A UNEP rapid response assessment.* Nellemann, C., MacDevette, M., Manders, T. Eickhout, B., Svihus, B., Prins, A.G., Kaltenborn, B.P., (Eds.). Arendal, Norway: UNEP, GRID-Arendal, `www.grida.no/files/publications/FoodCrisis_lores.pdf`.

UNEP (2012). *Global Chemicals Outlook (GCO): Towards Sound Management of Chemicals.* UNEP, `http://www.unep.org/chemicalsandwaste/Portals/9/Mainstreaming/GCO/Rapport_GCO_calibri_greendot_20131211_web.pdf`.

UNEP (2014). *The Emissions Gap Report 2014.* Nairobi, Kenya: United Nations Environment Programme (UNEP), `http://www.unep.org/publications/ebooks/emissionsgapreport/`.

UNESCO (2012). *Five Stylized Scenarios* (01 ed.). Global Water Futures 2050 (Gallopín,G.C., Ed.). Paris, France: UNESCO, `http://unesdoc.unesco.org/images/0021/002153/215380e.pdf`.

UNESCO (2014a). *The United Nations World Water Development Report (WWDR): Facing the Challenges,* Volume 2. Paris, France: UNESCO, `http://www.unesco.org/new/en/natural-sciences/environment/water/wwap/wwdr/2014-water-and-energy/`.

UNESCO (2014b). *The United Nations World Water Development Report (WWDR): Water and Energy,* Volume 1. Paris, France: UNESCO, `http://www.unesco.org/new/en/natural-sciences/environment/water/wwap/wwdr/2014-water-and-energy/`.

Uno, K. (2002). *Economy-Energy-Environment Simulation.* Economy and Environment. Dordrecht, The Netherlands: Kluwer Academic Publishers.

UNU-FLORES (2014). *Advancing a Nexus Approach to the Sustainable Management of Water, Soil and Waste.* Dresden: United Nations University (UNU), Institute for Integrated Management of Material Fluxes and of Resources (FLORES) (Huelsmann, S., Ardakanian, R., Eds.), `https://flores.unu.edu/publications/`.

Uusitalo, L. (2007). Advantages and challenges of Bayesian networks in environmental modelling. *Ecological Modelling 203,* 312–318.

Uzogara, S. (2000). The impact of genetic modification of human foods in the 21st Century: A review. *Biotechnology Advances 18*, 179–206.

Valbuena, D., P. Verburg, and A. Bregt (2008). A method to define a typology for agent-based analysis in regional land-use research. *Agriculture, Ecosystems & Environment 128*, 27–36.

Valin, H., R. Sands, D. Van der Mensbrugghe, G. Nelson, H. Ahammad, E. Blanc, B. Bodirsky, S. Fujimori, T. Hasaegawa, P. Havlik, E. Heyhoe, P. Kyle, D. Mason-D'Croz, S. Paltsev, S. Rolinski, A. Tabeau, H. Van Meijl, M. Von Lampe, and D. Willenbockel (2014). The future of food demand: Understanding differences in global economic models. *Agricultural Economics 45*, 51–67.

van Bilsen, A., G. Bekebrede, and I. Mayer (2010). Understanding complex adaptive systems by playing games. *Informatics in Education 9*(1), 1–18.

Van den Bergh, F. and A. Engelbrecht (2006). A study of particle swarm optimization particle trajectories. *Information Sciences 176*, 937–971.

van Ittersum, M., K. Cassman, P. Grassini, J. Wolf, P. Tittonell, and Z. Hochman (2013). Yield gap analysis with local to global relevance - A review. *Field Crops Research 143*, 4–17.

van Nes, E., M. Scheffer, V. Brovkin, T. Lenton, H. Ye, E. Deyle, and G. Sugihara (2015). Causal feedbacks in climate change. *Nature Climate Change 5*, 445–448.

van Vuuren, D., J. Edmonds, M. Kainuma, K. Riahi, A. Thomson, K. Hibbard, G. Hurtt, T. Kram, V. Krey, J.-F. Lamarque, T. Masui, M. Meinshausen, N. Nakicenovic, S. Smith, and S. Rose (2011). The representative concentration pathways: An overview. *Climatic Change 109*(1-2), 5–31.

van Vuuren, D., J. Lowe, E. Stehfest, L. Gohar, A. Hof, C. Hope, R. Warren, M. Meinshausen, and G.-K. Plattner (2011). How well do integrated assessment models simulate climate change? *Climate Change 104*(2), 255–285.

Vapnik, V. (2000). *The Nature of Statistical Learning Theory* (2nd. ed.). Information Science and Statistics. Berlin and Heidelberg: Springer Science+Business Media.

Varon, E. (1936). Alfred Binet's concept of intelligence. *Psychological Review 43*(1), 32–58.

Vasconcelos, V., F. Raischel, M. Haase, J. Peinke, M. Wächter, P. Lind, and D. Kleinhans (2011). Principal axes of stochastic dynamics. *Physical Review E 84*(3), 031103.

Vasconcelos, V., F. Santos, and J. Pacheco (2013). A bottom-up institutional approach to cooperative governance of risky commons. *Nature Climate Change 3*, 797–801.

Vassilev, N., M. Vassileva, and I. Nikolaeva (2006). Simultaneous p-solubilizing and biocontrol activity of microorganisms: potentials and future trends. *Applied Microbiology and Biotechnology 71*, 137–144.

Ver Hoef, J. and R. Barry (1998). Constructing and fitting models for cokriging and multivariable spatial prediction. *Journal of Statistical Planning and Inference 69*, 275–294.

Villa, F. (2001). Integrating modelling architecture: A declarative framework for multi-paradigm, multi-state ecological modelling. *Ecological Modelling 137*, 23–42.

Villa, F., K. Bagstad, B. Voigt, G. Johnson, R. Portela, M. Honzák, and D. Batker (2014). A methodology for adaptable and robust ecosystem services assessment. *PLoS ONE 9*(3), e91001.

Vincent, L. (1998). A technique for the identification of inhomogeneities in Canadian temperature series. *Journal of Climate 11*, 1094–1104.

Vincent, L. and D. Gullett (1999). Canadian historical and homogeneous temperature datasets for climate change analyses. *International Journal of Climatology 19*, 1375–1388.

Vincent, L., E. Milewska, R. Hopkinson, and L. Malone (2009). Bias in minimum temperature introduced by a redefinition of the climatological day at the Canadian synoptic stations. *Journal of Applied Meteorology and Climatology 48*, 2160–2168.

Vincent, L., X. Wang, E. Milewska, H. Wan, F. Yang, and V. Swail (2012). A second generation of homogenized Canadian monthly surface air temperature for climate trend analysis. *Journal of Geophysical Research: Atmospheres 117*, D18110.

Vincent, L., X. Zhang, B. Bonsal, and W. Hogg (2002). Homogenization of daily temperatures over Canada. *Journal of Climate 15*, 1322–1334.

Viseu, A. (2015). Integration of social science into research is crucial. *Nature 525*(291), 7569.

Viswanathan, R. and P. Varshney (1997). Distributed detection with multiple sensors: Part i: Fundamentals. *Proceedings of the IEEE 85*(1), 54–63.

Von Bertalanffy, L. (1968). *General System Theory: Foundations, Development, Applications*. New York, NY, USA: George Braziller.

Von Lampe, M., D. Willenbockel, H. Ahammad, E. Blanc, Y. Cai, K. Calvin, S. Fukimori, T. Hasegawa, P. Havlik, E. Heyhoe, P. Kyle, H. Lotze-Campen, D. Mason d'Croz, G. Nelson, R. Sands, C. Schmitz, A. Tabeau, H. Valin, D. Van der Mensbrugghe, and H. Van Meijl (2014). Why do global long-term scenarios for agriculture differ? An overview of the AgMIP Global Economic Model Intercomparison. *Agricultural Economics 45*, 3–20.

Von Neumann, J. (1928). Zur Theorie der Gesellschaftsspiele. *Mathematische Annalen 100*, 295–320.

Von Storch, H. and F. Zwiers (2002). *Statistical Analysis in Climate Research.* Cambridge, UK: Cambridge University Press.

Vorosmarty, C., P. McIntyre, M. Gessner, D. Dudgeon, A. Prusevich, P. Green, S. Glidden, S. Bunn, C. Sullivan, C. Reidy Liermann, and P. Davies (2010). Global threats to human water security and river biodiversity. *Nature 467*, 555–561.

Vörösmarty, C., C. Pahl-Wostl, S. Bunn, and R. Lawford (2013). Global water, the Anthropocene and the transformation of a science. *Current Opinions in Environmental Sustainability 5*, 539–550.

Wahba, G. (1990). *Spline models for observational data.* CBMS-NSF Regional Conference Series in Applied Mathematics. Philadelphia, PA, USA: Society for Industrial and Applied Mathematics (SIAM).

Waheed, B., F. Khan, and B. Veitch (2009). Linkage-Based Frameworks for Sustainability Assessment: Making a case for Driving Force-Pressure-State-Exposure-Effect-Action (DPSEEA) Frameworks. *Sustainability 1*, 441–463.

Walker, B. and S. D. (2006). *Resilience Thinking: Sustaining Ecosystems and People in a Changing World.* Washington, DC, USA: Island Press.

Walker, M. (1999). The unquiet voice of 'Silent Spring': The legacy of Rachel Carson. *The Ecologist 29*(5), 322.

Walrut, B. (2002). *Sea ranching and aspects of the common law: A proposal for a legislative framework.* PhD., diss. (University of British Columbia).

Walters, C. (2002). *Adaptive Management of Renewable Resources.* Caldwell, N.J., USA: Blackburn Press.

Wan, H., X. Zhang, and E. Barrow (2005). Stochastic modelling of daily precipitation for Canada. *Atmosphere-Ocean 43*(1), 23–32.

Wang, D., E. Carr, L. Gross, and M. Berry (2005). Toward ecosystem modeling on computing grids. *Computing in Science and Engineering 1*(September-October), 44–52.

Wang, N., N. Zhang, and M. Wang (2006). Wireless sensors in agriculture and food industry - Recent development and future perspective. *Computers and Electronics in Agriculture 50*, 1–14.

Wang, W. and A. Gelman (2014). Difficulty of selecting among multilevel models using predictive accuracy. *Statistics and Its Inference 7*(1), 1–8.

Wang, W., K. Ichii, and H. Hashimoto (2009). A hierarchical analysis of terrestrial ecosystem model Biome-BGC: Equilibrium analysis and model calibration. *Ecological Modelling 220*(17), 2009–2023.

Wang, Y. (1998). Smoothing spline models with correlated random errors. *Journal of the American Statistical Association 3*(441), 341–348.

Wang, Y., R. Law, and B. Pak (2010). A global model of carbon, nitrogen and phosphorus cycles for the terrestrial biosphere. *Biogeosciences 7*, 2261–2282.

Watkins, C. and D. P. (1992). Q-learning. *Machine learning 8*(3-4), 279–292.

Watson, L. and M. Spence (2007). Causes and consequences of emotions on consumer behaviour – A review and integrative cognitive appraisal theory. *European Journal of Marketing 41*(5-6), 487–511.

Watson, R. and D. Pauly (2001). Systematic distortions in world fisheries catch trends. *Nature 424*, 534–536.

WBCSD (2010). *Vision 2050: The New Agenda for Business.* Washington, DC, USA: World Business Council for Sustainable Development (WBCSD), `http://www.wbcsd.org/pages/edocument/edocumentdetails.aspx?id=219`.

WBG (2012). *Inclusive Green Growth: The Pathway to Sustainable Development.* Washington, DC, USA: World Bank Group (WBG), `http://siteresources.worldbank.org/EXTSDNET/Resources/Inclusive_Green_Growth_May_2012.pdf`.

WBG (2014). *Global Economic Prospects: Shifting Priorities, Building for the Future*, Volume 9. Washington, DC, USA: World Bank Group, `http://documents.worldbank.org/curated/en/2015/05/19653612/global-economics-prospects-shifting-priorities-building-future`.

WBG (2016). *World Development Report 2016: Digital Dividends*, Volume 9. Washington, DC, USA: World Bank Group, `http://www.worldbank.org/en/publication/wdr2016`.

Weber, M., N. Krogman, and T. Antoniuk (2012). Cumulative effects assessment: Linking social, ecological and governance dimensions. *Ecology and Society 17*(2), 22.

WEF (2012). *Global Agenda: Future of Government - Fast and Curious. How innovative governments can create public value by leading citizen-centric change in the face of global risks.* Global Agenda Council on the Future of Government, `www3.weforum.org/docs/WEF_GAC_FutureGovernment_2012.pdf`.

WEF (2013). *Building Resilience in Supply Chains.* Geneva, Switzerland: An Initiative of the Risk Response Network In collaboration with Accenture, `http://www.weforum.org/reports/building-resilience-supply-chains`.

Wei, G. (2000). A unified approach for the solution of the Fokker–Planck equation. *Journal of Physics A: Mathematical and General 33*(27), 4935–4953.

Weinzettel, J., E. Hertwich, G. Peters, K. Steen-Olsen, and A. Galli (2013). Affluence drives the global displacement of land use. *Global Environmental Change 23*, 433–438.

Weitzman, M. (2001). Gamma discounting. *American Economics Review 91*, 260–271.

West, P., H. Gibbs, C. Monfreda, J. Wagner, C. Barford, S. Carpenter, and J. Foley (2010). Trading carbon for food: Global comparison of carbon stocks vs. crop yields on agricultural land. *Proceedings of the National Academy of Sciences (PNAS) 107*(46), 19645–19648.

Westley, F., P. Olsson, C. Folke, T. Homer-Dixon., H. Vredenburg, D. Loorbach, J. Thompson, M. Nilsson, E. Lambin, J. Sendzimir, B. Banerjee, V. Galaz, and S. van der Leeuw (2011). Tipping toward sustainability: Emerging pathways of transformation. *AMBIO: A Journal of the Human Environment 40*(7), 762–780.

White, M. (2012). Sustainable approaches to advanced materials research. *Physics in Canada 68*(1), 27–29.

Whitehead, A. and B. Russell (1997). *Principia Mathematica to *56* (2nd. ed.). Cambridge Mathematical Library. Cambridge, UK: Cambridge University Press.

WHO (2006). *Overview of greywater management health considerations.* Amman, Jordan: World Health Organization (WHO).

Wiedmann, T., M. Lenzen, K. Turner, and J. Barrett (2007). Examining the global environmental impact of regional consumption activities-Part 2: Review of input–output models for the assessment of environmental impacts embodied in trade. *Ecological Economics 61*, 15–26.

Wilke, C. (2003). Hierarchical models in Environmental Science. *International Statistical Review 71*(2), 181–199.

Wilke, C. (2014). *Agent Based Models: Statistical Challenges and Opportunities.* Statistics Views. Hoboken, NJ, USA: John Wiley & Sons Ltd., http://www.statisticsviews.com/details/feature/6354691/Agent-Based-Models-Statistical-Challenges-and-Opportunities.html.

Wilke, C., L. Berliner, and N. Cressie (1998). Hierarchical Bayesian space-time models. *Environmental and Ecological Statistics 5*, 117–154.

Wilks, D. (2008). Effects of stochastic parameterization on conceptual climate models. *Philosophical Transactions of the Royal Society A 366*, 2477–2490.

Wilmers, C. (2007). Understanding ecosystem robustness. *TRENDS in Ecology and Evolution 22*(10), 504–506.

Wilson, A. and J. Silander (2014). Estimating uncertainty in daily weather interpolations: A Bayesian framework for developing climate surfaces. *International
Journal of Climatology 34*, 2573–2584.

Wilson, E. (1998). *Consilience: The Unity of Knowledge*. New York, NY, USA:
Alfred A. Knopf.

Wilson, K. (1975). Renormalization Group Methods. *Advances in Mathematics 16*,
170–186.

Wing, J. (2006). Computational Thinking. *Communications of the ACM 49*(3),
33–35.

Wissner-Gross, A. and C. Freer (2013). Causal Entropic Forces. *Physical Review
Letters 110*, 168702.

Wolpert, D. (1990). The relationship between Occam's Razor and convergent guess-
ing. *Complex Systems 4*, 319–368.

Wood, S. (2010). Statistical inference for noisy nonlinear ecological dynamic systems.
Nature (Letters) 446, 1102.

Wu, J. and J. David (2002). A spatially explicit hierarchical approach to modeling
complex ecological systems: Theory and applications. *Ecological Modelling 153*,
7–26.

WWF (2010). *Living Planet Report 2010: Biodiversity, biocapacity and development*. Glant, Switzerland: World Wildlife Fund (WWF) International, `http://wwf.panda.org/about_our_earth/all_publications/living_
planet_report/living_planet_report_timeline/2010_lpr2/`.

WWF (2012). *Living Planet Report 2012: Biodiversity, Biocapacity and Better
Choices*. Gland, Switzerland: World Wildlife Fund (WWF) International.

WWF (2014). *Living Planet Report: Species and Spaces, People and Places*. Gland,
Switzerland: World Wildlife Fund (WWF) International.

Xie, J., J. Vincent, and T. Panayotou (1996). *Computable General Equilibrium
Models and the Analysis of Policy Spillovers in the Forestry Sector*. United States
Environmental Protection Agency (EPA).

Yang, J. (2011). Convergence and uncertainty analyses in Monte-Carlo based sensitivity analysis. *Environmental Modelling and Software 26*(4), 444–457.

Yodzis, P. and K. McCann (2007). *Dynamic Signatures of Real and Model Ecosys-
tems*. From Energetics to Ecosystems: The Dynamics and Structure of Ecological
Systems (pp. 185-190).The Peter Yodzis Fundamental Ecology Series (Volume 1)
(Rooney, N., McCann, K.S., Noakes, D.L.G., Eds.). Dordrecht, The Netherlands:
Springer-Verlag.

Yu, A. and H. Huang (2014). Maximizing masquerading as matching in human visual search choice behavior. *Decision 1*(4), 275–287.

Yuan, M. and Y. Lin (2006). Model selection and estimation in regression with grouped variables. *Journal of the Royal Statistical Society: Series B (Methodology) 68*(1), 49–67.

Zecha, C., J. Link, and W. Claupein (2013). Mobile sensor platforms: Categorisation and research applications in precision farming. *Journal of Sensors and Sensor Systems 2*, 51–72.

Zeng, C., Q. Yang, and Y. Chen (2014). Lyapunov techniques for stochastic differential equations driven by fractional brownian motion. *Abstract and Applied Analysis 2014*, 9.

Zhang, C. and J. Kovacs (2012). The application of small unmanned aerial systems for precision agriculture: A review. *Precision Agriculture 13*, 693–712.

Zhang, X., H. Wan, F. Zwiers, G. Hegerl, and S.-K. Min (2013). Attributing intensification of precipitation extremes to human influence. *Geophysical Research Letters 40*(19), 5152–5257.

Zhang, Y., N. Hamm, N. Meratnia, A. Stein, M. van de Voort, and P. Havinga (2012). Statistics-based outlier detection for wireless sensor networks. *International Journal of Geographical Information Science 26*(8), 1373–1392.

Zheng, X., G. Wu, S. Zhang, X. Liang, Y. Dai, and Y. Li (2013). Using analysis state to construct a forecast error covariance matrix in ensemble Kalman filter assimilation. *Advances in Atmospheric Science 30*(5), 1303–1312.

Zhu, J. and M. Ruth (2013). Exploring the resilience of industrial ecosystems. *Journal of Environmental Management 122*, 65–75.

Zhu, W., S. Wang, and C. Caldwell (2012). Pathways of assessing agroecosystem health and agroecosystem management. *Acta Ecologica Sinica 32*, 9–17.

Zidek, J., G. Shaddick, and C. Taylor (2014). Reducing estimation bias in adaptively changing monitoring networks with preferential site selection. *The Annals of Applied Statistics 8*(3), 1640–1670.

Zidek, J., W. Sun, and N. Le (2002). Designing and integrating composite networks for monitoring multivariate gaussian pollution fields. *Journal of the Royal Statistical Society: Series C (Applied Statistics) 49*(1), 63–79.

Ziegler, B., H. Praehofer, and T. Kim (2000). *Theory of Modeling and Simulation: Integrating Discrete Event and Continuous Complex Dynamic Systems* (2nd. ed.). San Diego, CA, USA: Academic Press, Inc.

Zwart, S., W. Bastiaanssen, C. de Fraiture, and D. Molden (2010a). A global benchmark map of water productivity for rainfed and irrigated wheat. *Agricultural Water Management 97*(10), 1617–1627.

Zwart, S., W. Bastiaanssen, C. de Fraiture, and D. Molden (2010b). WATPRO: A remote sensing based model for mapping water productivity of wheat. *Agricultural Water Management 97*(10), 1628–1636.

Author Index

Subject Index